主编 ◎ 李 健

中国海水养殖模式

MARICULTURE MODES IN CHINA

Editor in Chief　Li Jian

·青岛·

本书得到国家重点研发计划项目"海水池塘和盐碱水域生态工程化养殖技术与模式（2019YFD0900400）"、农业国际交流与合作项目"'一带一路'热带国家水产养殖科技创新合作"和"'一带一路'海水养殖技术培训班（12200039）"资助

图书在版编目（CIP）数据

中国海水养殖模式 = Mariculture modes in China：英文 / Li Jian 主编 . —青岛：中国海洋大学出版社，2020.8
ISBN 978-7-5670-2565-3

Ⅰ. ①中… Ⅱ. ① L… Ⅲ. ①海水养殖—模式—研究—中国—英文 Ⅳ. ① S967

中国版本图书馆 CIP 数据核字（2020）第 167781 号

出版发行	中国海洋大学出版社
社　　址	青岛市香港东路 23 号　邮政编码　266071
网　　址	http://pub.ouc.edu.cn
出 版 人	杨立敏
责任编辑	董　超　孟显丽
电　　话	0532-85902533
电子信箱	1079285664@qq.com
印　　制	青岛海蓝印刷有限责任公司
版　　次	2020 年 11 月第 1 版
印　　次	2020 年 11 月第 1 次印刷
成品尺寸	185 mm × 260 mm
印　　张	36.5
字　　数	732 千
印　　数	1—1500
定　　价	160.00 元
订购电话	0532-82032573（传真）

发现印装质量问题，请致电 0532-88785354，由印刷厂负责调换。

编委会

主　编　李　健
副主编　刘志鸿　徐甲坤
编　委　（按姓氏首字母排序）

白昌明	曹　荣	陈松林	崔　勇	崔正国	董　宣	高保全
关长涛	江　涛	李吉涛	李佳琦	廖梅杰	刘宝良	刘福利
栾　生	吕建建	马爱军	邱　亮	史成银	隋　娟	万晓媛
汪文俊	王新安	王秀华	卫育良	吴　彪	吴文广	徐后国
徐文腾	徐永江	许　华	杨　冰	于　宏	张继红	张庆利
赵付文						

Editorial Committee

Editor in Chief　　　　Li Jian
Deputy Editors in Chief　Liu Zhihong　Xu Jiakun
Editors　　　(in alphabetical order by pinyin of family name)

Bai Changming	Cao Rong	Chen Songlin
Cui Yong	Cui Zhengguo	Dong Xuan
Gao Baoquan	Guan Changtao	Jiang Tao
Li Jitao	Li Jiaqi	Liao Meijie
Liu Baoliang	Liu Fuli	Luan Sheng
Lv Jianjian	Ma Aijun	Qiu Liang
Shi Chengyin	Sui Juan	Wan Xiaoyuan
Wang Wenjun	Wang Xin'an	Wang Xiuhua
Wei Yuliang	Wu Biao	Wu Wenguang
Xu Houguo	Xu Wenteng	Xu Yongjiang
Xu Hua	Yang Bing	Yu Hong
Zhang Jihong	Zhang Qingli	Zhao Fuwen

PREFACE

Peace and development are the common aspiration of people all over the world and the unremitting pursuit of mankind. After the Industrial Revolution, the ability of human to exploit and utilize natural resources has been greatly improved. However, human being is still facing many severe challenges such as food crisis, deterioration of the ecological environment and depletion of natural resources. Aquaculture not only provides human with high quality animal protein food economically, but also is one of the effective ways to alleviate world food crisis. A mutually beneficial and win-win "Belt and Road" (B&R) has spread with railways and sea lanes to other countries with China as the center. Tropical countries involved in the B&R project have similar aquatic resources with China, however, the aquaculture techniques in these tropical countries and regions are relatively weak. China has a long history in aquaculture. As early as in the Spring and Autumn Period, the earliest work extant in China named *Fanli's Scripture of Pisciculture* came out, which also has been viewed as the earliest technical document on fish farming in the world. Since the implementation of the reform and opening policy in China, numerous aquaculture technologies, farming modes, and new equipment have been developed and innovated by industrious and intelligent Chinese people, creating significant impacts on modern aquaculture. China has become the world's leading producer of aquaculture products and the B&R tropical countries and regions have a great demand for modern aquaculture technology. Under the background of economic globalization, it is of great significance for the application and promotion of China's aquaculture experience in these B&R countries and regions.

Since 2018, Chinese Academy of Fishery Sciences (CAFS) has been implementing the program of "Technology and Innovation Cooperation in Aquaculture with tropical countries along the 'Belt and Road'" supported by Ministry of Agriculture and Rural Affairs of the People's Republic of China. Scientists from CAFS and several tropical countries jointly carried out a series of surveys and assessment on genetic resource, and demonstrated agricultural techniques suited to the local ecosystem and needs of those tropical countries. This program has encouraged and promoted the application of Chinese aquaculture technology, and facilitated the development of aquaculture in the B&R countries.

There is an ancient Chinese proverb saying: "Give a man a fish, he eats for a day. Teach a man to fish and he eats for a lifetime." Yellow Sea Fisheries Research Institute (YSFRI), CAFS, shares successful aquaculture experiences with countries and regions along the B&R through training courses and on-site technical guidance to help train aquaculture technicians and management personnel in those countries and regions. In 2018 and 2019, over 300 trainees from more than 10 countries including Vietnam, Cambodia, Thailand, Malaysia, Indonesia, Brunei, the Philippines, Pakistan, Bangladesh, Egypt and China were trained in YSFRI. This book aims to provide technical information based on the training courses for improving aquaculture technology and management, facilitating investment, trade, and culture exchange, so as to achieve common development with countries and regions along the B&R.

Wang Qingyin (王清印)

August 18, 2020

Qingdao, China

CONTENTS

Chapter 1 Mariculture technology ·· 1

Fast-growing strain of turbot selected based on large-scale family breeding ······················ 2
Progeny production and culture technology of marine fish ··· 25
Molecular sex control and genomics of Chinese tongue sole ··· 49
History, status and prospect of shrimp farming in China··· 77
Quantitative genetic basis of shrimp selective breeding·· 95
Breeding program for Pacific white shrimp, *Penaeus vannamei* ··································· 110
Technical system of ecological culture of shrimp in China ··· 122
Selective breeding of swimming crab *Portunus trituberculatus* in China ······················· 146
Studies on techniques of the artificial propagation of *Portunus trituberculatus* ·············· 161
An overview of marine shellfish industry in China ·· 177
Seedling of gastropods with high economic value in China—taking the Pacific abalone as an example ··· 192
Breeding and culture of sea cucumber (*Apostichopus japonicus*) in China···················· 208
Theory and technology of kelp cultivation in China ··· 231
Theory and technology of *Pyropia* culture in China ··· 248

Chapter 2 Mariculture modes··· 263

Mariculture mode—Recirculating aquaculture system (RAS) ·· 264
Advances in the research of recirculating aquaculture systems (RAS) in China··············· 276
Development of coastal integrated multi-trophic aquaculture (IMTA) in China················ 287

MOM-B system and its application in IMTA ··310
Cage mariculture in China ···323
Research and application of marine ranching technology ··································340

Chapter 3 Maricultural organism disease control and molecular pathology ············ 359

Activities for WSD and IHHN OIE reference laboratory ··································360
Competence criteria of aquatic animal disease testing laboratory ·······················371
Rapid detection of aquaculture pathogens ··378
Acute hepatopancreatic necrosis disease (AHPND) in shrimp ··························387
Infection with decapod iridescent virus 1 (DIV1), an emerging disease in shrimps ···············398
Viral nervous necrosis of teleost fish ··412
Viral covert mortality disease and its pathogen ···426
Threats of concern to mollusk aquaculture in China ······································439

Chapter 4 Fishery environment and bio-remediation ·································· 447

Environmental monitoring for marine aquaculture ··448
Treatment of aquaculture wastewater with constructed wetlands ······················463

Chapter 5 Food engineering and nutrition ·· 477

Seafood products processing ··478
Best management practices for fish feed and feeding ······································496
Live food technology of marine fish larvae ···515
Protein nutrition for aquacultural fish ··536

Chapter 6 International science and technology cooperation in mariculture of Yellow Sea Fisheries Research Institute, Chinese Academy of Fishery Sciences ················ 551

International science and technology cooperation in mariculture of Yellow Sea Fisheries Research Institute, Chinese Academy of Fishery Sciences ·································552

Chapter 1
Mariculture technology

Fast-growing strain of turbot selected based on large-scale family breeding

Author: Ma Aijun, Wang Xin'an
E-mail: maaj@ysfri.ac.cn

1 A brief introduction to turbot

1.1 Turbot *Scophthalmus maximus*

Turbot *Scophthalmus maximus* (Linnaeus) (Fig. 1-1-1), a species of demersal marine flatfish, naturally inhabits the Baltic Sea, Black Sea and Mediterranean Sea (Lei et al., 1995)

Fig. 1-1-1   Turbot *S. maximus*

1.2 Development history of turbot industry in China

In 1992, turbot was first introduced into China (Ma et al., 2010).

In 1995, parent fish were successfully nurtured (Ma et al., 2010).

In 1997, breeding technique obtained significant breakthrough (Ma et al., 2015).

In 1999, breeding seedling reached large-scale production (Ma et al., 2012).

The maximum annual output is 50 thousand tons and the annual output value is 4 billion *yuan* (Lei et al., 2012).

1.3 Necessity of carrying out turbot selective breeding

The genetic improvement of turbot is necessary to sustain the development of the industry in a highly competitive aquaculture market. Otherwise, it will cause serious germplasm degeneration because of the inbreeding and the lack of effective long-term broodstock management programs.

High mortality and slow growth rate will lead to unstable production of farmed turbot and a gradual decline in total output. Fortunately, the companies we work with are also aware of possible problems.

1.4 Large-scale family breeding for turbot by marker assisted selection

Considering that germplasm degeneration of turbot would be resulted from inbreeding, Yellow Sea Fisheries Research Institute has initiated and run the genetic improvement program of turbot with the enterprise since 2004 by using large-scale family breeding.

Large-scale family breeding by electronic markers is also called classical breeding technology, because the technology itself uses classical methods such as weighing and quantifying, and the principle is classical genetic principle. Since the core of the technology is the modern statistical method and computer data processing technology, it is also called modern breeding technology (Ma et al., 2010).

Large family selection has the characteristics of high efficiency. The genetic progress of each generation can obtain the genetic progress of 20%–30%. Besides, it is the most effective and most practical technology to inhibit genetic decline.

At present, a fast-growing strain is obtained by large family selection.

2 Collection of breeding population and background analysis

2.1 Parental origin

Four turbot populations were imported from UK, France, Denmark and Norway (Table 1-1-1).

Table 1-1-1 Parental origin

Country	Number of males	Number of females	Date of import	Body-length at import/cm
UK	16	22	August 11, 2002	5
France	18	24	September 22, 2003	5
Denmark	15	29	April 4, 2003	5
Norway	19	27	June 10, 2002	5
Total	68	102	–	–

2.2 Genetic background analysis based on multivariate statistics

The morphologic characters of the four different populations were compared by using hierarchical cluster analysis (Fig. 1-1-2) and principal component analysis (Fig. 1-1-3).

The hierarchical cluster analysis showed that the morphological characteristics were similar between the populations of Denmark and Norway as well as UK and France, while large variation existed between the populations of Denmark and Norway as well as that of UK and France.

The principal component analysis showed that the morphological variations between the populations of Denmark and Norway as well as UK and France were not significant and that the morphological variations between the populations of Denmark and UK, Denmark and France, Norway and UK as well as Norway and France were significant.

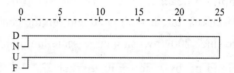

D, Denmark; N, Norway; U, UK; F, France.

Fig. 1-1-2　Hierarchical cluster analysis

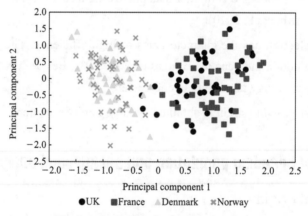

Fig. 1-1-3　Principal component analysis

2.3 Genetic background analysis based on molecular biology

A. Genetic diversities within and between four geographical stocks were assessed through 12 microsatellite loci.

B. The rank of heterozygosity values for the four stocks in an increasing order was Denmark < UK and < France < Norway.

C. The stocks were pooled into two groups by cluster analysis: one group of Norwegian and Danish stocks, the other group of France and UK stocks (Fig. 1-1-4).

D. Hardy-Weinberg equilibrium and genetic deviation index showed a reasonable difference within each stock.

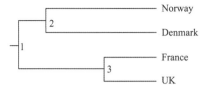

Fig. 1-1-4 Genetic background analysis based on molecular biology (UPGMA clustering graph)

3 Family construction and rearing

3.1 Mating design programme

The families were produced using a nested mating design with two females nested within a male, depending on the parent fish from four turbot populations (Fig. 1-1-5).

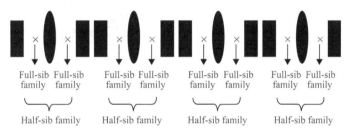

Fig. 1-1-5 Nested mating design with two females nested within a male

3.2 Standardization rearing of families rearing at the early stages

The families were reared in separate tanks until tagging. To obtain similar rearing conditions for all families at the early breeding stages, some effective measures were taken to standardize the stocking density of fish and the environment (Fig. 1-1-6).

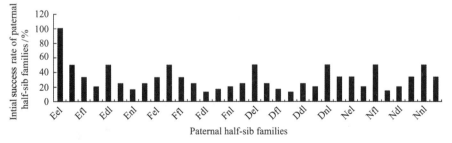

E and e, the individual from Uk population. F and f, the individual from France population. D and d, the individual from Denmark population. N and n, the individual from Norway population. Capital letter, male. Lower case Letter, female.

Fig. 1-1-6 Initial success rates of half-sib families

3.2.1 Environment standardization

All families from hatching to commune stocking were reared under same conditions according to the culture conditions of turbot, mainly including water temperature, salinity, illumination intensity, pH, values of dissolved oxygen, etc.

All families were placed in the same workshop for breeding management according to different developmental stages (Fig. 1-1-7).

At the early stages of rearing, water temperature, salinity, illumination intensity, pH and values of dissolved oxygen were 13–18 ℃, 30–40, 500–2,000 lx, 7.8–8.2 and >6 mg/L, respectively.

Fig. 1-1-7　The workshop for breeding management

3.2.2　Quantity standardization

Three-level quantity standardization rearing were carried out according to the development of turbot:

A. The first quantity standardization: at 15 d post-spawning, random samples of 10,000 young fish per full-sib families were transferred to 4 m×1.5 m×1 m concrete tanks separately (Fig. 1-1-8).

B. The second quantity standardization: at 30 d post-spawning, random samples of 5,000 young fish per full-sib family were transferred to 4 m×1.5 m×1 m concrete tanks separately (Fig. 1-1-9).

C. The third quantity standardization: at 45 d post-spawning, random samples of 2,000 young fish per full-sib family were transferred to 5 m×3 m×1 m concrete tanks separately (Fig. 1-1-10).

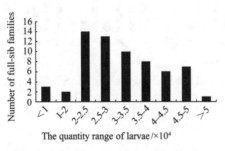

Fig. 1-1-8　First quantity standardization

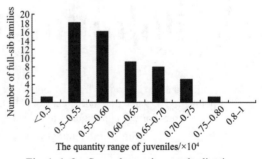

Fig. 1-1-9　Second quantity standardization

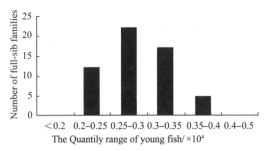

Fig. 1-1-10 Third quantity standardization

All families were constructed in Yantai Tianyuan Aquatic Limited Corporation, Yantai, China (Fig. 1-1-11).

In 2007, 56 full-sib families (including 28 paternal half-sib families) were constructed successfully.

In 2008, 38 full-sib families (including 8 paternal half-sib families) were constructed successfully.

In 2009, 20 full-sib families were constructed successfully.

In 2010, 57 full-sib families (including 26 paternal half-sib families) were constructed successfully.

Fig. 1-1-11 Yantai Tianyuan Aquatic Limited Corporation, Yantai, China

3.3　Standardization of families rearing at the later stage

3.3.1　Quantity standardization

At 2 months of age, random samples of 1,000 young fish per full-sib families were transferred to 12 m^3 concrete tanks separately.

When the fish were reared up to 3 months of age, samples of 250–300 fish were randomly selected from each tank for tagging using Visible Implant Elastomer (VIE) (for distinguishing different families) and then stocked communally.

50 fish from each full-sib family were randomly sampled and individually tagged with Passive Integrated Transponders (PIT) (for distinguishing families and individuals) when tagged fish with VIE were communally reared up to 9 months of age.

3.3.2 Environment standardization

During the larvae-cultured period and juvenile-cultured period, water temperature, salinity, illumination intensity, pH and values of dissolved oxygen were 13–18 ℃, 30–40, 500–2,000 lx, 7.8–8.2 and over 6 mg/L, respectively.

During the adolescent-culture period, the above five indexes were 15–18 ℃, 25–30, 500–1,500 lx, 7.6–8.2 and over 6 mg/L, respectively.

4 Tagging method for parent fish and offspring

4.1 Tag of base population parent fishes: PIT electronic tag and liquid nitrogen freeze-branding

All broodstocks from the four populations were individually tagged with PIT electronic tag (Fig. 1-1-12) and liquid nitrogen freeze-branding (Fig. 1-1-13).

A. PIT electronic tag: long time reservation.

B. Liquid nitrogen freeze-branding: early selection of parent fish.

Fig. 1-1-12 PIT electronic tag   Fig. 1-1-13 Liquid nitrogen freeze-branding

The survival rate of fish tagged with liquid nitrogen was 100% three months later and the label recognition rate was 100% one year later. The materials required for liquid nitrogen figure notation are only liquid nitrogen tanks and red copper and the cost is very low. The practicability of this method is obvious because the number of water marks is clearly visible (Fig. 1-1-14).

Although the label recognition rate of fish tagged with liquid nitrogen was 100% one year later, some marked numbers were not very clear. Considering persistence, security and practicality of tags, each parent fish was tagged with both PIT and liquid nitrogen freeze-branding.

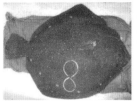 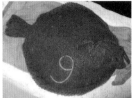  

Fig. 1-1-14 Fish tagged with liquid nitrogen

4.2 Offspring tag: VIE and PIT tagging

PIT electronic tagging is an important link of large-scale family selection. Thus, different families tagged with PIT are communally reared. The correlation between phenotypic values and genotypic values is greatly improved due to the elimination of environmental effects.

However, for turbot, if the tagging time is too early, the damage to the fish will be greater, which will cause a large amount of seedling mortality. If the tagging time is too late, the environmental impact will last much longer. Especially in the early stages of fish development, the impact of environmental factors is more obvious.

In order to solve this contradiction, VIE and PIT tagging methods were used. That is to say, VIE tagging method was used in the early stage and PIT markers were used when the individuals were suitable for PIT tagging.

When the fish were reared up to 3 months of age, individuals were tagged with VIE (for distinguishing different families) and then stocked communally.

The individuals were tagged with PIT (for distinguishing families and individuals) when tagged fish with VIE were communally reared up to 9 months of age. The fish tagged with PIT were still pooled together.

5 Genetic evaluation of growth traits in turbot

5.1 Estimation of genetic parameters and prediction of breeding value based on animal model BLUP method

5.1.1 Construction of pedigree database and data collection

A systematic pedigree database must be established for a complete breeding program.

A. Pedigree database mainly included family number, population origin, breeding generation, male number, female number, insemination time, fertilized egg volume, collection time of newly hatched larvae, birth batch, etc.

B. Pedigree database retained clear genealogical information for continuation breeding and provided direct basis for the selection of broodstock and the evaluation of breeding progress.

5.1.2 Construction of genetic evaluation model

The accuracy estimations of breeding values depend on the use of a correct model, a large sample size and reliable statistical methods.

During 12 months of age, the sex of turbot was unable to be distinguished and the difference between female and male turbot in growth was not significant, so that the effect was not taken into account in the model. The maternal effect was also not considered in the model because it was generally recognized that there was no evidence for the influence of maternal fish on the size of growth during the post-juvenile. Simulation studies revealed that the estimation value was more accurate for larger sample size. Our sample size was large enough for genetic evaluation of growth trait. In addition, the test fish would often be reared over a period of about 60 d, so that an adjustment for age at harvesting would also be desirable.

In the present study, an animal model including additive genetic effect, full-sib family effect and effect of age was used.

5.1.3 Equation form of the animal model

$$f_{ijk} = u + a_i + f_j + \beta_t d + e_{ijk} \qquad \text{(Equation 1)}$$

u is the mean population;

f_{ijk} is the measured value of body weight;

a_i is the additive genetic effect of individual as the random effect;

f_j is full-sib family random effect;

e_{ijk} is the random residual;

$\beta_t d$ is the linear coefficient of the age of individual (d) on body weight.

5.1.4 Matrix notation of the animal model

$$y = Xb + Zu + e \qquad \text{(Equation 2)}$$

y is the vector of observations of trait;

b is the vector of fixed effects;

u is the vector of random effects;

e is a vector of random errors;

X and Z are known design matrices assigning the observations to levels of b and u, respectively.

5.1.5 Genetic evaluation method

The usual methods for estimating genetic parameters include maximum likelihood (ML) (Hartley and Rao, 1967), restricted maximum likelihood (REML) (Patterson and Thompson, 1971), average information restricted maximum likelihood (AI-REML) (Jensen et al.,1997), minimum variance quadratic unbiased estimation (MIVQUE) (Swallow and Searle, 1978), minimum norm quadratic unbiased estimation (MINQUE) (Rao, 1971) and Bayesian methods (Bayesian) (Harville, 1974; Gianola and Fernando, 1986).

For genetic evaluation of turbot, previous data were analyzed by REML using MTDFREML software, and current data were analyzed by AI-REML using AI-REML: R software.

5.2 Selection response prediction

In this study, genetic progress for body weight in turbot was evaluated using the following parameters:

A. Heritability of body weight: 0.64 ± 0.07; common environmental coefficient: 0.00016; phenotypic variance: 6,194.76.

B. Broodstock selected per generation: male 25; female 50.

C. Individual number selected each family (tagged individuals): male 25; female 25.

D. Male fraction selected: 0.02; female fraction selected: 0.04.

Prediction formula:

$$R = \frac{1}{2} i_m \sigma_A r_{APm} + \frac{1}{2} i_f \sigma_A r_{APf} \qquad \text{(Equation 3)}$$

i_m and i_f are selection intensities of male and female, respectively;

r_{APm} and r_{APf} are accuracies of breeding value prediction of male and female, respectively;

σ_A is genetic variation degree of body weight.

According to genetic evaluation of F1 breading families, predicted response to selection was as follows: male 46.998 g, female 41.964 g.

The predicted selection response was 88.962 g and the predicted genetic progress was 28.421% (Table 1-1-2).

Table 1-1-2 Selection response prediction

	Selection response prediction	Data
Parameters	Heritability of body weight	0.64 ± 0.07
	Common environmental coefficient	0.00016 ± 0.115
	Phenotypic variance	6,194.76
	Male broodstock selected per generation	25
	Female broodstock selected per generation	50
	Male individuals selected each family	25
	Female individuals selected each family	25
	Male fraction selected	0.02
	Female fraction selected	0.04

(to be continued)

Conclusions	Selection response prediction	Data
	Selection response of male	46.998
	Selection response of female	41.964
	Total selection response	88.962
	Predicted genetic progress	28.42137%

5.3 Mating design of F2 families and construction

The F2 breeding program was developed using the individual breeding values and inbreeding coefficients of the F1 animals based on linear programming principle (Fig. 1-1-15).

♂		♀
20037862999	×	20037862945
20037862905	×	20037862948
20037862948	×	20037862903
20037862947	×	20037867181
20037862905	×	20037867114
20037862940	×	20037862967
20037862974	×	20037862947
20037862974	×	20037862964
20037862916	×	20037862919
20037862940	×	20037862910

Fig. 1-1-15 Mating design of F2 families (PIT number of fish)

Selection line including 57 full-sib families was created with high breeding value and control line including 10 full-sib families was created with average breeding values.

5.4 Response to selection, realized heritability and genetic gain of F2

In order to evaluate the early breeding effect of F2, three methods were used to evaluate response to selection, realized heritability and genetic gain of live weight at 6 months.

A. Comparing the least mean squares of the second-generation of selected line and control line.

B. Comparing the estimated breeding values of live weight between the first- and the second-generation selected line.

C. Comparing the estimated breeding values of the second-generation of selection line and control line, respectively.

Responses to selection, realized heritability and genetic gain estimated by three methods were different and their mean values were 3.1983 ± 0.5880, 0.2941 ± 0.0531 and 8.70 ± 1.60, respectively, which revealed that response to selection, moderate realized heritability and lower genetic gain were obtained (Table 1-1-3, Table 1-1-4).

Table 1-1-3 Variance components, heritability and common environmental effects of live weight at 6 months of *S. maximus*

Parameters	REML estimate
Additive genetic variance	70.6176
Common environment variance	9.0965
Phenotypic variance	118.13692
Heritability (mean ± S.E.)	0.5978 ± 0.1550
Common environment coefficient (mean ± S.E.)	0.0770 ± 0.7020

Table 1-1-4 Selection response, realized heritability and genetic gain at 6 months of age

Method	Selection response	Realized heritability	Genetic gain/%
Comparing the least mean squares for the selection and control lines of F2	2.6505 ± 0.3154	0.3471	7.21
Comparing the estimated breeding values of live weight between F1 families and F2 Selected line	3.1248 ± 0.3718	0.2944	8.50
Comparing the estimated breeding values of the selection and control lines of F2	3.8196 ± 0.4545	0.2409	10.39
Means	3.1983 ± 0.5880	0.2941 ± 0.0531	8.70 ± 1.60

Family selection is an effective approach to improve growth traits of turbot. At the same time, the population still had additive genetic variance to enable further improvement.

5.5 Developmental quantitative genetic analysis of growth traits

In order to elucidate the genetic mechanism of growth traits in turbot during ontogeny, developmental genetic analysis of the body weight, total length, standard length and body height of turbots was conducted.

During the breeding process, determining the optimum selection period is an important step to ensure successful selection.

5.5.1 Unconditional genetic model: static model

$$Y_{ij(t)} = u_{(t)} + A_{i(t)} + A_{j(t)} + D_{ij(t)} + e_{ij(t)} \qquad \text{(Equation 4)}$$

$Y_{ij(t)}$ is the unconditional phenotypic value;

$u_{(t)}$ is the mean unconditional population at time t;

$A_{i(t)}$ and $A_{j(t)}$ are the unconditional additive effects;

$D_{ij(t)}$ is the unconditional dominance effect;

$e_{ij(t)}$ is the unconditional residual error.

5.5.2 Conditional genetic model: dynamic model

$$Y_{ij(t|t-1)} = u_{(t|t-1)} + A_{i(t|t-1)} + A_{j(t|t-1)} + D_{ij(t|t-1)} + e_{ij(t|t-1)} \quad \text{(Equation 5)}$$

$Y_{ij(t|t-1)}$ is the conditional phenotypic value;

$u_{(t|t-1)}$ is the mean conditional population at time t;

$A_{i(t|t-1)}$ and $A_{j(t|t-1)}$ are the conditional additive effects;

$D_{ij(t|t-1)}$ is the conditional dominance effect;

$e_{ij(t|t-1)}$ is the conditional residual error.

5.5.3 Genetic dynamic analysis of body weight and total length

Genetic dynamic analysis of body weight and total length is shown as Figs. 1-1-16 to 1-1-19.

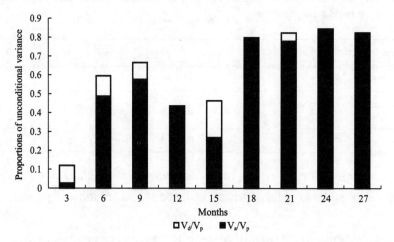

Fig. 1-1-16 Proportions of unconditional variance components for body weight

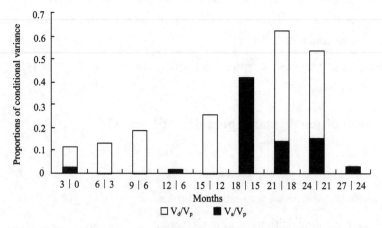

Fig. 1-1-17 Proportions of conditional variance components for body weight

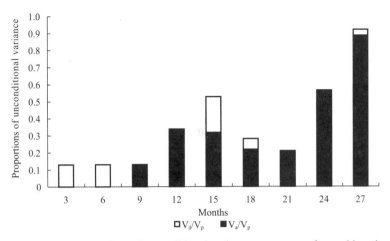

Fig. 1-1-18 Proportions of unconditional variance components for total lengths

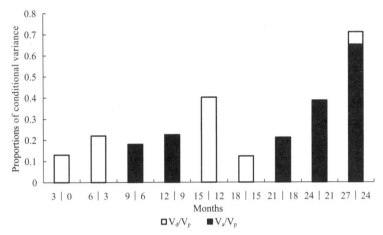

Fig. 1-1-19 Proportions of conditional variance components for total length

5.5.4 Genetic dynamic analysis of standard length and body height

Genetic dynamic analysis of standard length and body height is shown as Figs. 1-1-20 to 1-1-23.

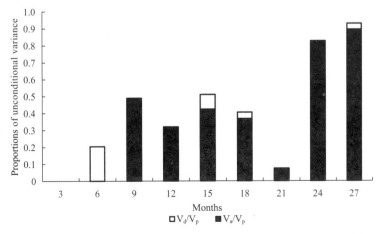

Fig. 1-1-20 Proportions of unconditional variance components for standard length

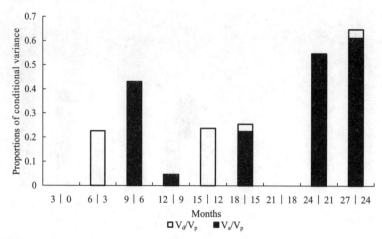

Fig. 1-1-21   Proportions of conditional variance components for standard length

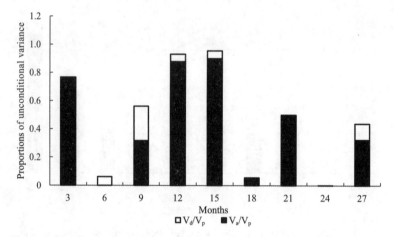

Fig. 1-1-22   Proportions of unconditional variance components for body height

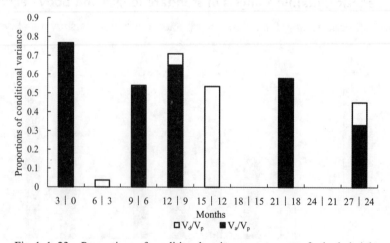

Fig. 1-1-23   Proportions of conditional variance components for body height

5.5.5   Conclusions of genetic dynamic analysis

The unconditional additive effects for four traits were more significant than unconditional dominance effects, meanwhile, the alternative expressions were also observed between the additive and dominant effects for body weight, total length and standard length (Fig. 1-1-16, Fig. 1-1-18, Fig. 1-1-20, Fig. 1-1-22).

Conditional analysis showed the developmental periods with active gene expression for body weight (15 months of age-18 months of age), total length (15 months of age and 12-24 months of age), standard length (15 months of age and 24 months of age) and body height (21 months of age and 27 months of age), respectively (Fig. 1-1-17, Fig. 1-1-19, Fig. 1-1-21, Fig. 1-1-23).

The accumulative effects of genes controlling the four quantitative traits were mainly additive effects, suggesting that the selection was more efficient for the genetic improvement of turbots.

The optimum selection periods of body weight (15-18 months of age), total length (15 months of age and 12-24 months of age), standard length (15 months of age and 24 months of age) and body height (21-27 months of age) respectively, were constructed based on the developmental periods of active gene expression for four traits.

5.6 Analysis on genotype-environment interaction

Quantitative traits include interaction effects between genotype and environment (G×E), which can change the relative rank or difference among under different environmental conditions.

Analysis on genotype-environment interaction is of great significance for identification of variety, extension of varieties, geographic division of crop breeding and formulating of breeding goal.

To breed superior families with universality and excellent turbot families suitable for specific environments, ten families from turbot selective breeding F2 were dispatched in randomized block design to five farms in different locations. AMMI model was used to analyze the data collected.

5.6.1 Model expression

$$y_{ge} = \mu + \alpha_g + \beta_e + \sum_{i=1}^{N} \lambda_n r_{gn} \delta_{gn} + \theta_{ge} \qquad \text{(Equation 6)}$$

y_{ge} is the body weight of genotype g in environment e;

μ is the mean body weight;

α_g is the deviation of genotype;

β_e is the environment main effect as deviation;

λ_n is the singular value for the IPC axis n;

r_{gn} and δ_{gn} are the genotype and environment IPC scores;

θ_{ge} is error.

5.6.2 Stability parameters

$$D_{g(e)} = \sqrt{\sum_{i=1}^{N} IPC\, A_{g\,(e)k}^{2}} \quad (k=1, 2, \ldots, m) \qquad \text{(Equation 7)}$$

$D_{g(e)}$ is AMMI stability value.

5.6.3 Variance and AMMI analysis on body weight

Variance analysis showed that sum of square between test locations, sum of square of interaction effects between families and sites, and sum of square of family were 76.48%, 10.82% and 7.50% of the total sum of square, respectively. The environment had a great influence on the growth of turbot. The next was the interaction between the family and the environment. The two effects were far greater than the genetic differences among families.

AMMI analysis showed that the first three components IPCA1, IPCA2 and IPCA3 account for 76.74%, 16.59% and 6.63% of interactions, respectively, with a cumulative explanation of 99.96%, which indicated that the AMMI model could analyze the interaction between genotype and environment more effectively (Table 1-1-5).

Table 1-1-5 Results of variance and AMMI analysis on body weight

Source of variation	df (Degree freedom)	SS (Sum of squares)	MS (Mean square)	F (F-statistics)	Proportion of total sum of square/%
Total variation	258	140,028,128.50	542,744.70		
Genotypes	10	10,507,454.39	1,050,745	60.58**	7.50
Environments	4	107,093,597.40	267,733,99	1,543.68**	76.48
G×E interaction	37	15,152,650.81	409,531.10	23.61**	10.82
IPCA1	13	11,628,950.55	894,534.70	51.57**	76.74
IPCA2	11	2,513,921.68	228,538.30	13.17**	16.59
IPCA3	9	1,005,538.45	111,726.50	6.44**	6.63
Residual	6	104,062.86	17,343.81		
Error	207	134,294,999.40	648,768.10		

Note: **, it is significant at 0.01 level.

5.6.4 Stability of families and discrimination of environment

In AMMI biplot, the closer the distance from the origin of coordinates is, the more stable the family and the worse the environmental discrimination is.

Among the ten families, the best family in terms of growth stability was G5 (6# family), followed by G8 (18# family) and G7 (26# family).

Among the five test locations, environmental discrimination of Zhifu, Yantai, Shandong is far greater than the other four test locations (Fig. 1-1-24).

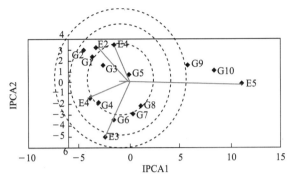

G1-G10, the serial number of families; E1-E5, the serial number of test locations.
Fig. 1-1-24　AMMI$_2$ biplots (IPCA1, IPCA2)

5.6.5　Relationship between family and environment

Among the ten families, families that adapt to specific environments need to be studied using AMMI biplot.

Five test locations were divided into three environment groups, *i.e.*, Group E3, Group E5 and Group E1-E2-E4.

Families G6, G10 and G2 should be dispatched to the environment Groups E3, E5, E1-E2-E4, respectively, based on "Which-won-where" view of the AMMI biplot (Fig. 1-1-25).

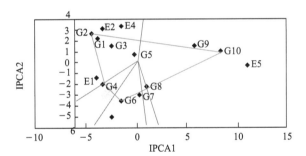

G1-G10, the serial number of families; E1-E5, the serial number of test locations.
Fig. 1-1-25　"Which-won-where" view of the AMMI$_3$ biplot

5.6.6　High yield and stability of families

A. Family G5 was a family with both high yield and stability.

B. The yield of family G6 was 851.768 g. The family leaded to the highest yield under environmental E3 condition, and was suitable for popularizing in E3.

C. The yield of family G10 was 911.664 g. The family leaded to the highest yield under environmental E5 condition, and was suitable for popularizing in E5.

D. The yield of family G2 was 784.764 g. The family leaded to the highest yield under environmental E1-E2-E4 condition, and is suitable for popularizing in E1-E2-E4 (Fig. 1-1-26).

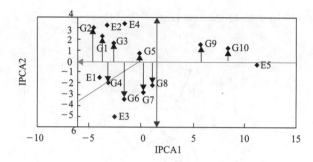

G1-G10, the serial number of families; E1-E5, the serial number of test locations.

Fig. 1-1-26 High yield and stability of families

6 Selection of fast-growing strain

6.1 Evaluation of selection progress

For F2 and F2 breeding families, body weight of each family at 3, 6, 9 and 18 months of age was measured. The first sixteen families were selected as excellent families of G1 and G2 based on average breeding values of families.

Selection progresses of F2 body weight were 69.47%, 18.12%, 31.43% and 29.76%, respectively, at 3, 6, 9 and 18 months of age. Mean selection progress of four stages was 37.20% (Table 1-1-6).

Table 1-1-6 Selection progresses of four different growing stages

Months of age	G1 body weight/g	G2 body weight/g	Selection progress/%
3	1.43	2.42	69.47
6	36.76	43.42	18.12
9	108.30	142.35	31.43

6.2 Selection of fast-growing strain

The first six families 14 ②, 14 ①, 13 ②, 2 ②, 12 ① and 1 ①, were selected based on average breeding values of F2 families (18 months of age).

The mean breeding values of the six families were shown in Table 1-1-7. The six families were selected and designated as fast-growing strain of turbot.

Table 1-1-7 The first six families selected based on average breeding values of F2 families

Families	Rank	Mean breeding values/g	Mean body weight/g
14 ② (F♂×F♀)	1	145.29	926.22
14 ① (F♂×F♀)	2	118.71	1,001.06
13 ② (E♂×E♀)	3	98.46	860.38

(to be continued)

Families	Rank	Mean breeding values/g	Mean body weight/g
2 ② (F♂×E♀)	4	74.23	764.02
12 ① (N♂×D♀)	5	65.48	749.10
1 ① (E♂×F♀)	6	63.94	762.22

6.3 Comparison of growth performance of fast-growing strain, common cultured strain and control line

Mean body weight of fast-growing strain, common cultured strain and control line were 843.8333 g, 676.0127 g and 711.6752 g respectively at 18 months of age (Fig. 1-1-27).

Mean body weight of fast-growing strain was 24.82% higher than that of common cultured strain and 18.57% higher than control line.

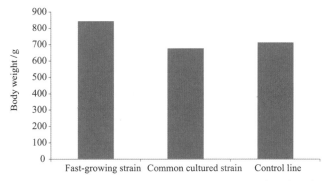

Fig. 1-1-27 Growth performance of fast-growing strain, common cultured strain and control line

6.4 Morphological characters of fast-growing strain

The dynamics of changes in body shape of fast-growing and slow-growing strains of turbot, and the differences in body shape between the two strains, were evaluated. The ratios of total length to body length (TL/BL), body width to body length (BW/BL) and total length to body width (TL/BW) were used as morphometric indices (Fig. 1-1-28).

The fast-growing and slow-growing strains were selected according to mean values of body weight of different families.

The two strains exhibited different temporal trends in TL/BL but similar trends in BW/BL and TL/BW (Figs. 1-1-29 to 1-1-31).

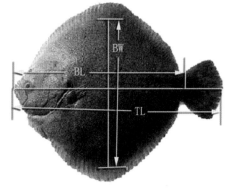

BL, Body length; TL, total length; BW, body width.

Fig. 1-1-28 Morphometric characteristics measured in turbot

Generally, BE/BL of the two strains increased with time while total length/body width decreased. Thus, the bodies of both fast-growing and slow-growing strains of turbot changed from a narrow to a rounded shape.

The ratio of total length to body length was generally lower, the ratio of body width to body length was higher, and the ratio of total length to body width was consistently lower in the fast-growing strain than in the slow-growing strain.

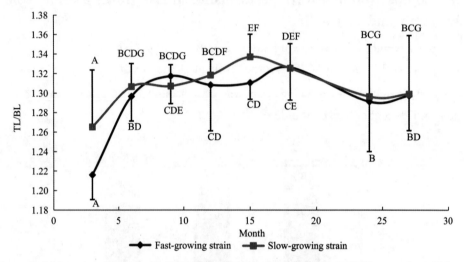

For each shrain, valves labeled with different capital letters are significantly different at the 0.05 level.

Fig. 1-1-29　The time course of TL/BL in the fast- and slow-growing strains

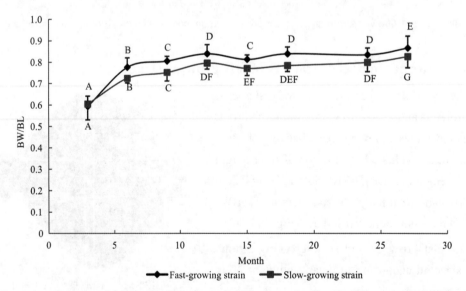

For each strain, valves labeled with different capital letters are significantly different at the 0.05 level.

Fig. 1-1-30　The time course of BW/BL in the fast- and slow-growing strains

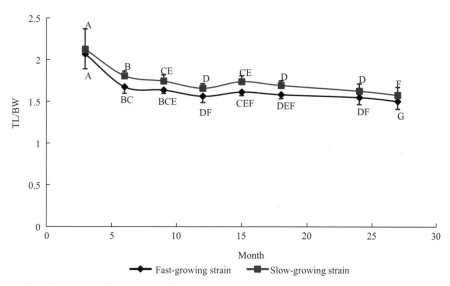

For each strain, valves labeled with different capital letters are significantly different at the 0.05 level.

Fig. 1-1-31　The time course of TL/BW in the fast- and slow-growing strains

Correlation analysis of the three shape ratios showed that TL/BL and TL/BW were unsuitable, and BW/BL was suitable.

The conclusions indicated that BW/BL could be used as a phenotypic marker for selective breeding of turbot (Table 1-1-8).

BW/BL showed an approximately parallel pattern on the same time series, and the correlation between BW/BL and body weight was significant, which indicated that early selection could be achieved with the help of "BW/BL" in turbot breeding.

Table 1-1-8　The correlation coefficients among body weight and shape ratio of turbot in different month

Months of age	TL/BL	BW/BL	TL/BW
3	−0.436 ± 0.000**	−0.061 ± 0.511	−0.144 ± 0.117
6	−0.232 ± 0.011*	0.555 ± 0.000**	−0.672 ± 0.000**
9	0.180 ± 0.050*	0.645 ± 0.000**	−0.670 ± 0.000
12	−0.316 ± 0.000**	0.387 ± 0.000**	−0.546 ± 0.000**
15	−0.572 ± 0.000**	0.592 ± 0.000**	−0.731 ± 0.000**
18	0.030 ± 0.767	0.652 ± 0.000**	−0.692 ± 0.000**
24	−0.042 ± 0.674	0.317 ± 0.000**	−0.308 ± 0.000**
27	−0.058 ± 0.531	0.317 ± 0.000**	−0.355 ± 0.000**

Notes: Data in the table were from the correlation analysis and tested by two-tailed t-test. * means significant difference ($P < 0.05$), and ** means very significant difference ($P < 0.01$).

References

Lei J L, Liu X F. Primary study of turbot introduction in China [J]. Modern Fisheries Information, 1995, 10(11): 1-3.

Lei J L, Liu X F, Guan C T. Turbot culture in China for two decades: achievements and prospect [J]. Progress in Fishery Sciences, 2012, 33(4): 123-130.

Ma A J, Wang X A, Xue B G, et al. Investigation on family construction and rearing techniques for turbot (*Scophthalmus maximus* L.) family selection [J]. Oceanologia et Limnologia Sinica, 2010, 41(3): 301-306.

Ma A J, Wang X A. Development of turbot *Scophthalmus maximus* seed industry and application of advanced technology [J]. Oceanologia et Limnologia Sinica, 2015, 46(6): 1461-1466.

Progeny production and culture technology of marine fish

Author: Xu Yongjiang
E-mail: xuyi@ysfri.ac.cn

1 Overview of marine fish culture in China

China is one of the first countries in the world to culture marine fish. The modern China marine fish culture history could be divided into four relatively obvious stages since the 1950s (Fig. 1-2-1).

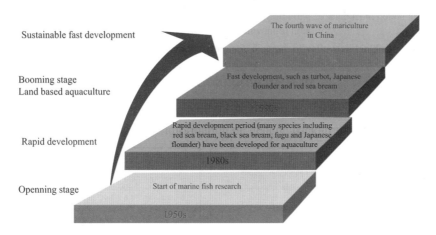

Fig. 1-2-1 The history of marine fish aquaculture in China

In 2018, the total mariculture yield in China was 20.31 million tons, wherein marine fish yield was 1.49 million tons (Fig. 1-2-2), accounting for 7.3%.

1.1 Pond culture

In the 1960s, the prevalent pattern was earthen pond culture in China, which is rough with low efficiency. Since the 1980s, more modern equipments have been employed, and the pond culture has become more profitable and efficient. More species were developed and researchers paid more attention on mix culture of fish and other species (Fig. 1-2-3).

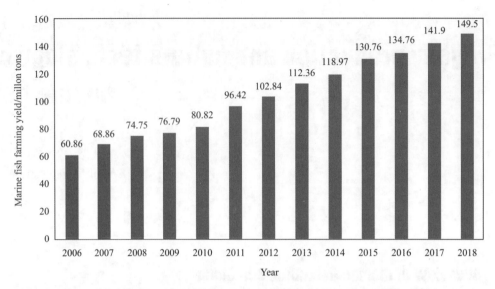

Fig. 1-2-2　The annual yield of farmed marine fish in China from 2006 to 2018

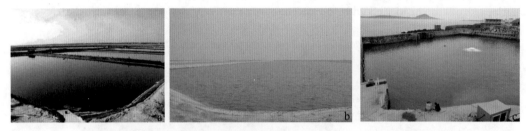

a. Intensive culture pond; b. Single large culture pond; c. Engineering culture pond.
Fig. 1-2-3　Mix culture

1.2　Recirculating aquaculture system (RAS) pond culture system

In RAS pond culture system, the traditional large pond is divided into small ponds of 0.2–0.4 hm², and then reunion in a parallel combination pattern. The separate inflow and effluent system was constructed. The culture water was collected in a treatment pond and treated with plant and probiotics, and then was reused in the culture system (Fig. 1-2-4).

1.3　Cage culture

Since the 1980s, inshore cage culture of marine fish has developed quickly. In the 1990s, succeeding in scaled production of juveniles of red seabream accelerated the cage culture industry. Now, there are about 1 million cages in China, and about 30 species are cultured. The offshore deep sea cage culture has been developing since 2005, and becomes important platforms for open ocean aquaculture (Fig. 1-2-5) (Guan, 2017).

1.4　RAS culture

China began to develop the RAS for marine fish culture in the the 1990s. RAS culture technology experienced sustainable development (Fig. 1-2-6), and now this stable, cost-efficient and environment-friendly mode has been widely accepted by government and farmers in China.

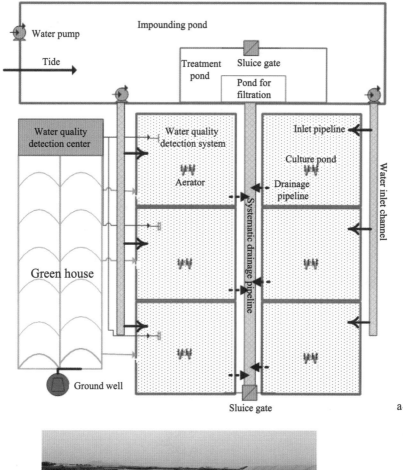

a. Schematic diagram of recirculating aquaculture pond system; b. Recirculating aquaculture pond system for marine fish.

Fig. 1-2-4 RAS pond culture system

1.5 Open ocean aquaculture

Open ocean aquaculture is the most promising way to accelerate large-scale food and protein production for human beings. It is developing fast now in China.

The safe gears and platforms and suitable species are key to successful open ocean aquaculture. China has built several heavy gears including the "Ocean Farm 1" platform, large cages, aquaculture craft, etc. for open ocean aquaculture (Fig. 1-2-7).

a. Quadrate metal culture cage; b. Circular HDPE culture cage.

Fig. 1-2-5　Cage culture modes

a. Sand filter tank; b. Schematic diagram of RAS; c. Protein separator; d. Computer monitor system;
e. Ultraviolet sterilizer; f. Microfiltration machine; g. Ozone generator.

Fig. 1-2-6　RAS culture equipments

a. Ocean recreational fishery platform; b. SalMar's "Ocean Farm 1" platform; c. Aquaculture vessel.

Fig. 1-2-7　Open ocean aquaculture gears

1.6 Sea ranching

Many countries are developing the sea ranching to conserve and maintain inshore marine fishery resources and environment (Fig. 1-2-8).

A. Five types: resources enhancement type, ecological restoration type, recreation type and sightseeing type, and germplasm protection type.

B. Key technologies: bioenvironmental construction, breeding and accommodation of target species, management strategy and technological development.

a. Schematic diagram of sea ranching; b. Typical sea ranching in China.

Fig. 1-2-8 Sea ranching

2 Introduction of several marine fishes in China

2.1 Turbot (*Scophthalmus maximus*)

Yellow Sea Fisheries Research Institute introduced turbot (Fig. 1-2-9) in 1992, and got success in artificial and massive spawning control and seedlings production in 2001.

The stable seedlings production and grow-out technology have been developed. Turbot supports the flatfish farming industry in northern China.

Fig. 1-2-9 Turbot (*S. maximus*)

Now, the large-scaled farming industry could annually produce about 70,000 t of turbot of market size in China, which is No.1 in flatfish culture yield in China and No.1 in the world.

Genetic breeding, market strategy, resource management, etc. have aroused the attention of the government and scientists, in order to maintain the sustainable development of farming industry (Daniels and Watanable, 2010).

2.2 Tongue sole (*Cynoglossus semilaevis*)

The interest to develop tongue sole for aquaculture has been raised since the 1980s.

In 2003, key breakthrough has been made in spawning of tongue sole broodstocks and seedlings production technologies, and the farming industry has developed quickly (Fig. 1-2-10).

Now, the annual aquaculture yield of tongue sole is about 5 000 t.

Tongue sole is ranked the three flatfish species supporting flatfish farming industry in northern China (Liu and Zhuang, 2014).

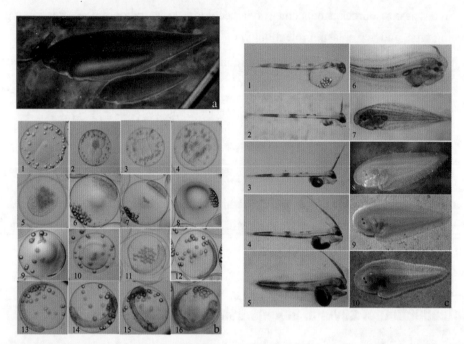

a. Tongue sole female fish (above), male fish (below). b. Embryonic development of tongue sole. 1. 2-cell stage; 2. 4-cell stage; 3. 8-cell stage; 4. 16-cell stage; 5. Multicellular stage; 6. Morula stage; 7. High blastula stage; 8. Low blastula stage; 9. Early gastrula stage; 10. Mid-gastrula stage; 11. Telo-phase of gastrula stage; 12. Neurula stage; 13. Embryo encircling 1/2 of yolk sac; 14. Embryo encircling 2/3 of yolk sac; 15. Embryo encircling 3/4 of yolk sac; 16. Hatching stage. c. Larval and juvenile development of tongue sole. 1. Newly hatched larvae; 2. Larvae of 4 DAH (day after hatching); 3. Larvae of 6 DAH; 4. Larvae of 16 DAH; 5. Larvae of 25 DAH; 6. Juvenile of 30 DAH; 7. Juvenile of 50 DAH; 8. Juvenile of 60 DAH; 9. God juvenile; 10. Young fish of 100 DAH; DAH, day after hatching.

Fig. 1-2-10　The early life cycle of tongue sole (*C. semilaevis*)

2.3　Japanese flounder (*Paralichthys olivaceus*)

Japanese flounder is a member of Family Bothidae. It has oval body shape, eyes on left side of head (Fig. 1-2-11). It is endemic in northeast China. In China, most capture fishery production of it occurs in the Yellow Sea and the Bohai Sea.

Fig. 1-2-11　Japanese flounder (*P. olivaceus*)

It is a cold-water benthic fish, and the suitable water temperature range is 15-25 ℃. It is euryhaline, and can live in high salinity areas offshore, as well as in the low salinity estuarine areas, and has a strong ability to tolerate hypoxia, with oxygen concentrations of 0.6-0.8 mg/L being lethal. The growth rate of female is significantly higher than that of male.

It takes two and three years for male and female to reach sexual maturity respectively. The spawning season under natural conditions is from April to June. It is a multiple spawner.

2.4 Groupers

Sevenband grouper (*Epinephelus septemfasciatus*), arrowtooth eel (*Epinehelus moara*) and *Epinephelus* hybrids (blotchy rock cod, *Epinephelus fuscoguttatus*♀× giant grouper, *Epinephelus lanceolatus* ♂) (Fig. 1-2-12) are new candidates for farming industry in China recently.

Market consumption demand is high, and farming industry has developed quickly. Millions of juveniles can be produced annually in land-based green house in northern China, and the main culture modes include land-based RAS culture, pond culture and cage culture (Meng, 1995).

a. Blotchy rock cod, *E. fuscoguttatus*; b. Hybrid from blotchy rock cod, *E. fuscoguttatus*♀ × giant grouper, *E. lanceolatus*♂; c. Sevenband grouper, *E. septemfasciatus*.

Fig. 1-2-12　Three kinds of *Epinephelus*

2.5　Large yellow croaker (*Larimichthys crocea*)

Large yellow croaker is a species of marine fish in the croaker family (Family Sciaenidae, Genus *Larimichthys*) native to the northwestern Pacific, generally in temperate waters (Fig. 1-2-13).

It lives in coastal waters and estuaries, and is found on muddy-sandy bottoms. It feeds on crustaceans and fishes. Body elongates and is compressed. Body colour is yellow, darker above and golden sheen below. Common size is 60 cm.

Its population collapsed in the 1970s due to overfishing. This species is now widely cultured in China, and the production has grown to about 200,000 t in 2018.

Fig. 1-2-13 Large yellow croaker (*L. crocea*)

2.6 Japanese sea bass (*Lateolabrax japonicas*)

L. japonicas belongs to Family Serranidae, Genus *Lateolabrax*. It is characterized by many black dots on the lateral body region (Fig. 1-2-14). It is a species of Asian sea bass native to the Western Pacific Ocean, where it occurs from the Sea of Japan to the South China Sea.

This species is important commercially, popular as a game fish, and farmed in China. *L. japonicas* inhabits fresh, brackish and marine waters of inshore rocky reefs and in estuaries at depths of at least 5 m.

This species is catadromous, with the young ascending rivers and then returning to the sea to breed. The culture modes in China include pond culture, cage culture and land-based industrial culture.

Fig. 1-2-14 Japanese sea bass (*L. japonicas*)

2.7 Golden pompano (*Trachinotus ovatus*)

T. ovatus belongs to Family Carangidae, Genus *Trachinotus*. It has large, strong fins and is common in the Mediterranean Sea and in the Atlantic Ocean from the Bay of Biscay to Guinea (Fig. 1-2-15).

Its coloration is back greenish-grey, sides silvery with 3-5 vertically elongate black spots on anterior half of lateral line, dorsal-, anal- and caudal-fin lobes black-tipped.

Adults feed on small crustaceans, mollusks and fishes. Eggs are pelagic.

It is widely farmed in China, and the main culture mode is cage. Its annual yield is about 100,000 t.

Fig. 1-2-15   Golden pompano (*T. ovatus*)

2.8   Tiger puffer (*Takifugu rubripes*)

It is a pufferfish that belongs to the Family Tetraodontidae, Genus *Takifugu*. It has a lovely oval body shape. There are about 18 species in genus *Takifugu* in China coastal waters. China has a long history for consuming tiger puffer for delicious dishes (Fig. 1-2-16).

It is distributed from the Sea of Japan, the East China Sea and the North Yellow Sea to Southern Sakhalin, at depths of 10-135 m. It is a demersal species.

Tiger puffer is classified as near threatened species by the IUCN. The current catches are small. Gear restrictions and adjustments of fishery seasons to protect juveniles have been recommended to aid recovery. The species is extensively raised in aquaculture in China.

Fig. 1-2-16   Tiger puffer (*T. rubripes*)

2.9   Yellowtail kingfish (*Seriola lalandi*)

Seriola l. is one of the largest members of the Family Carangidae. It is globally distributed, which is characterized as the big body size, a clear golden yellow longitudinal band from the snout to the tail, and the yellow color of pelvic fin, buttocks and caudal fin edge (Fig. 1-2-17).

It is a pelagic temperate teleost, and its fresh muscle has high content of EPA and DHA, which makes it rival tuna and salmon. It is particularly suitable for sashimi, with high economic value, and high domestic and international consumption market demand.

Its suitable temperature range for growth is 18-29 ℃, and the suitable salinity ranges from 25 to 34.

We have developed its culture technology since 2013, and got breakthrough in seedlings production in 2017. Farming industry has developed in process.

Fig. 1-2-17 Yellowtail kingfish (*S. lalandi*)

2.10 Humphead wrasse (*Cheilinus undulates*)

C. undulates is the largest member of Genus *Cheilinus*, which can be easily identified by its large size, thick lips, two black lines behind eyes and the hump that appears on the forehead of larger adults. It is located within the east coast of the Africa and the Red Sea, as well as in the Indian Ocean to the Pacific Ocean (Fig. 1-2-18).

The males are typically larger than the females. It is long-lived, but has a very slow breeding rate. Individuals become sexually mature over six years. They are protogynous hermaphrodites, with some members of the population becoming male at about 9 years old. In 1996, the humphead wrasse was listed as a vulnerable species on the IUCN Red List.

Since 2005, China has implemented studies to protect this species by artificially domesticating the broodstocks and controlling its reproductive cycles. The natural spawning and early seedlings have been obtained.

Fig. 1-2-18 Humphead wrasse (*C. undulates*)

2.11 Rock bream (*Oplegnathus fasciatus*)

Rock bream is a coral fish species distributed in China coastal waters, especially in the Yellow Sea (Fig. 1-2-19). It has beautiful shape and external morphology, and is also an ornamental fish and delicious fish in China.

Since 2004, the massive production of juveniles of rock bream has been achieved. The farming industry has developed in coastal areas in China.

Fig. 1-2-19 Rock bream (*O. fasciatus*)

3 Progeny production technology of marine fish in China

3.1 Broodstock breeding

3.1.1 Broodstock breeding facilities

The hatchery site should be located at coastal areas. Use indoor flow-through seawater systems to hold broodstock. Tanks are constructed of concrete and range from 10 to 100 m^3, and 0.6-1.2 m in depth. They can be circular, square or octagonal in shape. With the inlet water pipes on the top of the tank wall and drainpipes on the center of tank bottom, a circular water flow in the tank is formed easily to flush out feces and uneaten feed. Water level in the tank is controlled with an outside standpipe.

3.1.2 Broodstock source

Broodstocks are ususlly from wild population or artificial cultured population.

3.1.3 Broodstock selection criteria

Normal body shape and coloration; intact body surface; healthy and no-injured; active in feeding and swimming. As for grouper, the potential virus should be detected. It should reach the biological minimum size, which usually means that the females should be over 3 years old, and the males should be over 2 years old.

3.1.4 Reproduction population construction

Genetic diversity should be investigated using molecular markers, and the individuals with high genetic diversity could be selected and mated for spawning, which is beneficial for the quality control of progenies.

3.1.5 Broodstock breeding density

Sex-matured broodstocks are assigned into the breeding tanks, and are held at 2 to 4 individuals per square meter at a male to female sex ratio between 1 : 1 and 1 : 2 with the body weight of 2-3 kg/m^2.

3.1.6 Environment conditions

Suitable culture conditions should be determined according to the ecological and reproductive characteristics of target fish species. Simulative natural environmental conditions (mainly temperature and photoperiod) have been proved to be effective in maturing broodstock and regulating spawning in many fish species under artificial conditions.

3.1.7 Nutrition regulation

Chopped trash fish or moist pellets are fed to broodstock, including sand eel, horse mackerel, redlip croaker, white croaker and Japanese Spanish mackerel. Additives such as vitamins and lecithin are added to the chopped fish with boiled starch whenever it is necessary. Moist pellets are manufactured on-site by mixing a powdered commercial premix with the trash fish prior to extrusion through feed processor. To develop formulated dry pellet is important. The feeding frequency is usually twice per day. The feeding rate is 3%–5% of the body weight at non-spawning season and 1%–3% at the spawning season.

3.1.8 Water renewal

Water renewal is maintained at a rate of three to five exchanges per day.

3.2 Reproductive cycle control and spawning

Natural spawning and arificial propagation are two main ways to obtain viable eggs from marine fish broodstocks. Nutriton enhancement is important for general maturation (Fig. 1-2-20) (Lei, 2005).

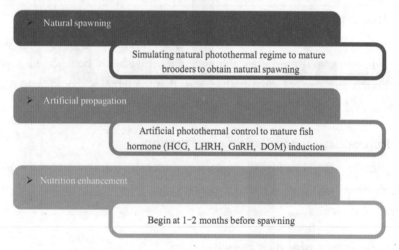

Fig. 1-2-20 Reproductive cycle control and spawning

3.2.1 Natural spawning

Before the spawning season, black plastic sheeting is used as a curtain to block outside light for the photothermal control of maturation. Four 60 W tungsten bulbs are located 1.0–1.5 m above the water surface to provide artificial light. Light intensity at the water surface is 200–600 lx and 0 lx during the light and dark phase, respectively (Fig. 1-2-21).

a. Photothermal control of fish broodstocks; b. Outlook of tank photothermal control of marine fish.

Fig. 1-2-21　Natural spawning

Boilers and chillers are used to maintain temperature within the optimum range for the broodstock. Measures such as mixing deep well water and natural seawater, shifting the timing of the breeding season are also adopted by some hatcheries to reduce energy cost.

Broodstock are kept in a quiet area insulated from the growout fish to minimize noise and the spread of infectious diseases. Rearing water temperature and photoperiod are manipulated to control the maturation and ovulation of the broodstock.

Many hatcheries will keep three to four batches of broodstocks under different photothermal protocols to obtain fertilized eggs year round.

3.2.2 Artificial propagation

Some fish species can not naturally spawn under artificial culture conditions. In order to supply fertilized eggs according to the timetable required by the customers, these hatcheries now tend to induce synchrony ovulation by hormones treatment to obtain viable eggs rather than natural ovulation.

Maturation and ovulation of the broodstock are monitored largely by observation of morphological changes of the gonad and the behavior of the broodstock. In turbot, abrupt bulging of the gonad and restless swimming of the broodstock signal the impending ovulation of the females. The appearance of milt-running male by hand stripping is also a strong indication that the breeding season has begun.

3.3　Eggs hatching

Most fish spawn buoyant eggs, whereas some spawn demersal eggs such as spotted halibut and Pacific cod. Eggs are usually incubated in a hatching system with water flowing through, or special hatching containers.

The stocking density of eggs is 10-20 eggs per liter. One day before hatching, buoyant eggs are disinfected with effective iodine concentration of 20-30 mg/L for 10 min or glutaraldehyde concentration of 200-400 mg/L for 2.5-10 min, then transferred into the rearing tank after volumetric enumeration (Fig. 1-2-22).

a. Fertilized eggs collection; b. Eggs hatching; c. Hatching system for marine fish eggs.

Fig. 1-2-22 Eggs hatching

3.4 Seedlings production technology

3.4.1 System design and requirements

Green water system is used for the larviculture of marine fish. Tank size varies among hatcheries, but is commonly 7-25 m³ in volume. Most larviculture tanks are circular, square or octagonal in shape, and constructed of concrete.

Sand-filtered seawater is adjusted to optimum temperature by boiler and cooler. Aeration is supplied by regenerative blowers to maintain dissolved oxygen and assist in water circulation. Additional rooms are also required for the culture of microalgae and rotifers, hatching of *Artemia* nauplii, and enrichment of live feed (Fig. 1-2-23).

a. Tank system for early larval culture of marine fish; b. Tank system for juvenile culture of marine fish.

Fig. 1-2-23 Tank system

3.4.2 Larval rearing

Fertilized eggs are stocked prior to hatching at densities of 15-30 eggs per liter in the tank. When eggs are stocked, the water temperature is the same as the incubation temperature.

Rearing tanks are illuminated by natural sunlight or artificial light for 14-16 h per day, and are left dark during the night. Light intensity at the water surface is 500-4,000 lx.

Water is not exchanged until 5 d post hatching. Exchange rates are increased gradually from 10% per day to 300% per day. Aeration is provided to mix gently the larvae and prey, and is adjusted according to the swimming ability of the larvae. From 10 d post hatching, the tank bottom is cleaned every 2–3 d by siphoning after stopping the aeration.

3.4.3 Water quality control

Natural sea water with a salinity of above 28 is preferred for the larval rearing. The dissolved oxygen (DO), pH, and ammonia nitrogen are monitored regularly twice a day. Pure oxygen and air stones generating micro-bubble are avoided in larval tank to prevent the occurrence of gas bubble disease, especially for first feeding larvae.

In practice, optimal water quality of pH range of 7.6 to 8.2 and ammonia below 0.02 mg/L could be maintained by a little adjustment of above water exchange protocol.

3.4.4 Food and feeding strategy

The feeding regime for larviculture is rotifer (*Brachionus plicatilis*), *Artemia* nauplii and micro-pellets feed.

Larvae are fed with rotifers as the initial live feed at a density of 5–10 individuals per milliliter, which happens normally 3 d after hatching. The feeding of rotifers lasts for 15–20 d. Microalgae are added to the rearing water at a concentration of 5×10^5 cells per milliliter to maintain green water environment.

Artemia nauplii are fed from 9 or 12 d post hatching to 40 d post hatching with an initial density of 0.1–0.2 individuals per milliliter then gradually increased to a final density of 0.5–1.0 individuals per milliliter. Micro-pellets are co-fed with *Artemia* from 20 d post hatching to 40 d post hatching; thereafter, juveniles are weaned to accept micro-pellets feed solely (Fig. 1-2-24).

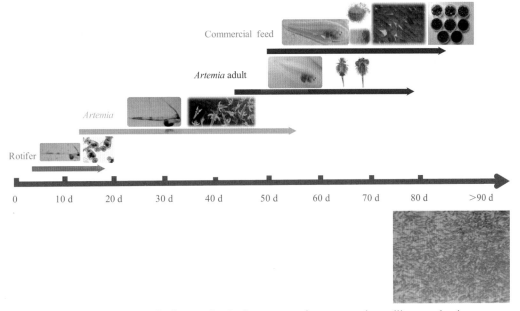

Fig. 1-2-24 Schematic diagram for feeding strategy for tongue sole seedlings production

3.4.5 Nursery culture

Juveniles can be transferred to nursery tanks when reach a total length of 3 cm. The juveniles are carefully collected with a small hand net, and transferred.

Size grading and removal of deformed or discolored juveniles are carried out during the transfer. Normal individuals are grouped into large, medium and small body sizes.

When the juveniles reach 5–6 cm in total length, they are selected and graded once more by body color, shape, activity and size, and then they can be sold. The seedlings production could attain one million in a single farm. Juveniles with abnormal body color, deformities, weak activity, lesion and symptom of diseases are rejected. Some hatcheries will volunteer to get the quarantine certificate issued by the state organization to guarantee the health condition of their juveniles.

4 Progress on yellowtail kingfish culture in China

Open ocean aquaculture development requirement: suitable fish species are urgently needed for deep sea net cage culture, etc.

Yellowtail kingfish is a globally distributed pelagic fish. Culture interest of which is raised for its high market value and tasty muscle in Japan, Australia, New Zealand, Chile, USA, China, etc.

Japanese farmers mainly rely on wild caught juveniles. Farmers in Australia and New Zealand produce juveniles from domestic broodstocks spawning.

Natural resources of yellowtail kingfish declined dramatically in China, and it is now a most promising candidate for culture and enhancement for its local availability and high market value.

4.1 Morphometric model

The measurement schematic diagram of yellowtail kingfish is determined (Fig. 1-2-25), to help identify ID and discriminate populations between China and international waters (Li et al., 2017).

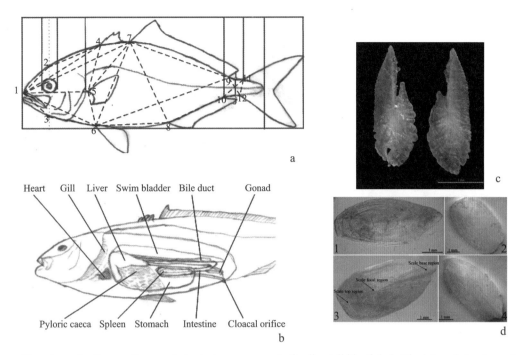

a. The schematic diagram for morphological measurement of yellowtail kingfish; b. The internal structure of yellowtail kingfish; c. The external morphology of otolith of yellowtail kingfish; d. The morphology of scales from different body areas of yellowtail kingfish (1. scale from head area; 2. scale from dorsal area; 3. scale from tail area; 4. scale from ventral area).

Fig. 1-2-25　The measurement schematic diagram of yellowtail kingfish

4.2　Karyotype and chromosome banding pattern

Karyotype and chromosome banding patterns were determined for yellowtail kingfish from China (Fig. 1-2-26), providing basic information on genetic background.

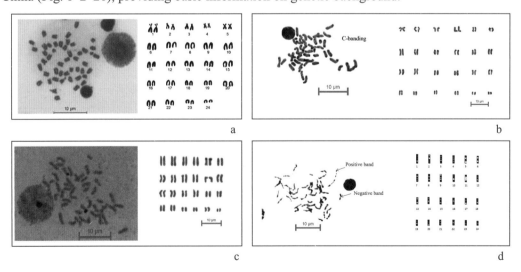

a. Karyotype of yellowtail kingfish; b. C-banding pattern of yellowtail kingfish; c. Ag-NORs pattern of yellowtail kingfish; d. G-banding pattern of yellowtail kingfish.

Fig. 1-2-26　Karyotype and chromosome banding pattern

4.3 Age determination

Methods for age determination by identifying vertebra and scales annual ring on which through staining have been constructed for yellowtail kingfish (Fig. 1-2-27).

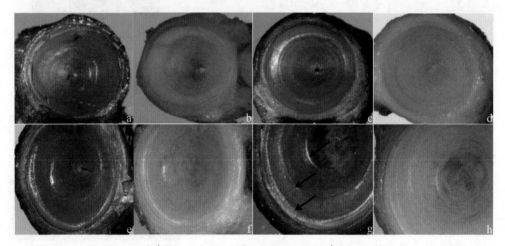

a, b. The second vertebrae of 0^+ fish; c, d. The second vertebrae of 1^+ fish; e, f. The second vertebrae of 2^+ fish; g, h. The second vertebrae of 3^+ fish.

Fig. 1-2-27　Third unstained and staining vertebrae with 0^+, 1^+, 2^+ and 3^+ age of yellowtail kingfish

4.4 Muscle nutrition evaluation

Yellowtail king fish is rich in crude protein, especially essential amino acids and delicious amino acids. Content of unsaturated fatty acids (DHA, EPA) is high. Its muscle nutrition quality is similar with salmon and tuna, and the flesh is excellent for sashimi (Tables 1-2-1 to 1-2-3).

Table 1-2-1　Muscle nutrition composition

Nutritional components	Farmed S. lalandi	Wild S. lalandi
Moisture/%	56.10	68.10
Crude protein/%	20.3	21.9
Crude fat/%	21.40	8.80
Ash/%	1.10	1.10
Total sugar/%	1.10	0.10
Energy/(kJ/g)	13.44	8.68
EP/%	66.21	39.64

Table 1-2-2　Muscle amino acid contents (%, wet weight)

Amino acids	Farmed S. lalandi	Wild S. lalandi
*Thr	0.78	0.87
*Val	0.96	1.07

(to be continued)

Amino acids	Farmed *S. lalandi*	Wild *S. lalandi*
*Met	0.60	0.66
*Ile	0.90	0.98
*Leu	1.65	1.82
*Phe	0.84	0.93
*Lys	1.79	1.96
#Asp	1.65	1.82
#Glu	1.33	1.47
#Gly	1.09	1.23
#Ala	1.68	1.89
Tyr	0.62	0.70
Pro	0.51	0.55
Ser	1.07	1.18
&His	1.18	1.28
&Arg	1.18	1.30
Essential amino acids (EAA)	7.52	8.29
Nonessential amino acids (NEAA)	7.95	8.84
Smi-essential amino acids (SEAA)	2.36	2.58
Delicious amino acids (DAA)	5.75	6.41
Total amino acids (TAA)	17.83	19.71
WEAA/WTAA/%	42.20	42.10
WEAA/WNEAA/%	94.60	93.80
WDAA/WTAA/%	32.20	32.50

Notes: *, indicate essential amino acids (EAA); &, indicate semi-essential amino acids (SEAA); #, indicate delicious amino acids (DAA); WEAA, whole essential amino acids; WTAA, whole total amino acids; WNEAA, whole non-essential amino acids; WDAA, whole delicious amino acids.

Table 1-2-3 Muscle fatty acid contents (%, wet weight)

Fatty acids	Farmed *S. lalandi*	Wild *S. lalandi*
*C14:0	2.88	2.51
*C15:0	0.42	0.5
*C16:0	22.25	21.23
*C17:0	0.43	0.84
*C18:0	4.82	7.69
&C16:1	6.58	5.52

(*to be continued*)

Fatty acids	Farmed *S. lalandi*	Wild *S. lalandi*
&C17:1	0.34	0.53
&C18:1	22.15	27.82
&C22:1n9	3.64	0.46
#C18:2n6	1.61	0.84
#C18:3n3	1.00	0.48
#C18:4n3	2.21	0.54
#C20:4n6	–	1.32
#C22:5n3	1.09	2.10
#EPA	6.60	4.10
#DHA	13.63	16.67
Saturated fatty acids (SFA)	30.8	32.77
Monounsaturated fatty acids (MUFA)	32.71	34.33
Polyunsaturated fatty acids (PUFA)	26.14	26.05
UFA/SFA	1.91	1.84

Notes: *, indicate essential amino acids (EAA); &, indicate semi-essential amino acids (SEAA); #, indicate delicious amino acids (DAA).

4.5 GH/IGF system in kingfish

Key genes including *GH*, *IGF-I*, *IGF-II* along GH axis were cloned and sequenced from yellowtail kingfish, and their differential expression patterns were determined in early life stages (Fig. 1-2-28).

4.6 Spawning manipulation

Yellowtail kingfish broodstocks (over 4 years old and 5 kg body weight) were exposed to artificial photothermal regime, and the natural spawning occurred in tanks after 70 d of regulation (Fig. 1-2-29).

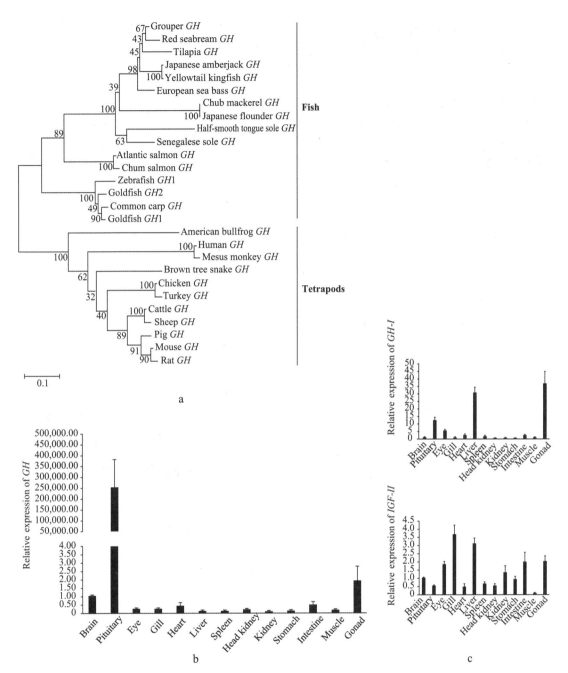

a. Phylogenetic analysis of IGF proteins in vertebrates. The phylogenetic tree was constructed by MEGA 6.06 using the neighbor-joining method with 1,000 bootstrap replicates. The number shown at each branch indicates the bootstrap value (%). b. Relative expression of *GH* mRNA in various tissues of yellowtail kingfish. c. Relative expression of *IGF-I* (1) and *IGF-II* (2) mRNAs in various tissues of yellowtail kingfish. Data were normalized to the abundance of 18S RNA expressed in the same tissue and presented as the mean ± SEM ($n = 3$).

Fig. 1-2-28　GH/IGF system in yellowtail kingfish

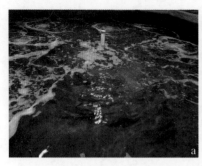

a. Yellowtail kingfish broodstocks under photothermal control; b. Brooder of yellowtail kingfish; c. The photothermal control equipment of yellowtail kingfish broodstocks.

Fig. 1-2-29　Spawning of yellowtail kingfish

4.7　Seedlings production technology

The morphometric characteristics of embryos, larvae, juveniles were studied, and the development and growth patterns during early life stage were clarified (Fig. 1-2-30).

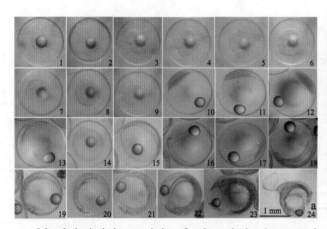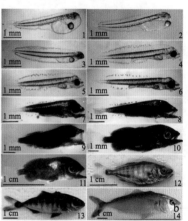

a. Morphological characteristics of embryonic development of yellowtail kingfish (1. unfertilized egg; 2. fertilized egg; 3. 2-cell stage; 4. 4-cell stage; 5. 8-cell stage; 6. 16-cell stage; 7. 32-cell stage; 8. 64-cell stage; 9. multicellular stage; 10. morula stage; 11. high blastula stage; 12. low blastula stage; 13. early gastrula stage; 14. early gastrula stage (show embryonic shield); 15. mid-gastrula stage; 16. ana-phase of gastrula stage; 17. telo-phase of gastrula stage; 18. neurula stage; 19. embryo encircling 1/2 of yolk sac; 20. embryo encircling 2/3 of yolk sac; 21. embryo encircling 4/5 of yolk sac; 22. embryo encircling the whole yolk sac; 23. hatching; 24. larva hatched out). b. Morphological characteristics of post-embryonic development of yellowtail kingfish (1. newly hatched larvae; 2. larvae of 1 DAH; 3. larvae of 2 DAH; 4. larvae of 3 DAH; 5. larvae of 5 DAH; 6. larvae of 7 DAH; 7. larvae of 10 DAH; 8. larvae of 15 DAH; 9. larvae of 20 DAH; 10. larvae of 25 DAH; 11. post-larvae of 35 DAH; 12. juvenile of 45 DAH; 13. juvenile of 60 DAH; 14. juvenile of 80 DAH).

Fig. 1-2-30　Morphology of early life stages of yellowtail kingfish

The suitable temperature, salinity, light regime and feed spectrum were established, and about 23,000 juveniles (total length 13-15 cm) were produced in 2017 (Fig. 1-2-31).

a. Indoor tanks for seedlings production of yellowtail kingfish; b. Size grading of yellowtail kingfish juveniles; c. Yellowtail kingfish juveniles.

Fig. 1-2-31 The production of juveniles in indoor tanks

4.8 Sea-land relay culture technology

"Cage culture + land greenhouse culture" has got fast growth (Fig. 1-2-32).

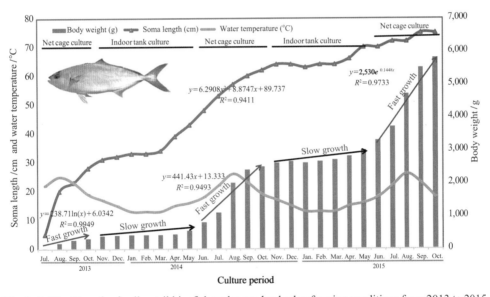

Fig. 1-2-32 Growth of yellowtail kingfish under sea-land relay farming conditions from 2013 to 2015

References

Daniels H V, Watanabe W O. Practical flatfish culture and stock enhancement [M]. Singapore: John Wiley & Sons Ltd, 2010: 185-204.

Guan C T. Annual report 2016 of China agriculture research system for flatfish culture industry [M]. Qingdao: China Ocean University Press, 2017: 1-450. (in Chinese)

Lei J L. Marine fish culture theory and techniques [M]. Beijing: China Agriculture Press, 2005: 1-1001. (in Chinese)

Li R, Xu Y J, Liu X Z, et al. Morphometric analysis and internal anatomy of yellowtail kingfish (*Seriola aureovittata*) [J]. Progress in Fishery Sciences, 2017, 38(1): 142-149. (in Chinese)

Liu X Z, Zhuang Z M. Reproductive biology and culture technology of tongue sole [M]. Beijing: China Agriculture Press, 2014: 1-394.

Meng Q W. Fish taxonomy [M]. Beijing: China Agriculture Press, 1995, 1-1158. (in Chinese)

Molecular sex control and genomics of Chinese tongue sole

Authors: Chen Songlin, Xu Wenteng

E-mail: chensl@ysfri.ac.cn

1 Background

1.1 Why do we choose Chinese tongue sole for investigation?

Chinese tongue sole (*Cynoglossus semilaevis*) is an economically important marine flatfish species in China, which is a kind of delicious sea food and shows big difference in growth rate between female and male (Figs. 1-3-1a, b).

Chinese tongue sole has a ZW sex determination system with a big W chromosome, and the ZW individual can be sex-reversed to neo-male/pseudomale (Zhou et al., 2005) (Fig. 1-3-1c, Fig. 1-3-2).

 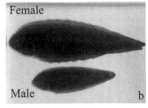

a. Chinese tongue sole is a kind of delicious seafood; b. Female and male show big difference in growth rate; c. The karyotype of Chinese tongue sole (ZW individual).

Fig. 1-3-1　Chinese tongue sole

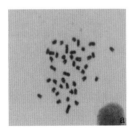 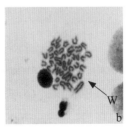

a. Male; b. Female (W chromosome).

Fig. 1-3-2　Karyotype of male and female Chinese tongue sole

1.2 The industrial challenge of Chinese tongue sole

Features or problems:

A. Why is male so small?

B. Does sex determining gene exist?

C. Why is male ratio up to 90%?

D. It is not only an economically important marine fish species, but also a promising model fish species for sex-specific gene screening and sex determination mechanism research.

2 Contents

2.1 Discovery of first generation female specific markers by AFLP markers (2005–2007)

Seven female specific AFLP markers were isolated from female genome (Figs. 1-3-3 to 1-3-4). Five female-specific AFLP markers were cloned and sequenced (Chen et al., 2007).

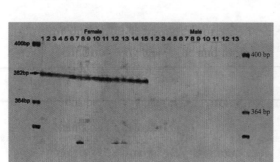

Fig. 1-3-3   Female-specific AFLP marker *CseF382*

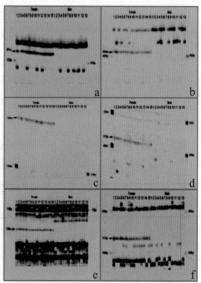

a. *CseF305*; b. *CseF464*; c. *CseF783*; d. *CseF575*; e. *CseF136*; f. *CseF618*.

Fig. 1-3-4   Six other female-specific AFLP markers

2.1.1 Female specific AFLP markers

One female specific AFLP marker, *CseF382*, was mapped on the W chromosome by FISH (Fig. 1-3-5a) and on LG5 of the 1st generation genetic linkage map of Chinese tongue sole (Fig. 1-3-5b).

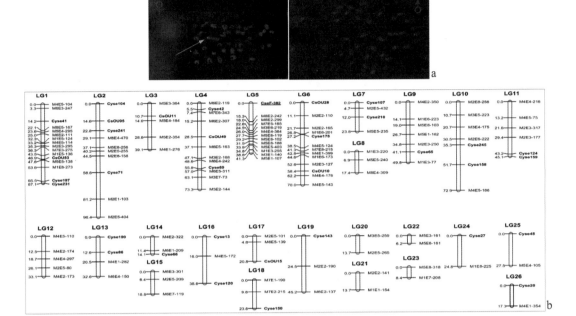

a. Mapping of *CseF382* on the W chromosome by FISH; b. Genetic linkage map of Chinese tongue sole.

Fig. 1-3-5　Identification of female specific marker

2.1.2　Development of genetic sexing technology

Specific primers were designed according to *CseF382* sequence and used for identifying genetic sex. A simple PCR method was developed for genetic determination for the tongue sole (Fig. 1-3-6) (Chen et al., 2007).

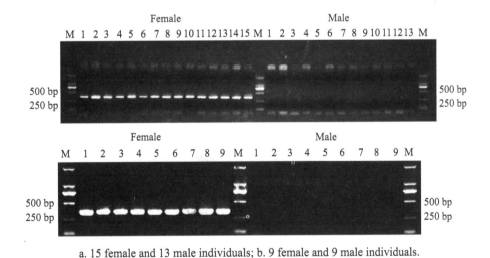

a. 15 female and 13 male individuals; b. 9 female and 9 male individuals.

Fig. 1-3-6　Genetic sex identification using *CseF382*

2.2 Cloning and expression analysis of sex-related genes

More than 10 sex-related genes were analyzed in Chinese tongue sole, including *Sox9a*, *Sox10*, *dmrt3*, *dmrt4*, *dax1*, *foxl2*, *P450 AromA*, *P450 AromB*, *gadd45g* and *FTZ-F1/SF-1*.

2.2.1 Expression pattern of *dax1*

Dax1 widely expresses in various tissues. Its expression level in male tissues is higher than that in female. Also, *dax1* expression is only existed in some development stages such as 7–18 months. These data show that *dax1* may play an important role in some stages of gonad development (Fig. 1-3-7).

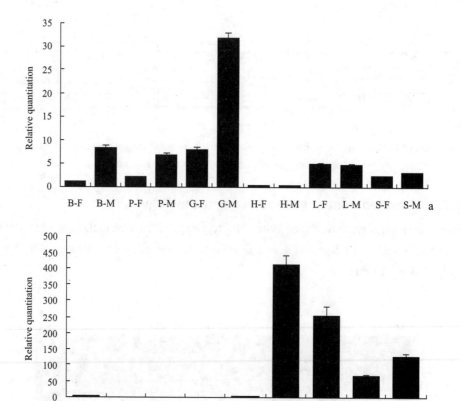

a. *Dax1* expression in various tissues; b. *Dax1* expression in different development stages. dph: day post hatching.

Fig. 1-3-7 *Dax1* expression pattern

Induced expression of *dax1* by high temperature and methyl testosterone in gonad of neo-male proves its role in maintaining the male traits and development (Fig. 1-3-8).

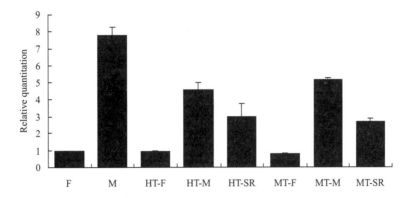

F: female control; M: male control;
HT-F, HT-M, HT-SR: the female, male and neo-male gonads treated with high temperature;
MT-F, MT-M, MT-SR: the female, male and neo-male gonads treated with methyl testosterone.

Fig. 1-3-8　The expression pattern of *dax1* in neo-male

2.2.2　*Sox9a* expression patterns at different sexes and stages of gonads

Sox9a expresses in the gonads of female and male, has no big difference between female and male during sex differentiation period from 28-90 dph (Fig. 1-3-9).

Sox9a is involved in gonad development, but has no feature as a sex determining gene in tongue sole (Dong et al., 2011).

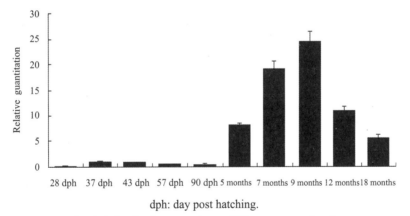

dph: day post hatching.

Fig. 1-3-9　*Sox9a* expression at different stages of testis

2.2.3　Cloning and expression analysis of *foxl2*

Foxl2 gene expresses in many tissues, and its expression level in brain-pituitary-gonad axis in female is significantly higher than that in male (Fig. 1-3-10) (Dong et al., 2011).

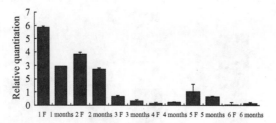

1. brain; 2. pituitary gland; 3. gonad; 4. heart; 5. liver; 6. kidney.

Fig. 1-3-10 *Foxl2* expression pattern in female (F) and male (M) tissues

2.2.4 *Foxl2* expression pattern in different development stages

In the course of female gonad development, *foxl2* expression level offers increase and then descending latter tendency. And the expression level is the highest in 12 months (oocytes gathered stage) (Fig. 1-3-11) (Dong et al., 2011).

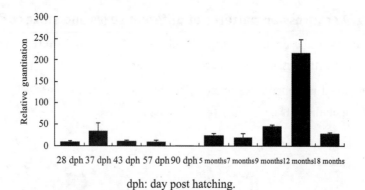

dph: day post hatching.

Fig. 1-3-11 *Foxl2* expression in different developmental stages of female tongue sole

2.2.5 *P450 AromA* (ovarian type)

P450 AromA cDNA is cloned and determined to be 2,266 bp long encoding 526 residues (Fig. 1-3-12a). The amino acid residues share 59%–77% identity with other fish (Fig. 1-3-12b) (Deng et al., 2009).

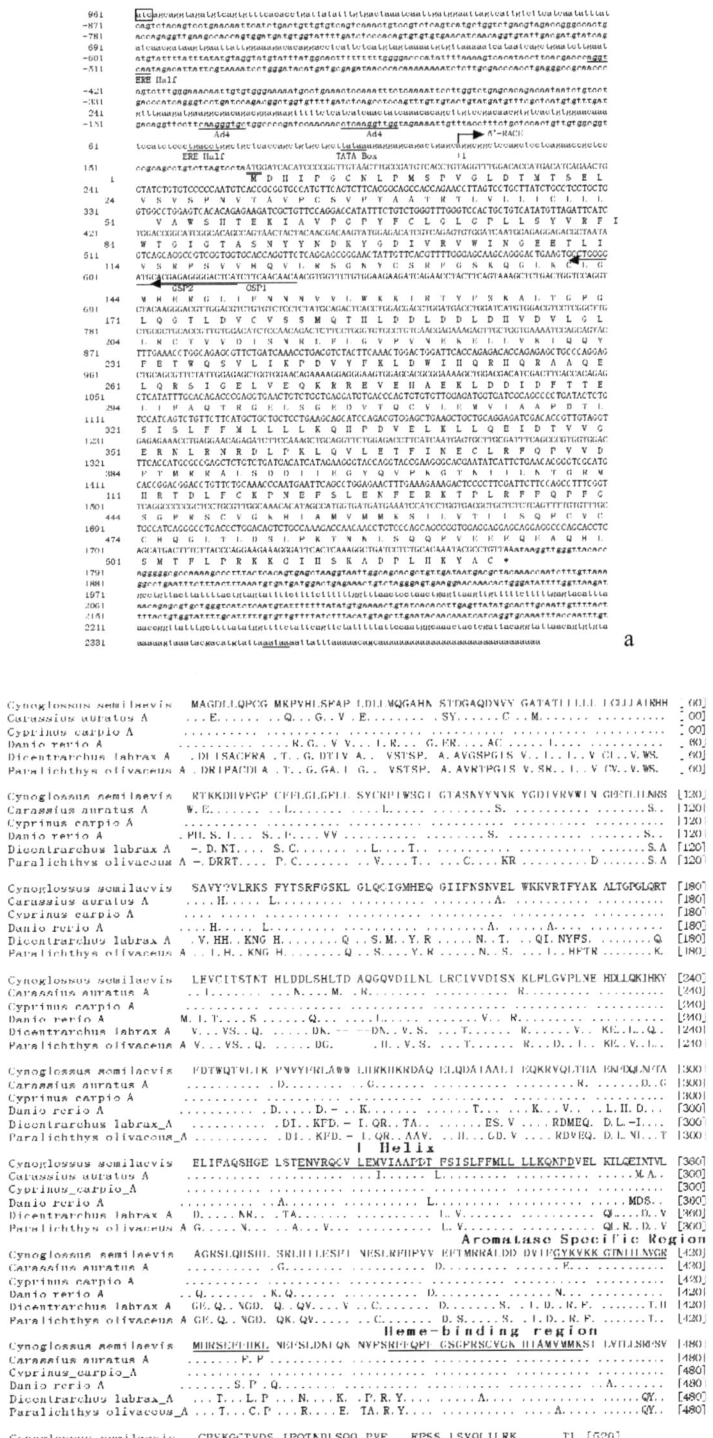

a. cDNA of P450 aromatase A (P450 AromA) and its deduced protein;
b. Alignment of tongue sole P450 AromA with other fish.

Fig. 1-3-12 Sequence analysis of P450 AromA

Expression of *P450 AromA* is detected only in gonads, and the expression level is much higher in female than in male (Fig. 1-3-13) (Deng et al., 2009).

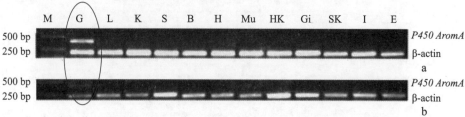

G: gland; L: liver; K: kidney; S: spleen; B: brain; H: heart; Mu: muscle; HK: head kidney; Gi: gill; Sk: skin; I: intestine; E: eye.

Fig. 1-3-13   *P450 AromA* expression in female (a) and male (b) tissues

P450 AromA expression level increases significantly along ovary development and maturation, but almost has no changes in testis in all stages (Fig. 1-3-14) (Deng et al., 2009).

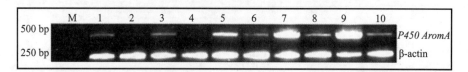

1. female of 7 month; 2. male of 7 month; 3. female of 12 month; 4. male of 12 month;
5. female of 18 month; 6. male of 18 month; 7. female of 30 month; 8. male of 30 month;
9. female of IV phase; 10. male of V phase.

Fig. 1-3-14   *P450 AromA* expression in different developmental stages

P450 AromA expression level decreases after sex-reversal from genetic female to neo-male. It implies that *P450 AromA* may play an important role in gonad differentiation and ovary development (Fig. 1-3-15) (Deng et al., 2009).

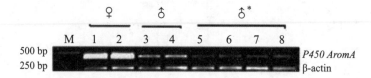

1. female control; 2. MT treated female (no sex-reversal); 3. male control;
4. MT treated male (no sex-reversal); 5-8. MT treated female (sex-reversal). *, pseudomale.

Fig. 1-3-15   *P450 AromA* expression in pseudomale gonad treated with methyltestosterone (MT)

2.2.6   Expression pattern of *P450 AromB* in Chinese tongue sole

P450 AromB (brain-type) is highly expressed in brain and gill, but very lowly expressed in gonad, showing a different role with *P450 AromA* in Chinese tongue sole (Fig. 1-3-16).

The sex determining gene should have the following features: sex-biased expression, early expression (before sex differentiation), and in right place (*e.g.*, gonadal part).

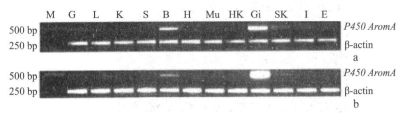

G: gonad; L: liver; K: kidney; S: spleen; B: brain; H: heart; Mu: muscle;
HK: head kidney; Gi: gill; Sk: skin; I: intestine; E: eye.

Fig. 1-3-16 *P450 AromB* expression in female (a) and male (b) tissues

According to these rules, a number of sex-related genes (mentioned above) are cloned and analyzed. They play important roles in sex differentiation, gonad development and sexual maturity, while they are not sex-determining genes.

So, in order to further screen sex-determining gene, we initiated whole genome sequencing for Chinese tongue sole.

2.3 Whole genome sequencing and assembly

The parameters of the two BAC libraries are shown in Table 1-3-1.

A. *Bam*HI library: 15,360 clones, empty ratio <1%, average insertion 160 kb, 3.9× genome coverage.

B. *Hin*dIII library: 39,936 clones, empty ratio <4%, average insertion 150 kb, 9.6× genome coverage.

C. Combined: 55,296 clones, empty clones <5%, average insertion 153 kb, 13.5× genome coverage.

Table 1-3-1 Parameters of the two BAC libraries

Library name	Clone vector	Restriction enzyme used	Total plates	Total clones	Insert empty clones/%	Average insert size /kb	Genome coverage
*Bam*HI library	pECBAC1	*Bam*HI	40	15,360	<1	160	3.9×
*Hin*dIII library	pECBAC1	*Hin*dIII	104	39,936	<4	150	9.6×
Combined library			144	55,296	<5	153	13.5×

The two BAC libraries of Chinese tongue sole provide a readily useable platform for genomic research, illustrated by the isolation of sex determination genes (Fig. 1-3-17).

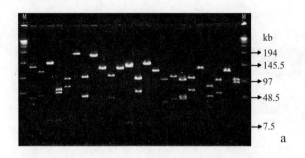
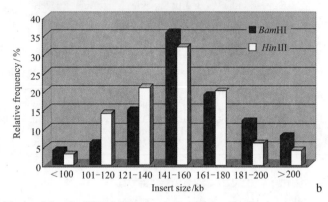

a. Insert size analysis of the *Hin*dIII BAC library; b. Insert size distribution in the two BAC libraries.

Fig. 1-3-17　Construction of BAC libraries in Chinese tongue sole

(Data from Shao et al., 2010)

Male genome is 495 Mb, and female genome is 545 Mb (Table 1-3-2).

Table 1-3-2　Whole genome sequencing data of female (ZW)

Library ID	Insert size /bp	Lanes	GC /%	Average read length /bp*	Raw reads /M	Raw bases /G	Average read length /bp**	Usable reads /M	Usable bases/G
BHSciu-RBFDBAAPEI-4	164	1	42.4	100	76.5	7.65	100	72.2	7.22
BHSciu-RAADBAAPE	165	1	40.7	100	56.2	5.62	100	52.6	5.26
BHSciu-RAODBAAPE	175	1	39.9	100	61.6	6.16	100	57.5	5.75
BHSciu-RAODBABPE	172	1	40.1	100	57.6	5.76	100	47.6	4.76
BHSciu-RADDIAAPE	471	2	39.8	100	119.9	11.99	100	103.2	10.32
BHSciu-RBFDIAAPEI-5	471	1	42.4	100	68.6	6.86	100	57.1	5.71
BHSciu-RACDMAAPE	765	2	39.2	100	87.7	8.77	100	68.8	6.88
BHSciu-RAODWAAPE	2,208	2	42.2	44	125.7	5.53	44	111.9	4.93

(*to be continued*)

Library ID	Insert size /bp	Lanes	GC /%	Average read length /bp*	Raw reads /M	Raw bases /G	Average read length /bp**	Usable reads /M	Usable bases/G
BHSciu-RAODWBBPE	2,424	1	43.3	44	55.9	2.46	44	49.7	2.19
BHSciu-RAODLAAPE	4,936	2	42.7	44	108.1	4.76	44	97.9	4.31
BHSciu-RAODTAAPE	8,546	1	43	44	47.8	2.1	44	35.6	1.57
BHSciu-RAADTAAPEI-1	9,104	1	44.1	49	159.1	7.8	49	27.5	1.35
BHSciu-RAADUAAPE	19,912	1	45.5	44	53	2.33	44	19.3	0.85
CYNcum-DAVDVAAPE	34,419	1	41.4	49	138.8	6.8	49	45.9	2.25
BHSciu-RAADVAAPE	34,467	1	39.3	49	138.3	6.78	49	10.5	0.51
All libraries	-	19	41.6	67	1,354.7	91.35	74	857.5	63.86

Notes: *, raw reads; **, clean reads.

110 Gb sequencing data are generated, which shows 220 genome coverage.

The scaffold N50 is 868 kb and assemble scaffold is 447 Mb, which covers 95% of the estimated genome (Table 1-3-3).

Table 1-3-3 Tongue sole genome assembly statistics

Type	Contig		Scaffold	
	Size/bp	Number	Size/bp	Number
N95	1,318	28,715	38,905	726
N90	4,419	20,032	272,602	526
N80	9,564	13,208	493,436	400
N70	14,668	9,388	627,811	315
N60	20,264	6,762	734,563	244
N50	26,524	4,806	867,956	185
N40	33,954	3,295	993,045	133
N30	43,078	2,108	1,132,664	88
N20	55,534	1,179	1,409,670	50
N10	75,365	467	1,818,425	20
Longest	194,815	1	4,694,140	1
Total	453,103,890	113,432	477,207,161	80,677

A total of 21,516 genes were annotated with average gene length of 8,575 bp (Fig. 1-3-18, Table 1-3-4) (Chen et al., 2014).

Table 1-3-4 Gene annotation

Gene set	Number	Average transcript length/bp	Average CDS length/bp	Number of exons per gene	Average exon length/bp	Average intron length/bp
Homology-based	18,284	9,252	1,595	9.4	169	910
RNA-seq	30,253	5,383	1,054	5.6	189	945
De novo	27,327	11,052	1,908	11.8	161	844
Reference	21,516	8,575	1,462	8.7	168	925

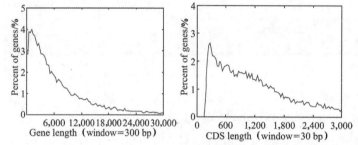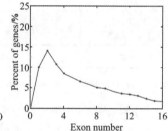

a. Gene length; b. CDS length; c. Exon number.

Fig. 1-3-18 The statistics of genes in Chinese tongue sole genome

2.4 Construction of high density genetic linkage maps and physical map in tongue sole

2.4.1 High density SSR genetic linkage map

From genome sequencing, over 80,000 SSR were identified (Table 1-3-5) (Song et al., 2012).

Table 1-3-5 Statistics of high density SSR genetic linkage map

Repeating base number of SSR/bp	Number of SSR
2	46,244
3	14,717
4	10,478
5	3,342
6	4,850

High density genetic linkage maps were constructed using 1,007 SSR markers with an average interval of 1.67 cM (Fig. 1-3-19) (Song et al., 2012).

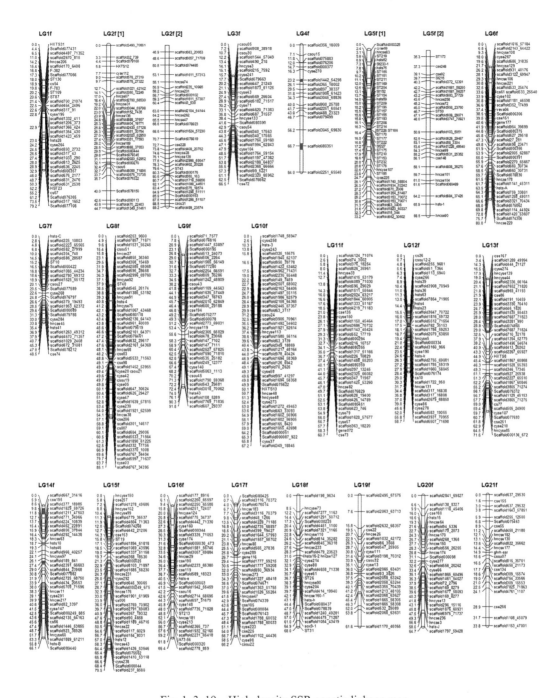

Fig. 1-3-19　High density SSR genetic linkage map

2.4.2　High density SNP genetic linkage map in Chinese tongue sole

12,142 SNP markers were used to construct a high density SNP genetic map with an average interval of 0.326 cM.

944 scaffolds representing 445 Mb genomic sequences were anchored to 20 autosomes, one Z chromosome and one W chromosome (Table 1-3-6).

Table 1-3-6 Chromosome construction

Chromosome	Number of SSR	Number of RAD-tag	Contig		Scaffold			Number Genes
			Number	Length/bp	Number	Length/bp	Source	
1	81	1,184	2,410	32,791,084	53	34,529,112	Female	1,487
2	40	810	1,227	19,259,417	29	20,052,734	Female	911
3	29	484	1,189	15,467,848	25	16,253,993	Female	596
4	85	323	1,263	19,377,156	31	20,014,501	Female	846
5	43	89	1,147	18,609,661	29	19,279,693	Female	706
6	30	825	1,270	18,113,957	29	18,841,016	Female	978
7	54	54	993	13,185,383	15	13,814,722	Female	613
8	53	642	2,144	28,615,567	37	30,153,790	Female	1,395
9	50	454	1,314	18,790,677	31	19,618,599	Female	1,029
10	46	777	1,507	20,081,642	33	21,015,569	Female	1,037
11	42	949	1,428	19,676,390	34	20,528,432	Female	1,022
12	40	517	1,349	17,485,432	35	18,398,590	Female	745
13	43	865	1,518	20,959,882	34	21,922,143	Female	892
14	50	1,288	1,782	27,668,722	47	28,847,931	Female	1,228
15	46	703	1,478	19,132,837	32	20,094,621	Female	761
16	40	430	1,252	17,874,443	29	18,785,820	Female	809
17	38	246	1,333	15,583,495	25	16,472,647	Female	984
18	28	639	1,092	14,404,870	22	15,207,555	Female	783
19	33	553	1,108	17,115,378	24	17,747,288	Female	847
20	34	226	1,036	14,355,002	18	15,234,830	Female	881
Z	37	53	2,044	20,757,346	26	21,915,962	Male	930
W	NA	NA	2,436	13,020,023	306	16,461,726	Female	320
Total	942	12 111	32,320	422,326,212	944	445,191,274	NA	19,800

The constuction of physical map is shown as Table 1-3-7.

Table 1-3-7 Construction of physical map (Zhang et al., 2014)

Item name	Number	Coverage of genome
BAC Number in experiment	33,575	8.3 times genome coverage
Assembled BAC Number	30,294	7.5 times genome coverage
Assembled BAC in contig	29,709	7.5 times genome coverage

(*to be continued*)

Item name	Number	Coverage of genome
Contig Number	1,485	–
Singleton number	585	1.93%
Clone number per contig	20	–
Contig Size/kb	537	–
N50/kb	664	–
Q-contig number	101	6.80%
Physical map size/Mb	797	1.27 times genome coverage

2.5 Evolutionary and structural analysis of Chinese tongue sole sex chromosome

2.5.1 Evolution analysis of Chinese tongue sole sex chromosome

The sex chromosomes of tongue sole and chicken were originated from the same ancestor (Fig. 1-3-20).

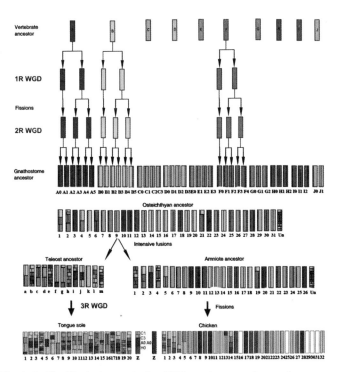

Fig. 1-3-20 Evolution analysis of Chinese tongue sole sex chromosome

2.5.2 Structural analysis of sex chromosome

A. Z chromosome: 937 genes with 382 Z-specific genes.

B. W chromosome: 395 genes with 2 W-specific genes.

297 genes are Z and W homologous (Fig. 1-3-21, Table 1-3-8).

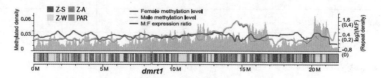

Fig. 1-3-21　Structural analysis of sex chromosome

Table 1-3-8　Classification of Z and W genes in non-PAR region

Type	Z (non-PAR)			W (non-PAR)		
	Functional genes	Pseudo genes	Total	Functional genes	Pseudo genes	Total
Z-W	286	11	297	272	67	339
Z-A	248	10	258	NA*	NA	NA
Z-S	370	12	382	NA	NA	NA
W-Z_random	NA	NA	NA	17	7	24
W-A	NA	NA	NA	26	4	30
W-S	NA	NA	NA	2	0	2
Total	904	33	937	317	78	395

A. Autosome: 46 genes/Mb, average size 8,876 bp, repeat sequence 4.3%.

B. Z chromosome: 42 genes/Mb, average size 9,857 bp, repeat sequence 13%.

C. W chromosome: 19 genes/Mb, average size 12,156 bp, repeat sequence 30% (Table 1-3-9).

Table 1-3-9　Gene density analysis of sex chromosome

Chromosome	Genes per megabase /%	Total Tes /%	DNA Tes /%	LINE /%	LTR/%	SINE /%	Unclassified Tes/%	Average gene size /bp
1	43	5.11	2.41	0.69	0.01	0.32	1.68	8,930
2	45	4.39	2.15	0.45	0.01	0.23	1.55	9,101
3	37	4.97	2.15	0.83	0.06	0.17	1.76	10,650
4	44	4.82	2.17	0.51	0.02	0.20	1.92	9,351
5	37	3.35	1.55	0.33	0.01	0.21	1.25	10,264
6	52	4.10	1.73	0.31	0.01	0.18	1.87	8,039
7	47	4.31	1.90	0.62	0.04	0.12	1.63	8,556
8	49	5.05	2.42	0.68	0.05	0.22	1.68	9,428
9	52	4.65	2.16	0.55	0.02	0.24	1.68	7,833
10	49	4.72	2.34	0.57	0.03	0.24	1.54	8,500
11	51	3.42	1.42	0.40	0.01	0.22	1.37	8,612

(to be continued)

Chromosome	Genes per megabase /%	Total Tes /%	DNA Tes /%	LINE /%	LTR/%	SINE /%	Unclassified Tes/%	Average gene size /bp
12	40	3.46	1.70	0.29	0.01	0.14	1.32	10,820
13	43	4.30	2.15	0.52	0.03	0.20	1.40	9,080
14	43	4.49	1.93	0.64	0.02	0.28	1.62	9,550
15	39	5.08	2.04	0.58	0.08	0.20	2.18	8,594
16	43	4.99	2.42	0.53	0.02	0.19	1.83	9,839
17	60	4.56	2.23	0.50	0.01	0.17	1.65	7,061
18	51	3.55	1.62	0.27	0.01	0.13	1.52	7,630
19	48	3.05	1.32	0.33	0.03	0.18	1.19	8,782
20	58	2.32	0.91	0.21	0.01	0.08	1.11	7,733
Autosomes	46	4.33	1.99	0.51	0.02	0.21	1.60	8,876
chrZ	42	13.13	4.74	3.95	0.23	0.43	3.78	9,857
chrW	19	29.94	8.74	9.39	1.09	0.46	10.26	12,156

Chinese tongue sole was divergent from other fish at 197 millon years ago.

Z and W were divergent at 31 million years ago, indicating that Chinese tongue sole has very young sex chromosomes (Table 1-3-10, Fig. 1-3-1c).

Table 1-3-10 Evolution time of sex chromosome

Type	Synonymous mutation frequency/Ks	Evolution and time	Min	Mean	Max
Lineage	0.47	Rate (/Site/Year)	0.00000000276	0.00000000239	0.00000000214
		Time (MY)	170	197	220
Z-W	0.15	Rate (/Site/Year)	0.00000000553	0.00000000477	0.00000000427
		Time (MY)	27	31	35

Note: MY, million years.

2.6 Identification and functional analysis of sex determining gene in Chinese tongue sole

2.6.1 Sex-related genes in Z chromosome of Chinese tongue sole

Four sex-related genes (*dmrt1*, *patched1*, *SF-1*, Follistatin) were detected on Z chromosome.

Dmrt1 expressed specifically in gonad in the key stages of differentiation of ZZ male, but not in female, indicating that *dmrt1* is a promising candidate for sex determining gene (Fig. 1-3-22).

No sex specific expression was detected in *Patched1*, *SF-1* and Follistatin genes.

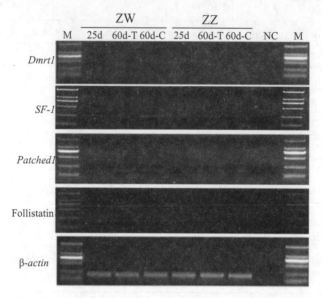

25d, day 26; 60d-T, day 60 with temperature treatment; 60d-T, day 60 control; NC, negative control; M, marker.

Fig. 1-3-22　Expression of Z-linked genes in gonads at the key stages of differentiation

2.6.2　*Dmrt1* was mapped on Z chromosome

It indicates that *dmrt1* signal only exist on Z, but not in W chromosome (Fig. 1-3-23).

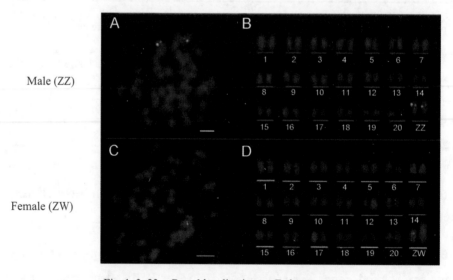

Fig. 1-3-23　*Dmrt1* localization on Z chromosome

2.6.3　*In situ* hybridization (ISH) of gonads with *dmrt1*

ISH showed that *dmrt1* expressed at 56 dph and increased with development of testis, while no signal was detected in ovary (Fig. 1-3-24).

qPCR demonstrated that *dmrt1* expressed at 48 dph and increased to peak at 1 year post hatching, while no expression was detected in ovary (Fig. 1-3-25) (Chen et al., 2014).

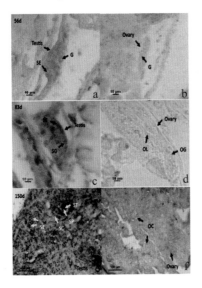

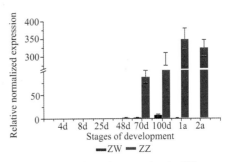

Fig. 1-3-25 *Dmrt1* expression in different developmental stages

a, c, e. Testes; b, d, f. Ovaries.

a. Male at 56 dpf; b. Female at 56 dpf; c. Male at 83 dpf; d. Female at 83 dpf; e. Male at 150 dpf; f. Female at 150 dpf.

dph: day post hatching.

Fig. 1-3-24 Localization of *dmrt1* in gonads by ISH

2.6.4 Development of genome editing in Chinese tongue sole

A. Breakthrough on embryo microinjection technique.

B. Established TALEN and CRISPR.

C. Genome editing platform for functional study of sex determining gene (Fig. 1-3-26).

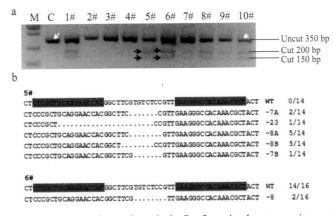

a. Screening by electrophoresis; b. Confirmation by sequencing.

Fig. 1-3-26 TALEN-mediated knockout of *dmrt1*

Since the chorion of Chinese tongue sole embryo is very tough, thick pointed injection needles were prepared using a P-97 Flaming/Brown micropipette puller (Fig. 1-3-27, Table 1-3-11) (Cui et al., 2017).

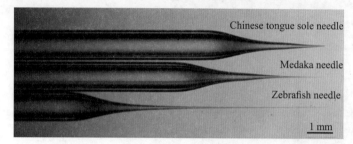

Fig. 1-3-27 Comparison of Chinese tongue sole, medaka and zebrafish needles

Table 1-3-11 Statistics of TALEN-mediated knockout technique in Chinese tongue sole

Vector type	Injected embryo number	Survival number	Mutation rate
TALEN mRNA1	5,000	About 50	20%–70%

Dmrt1 knockout led to abnormal development and degeneration of testis (Figs. 1-3-28 to 1-3-30). An ovary-like cavity was observed in *dmrt1* knockout male.

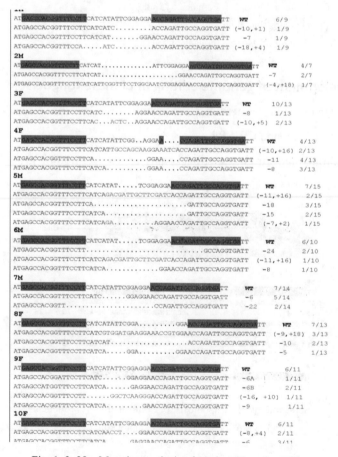

Fig. 1-3-28 Mutation analysis of 10 *dmrt1* knockout fries

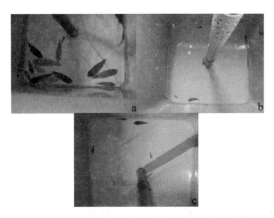

Fig. 1-3-29 Cultivation of *dmrt1* knockout fries

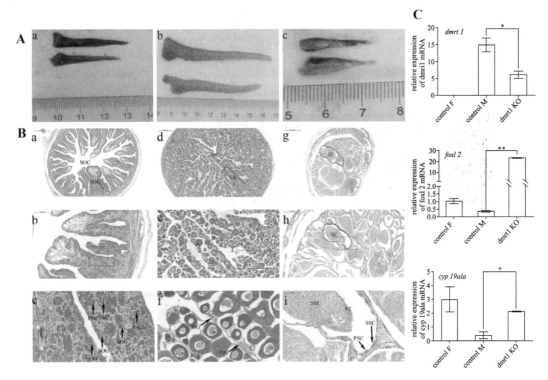

A. Gross morphology of gonads from approximately one year old fish. a. *Dmrt1*-deficient testis; b. Wild-type ovaries; c. Wild-type testes. B. Histology of gonads from approximately one year old fish. a,b. *Dmrt1*-deficient testis. The development of testis is ceased. The shape of the *dmrt1*-deficient testis in transverse sections is similar to control ovaries, and there are structures resembling ovarian cavity and ovarian lamella in the gonad of the mutant male fish. Ovarian cavity-like (OCL), ovarian lamella-like (OLL). c. Large magnification of frame area in (b). No secondary spermatocytes, spermatids and sperm are observed. Oogonia-like (OGL), spermatogonia (SG) and primary spermatocytes (PSC). d,e. Ovary of control female, including ovarian cavity (OC), ovarian lamella (OL). f. Large magnification of frame area in (e). Four stages of oocytes: stage I - IV and oogonia (OG). g,h. Testis of control male. Seminiferouslobuli (SL), seminiferous cyst (SC); i. Larger magnification of frame area in (e). Secondary spermatocytes (SSC), spermatids (ST) and sperm (SM). C. Gene expressions of sex differentiation markers on *dmrt1*-deficient gonad. *, Significant difference; **, Very significant difference.

Fig. 1-3-30 Analysis of *dmrt1*-knockout fish

Hermaphrodite was generated by *dmrt1* TALEN in Chinese tongue sole.

In hermaphrodite, *dmrt1* knockout led to expression increasing of *foxl2* and *cyp19a1a*, which usually showed high expression level in female (Figs. 1-3-31 to 1-3-32).

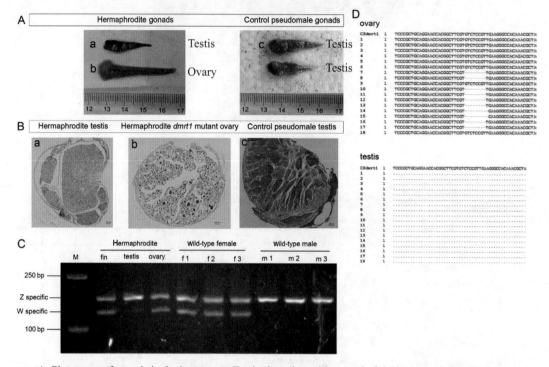

A. Phenotype of gonads in the intersex. a. Testis shaped up-side gonad of the intersex; b. Ovary shaped down-side mutant gonad of the intersex; c. Testes of a pseudomale. B. Histology of the gonads. a. Testis of the intersex showing normal male structures; b. *Dmrt1* mutant gonad (ovary) of the intersex; c. Testis of a pseudomale. Scale bar, 100 μm. C. Determination of the genetic sex of the intersex by SSR PCR yielding different sized products for the Z (169 bp) and W (134 bp) chromosomes. D. Sequences of wild-type and mutated target sites of *dmrt1* retrieved from ovary, partial, and only wild-type *dmrt1* in testis, partial.

Fig. 1-3-31　Effect of *dmrt1* disruption on intersex gonad phenotype, sex differentiation and gene expression

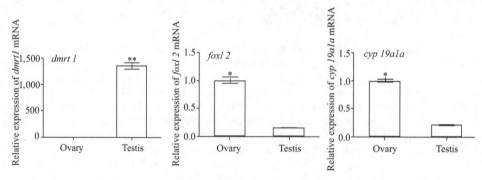

Fig. 1-3-32　Relative mRNA expression of *dmrt1*, *foxl2* and *cyp19a1a* in the *dmrt1*-deficient gonad at one year of age

The average weight of *dmrt1* knockout male was 241 g, 2.2 times as normal male (Table 1-3-12, Fig. 1-3-33). Genome editing has great application potential in sex control and breeding (Cui et al., 2017).

Table 1-3-12 Growth parameter of *dmrt1* knockout individual

Fish type (3 individuals)	Average weight/g	Average length/cm	Average width/cm	TALEN time
Knockout male	241.2	33.7	9.7	May-14
Normal male	110.3	26.1	7	May-14
Normal female	296.8	35.9	10	May-14

a. *Dmrt1* knockout; b. Normal male; c. Normal female.

Fig. 1-3-33 Accelerated growth of *dmrt1* knockout individual

2.6.5 *Dmrt1* is the male determining gene in Chinese tongue sole

These data suggest that *dmrt1* is Z-linked, male-specific expression and necessary for testis development. *Dmrt1* is male determining gene and its expression determines Chinese tongue sole to be phenotypic male (Fig. 1-3-34).

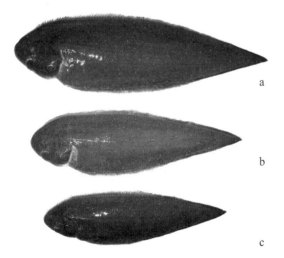

a. ZW female; b. ZW male; c. ZZ male.

Fig. 1-3-34 *Dmrt1* expression determines Chinese tongue sole to be phenotypic male

2.7 Identification of sex specific SSR markers and development of technique for fry with high female ratio

2.7.1 Screening of sex specific SSR markers in Chinese tongue sole

Sex specific SSR markers were identified after comparison of male and female genome (Fig. 1-3-35).

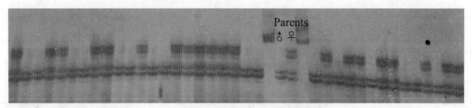

Fig. 1-3-35 Screening of sex specific SSR marker

2.7.2 Development of technique for WW super-female identification

Molecular technique (Patent: ZL201010602905.8, ZL201010239729.6.) is established to distinguish male/ZZ, female/ZW, neo-male/ZW and superfemale/WW (Fig. 1-3-36) (Chen et al., 2012).

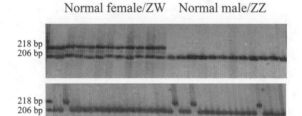

Lower panel: identification of WW superfemale embryos in mitogynogenetic embryos.

Fig. 1-3-36 Technique for WW super-female identification

2.7.3 Why the ratio of phenotypic male is about 70%–90% in common stock in tongue sole? —Discovery of reasons for high male proportion in tongue sole

A number of male and neo-male families were established. The offspring of neo-male exhibited 96% phenotypic male, 38% higher than that of normal male, indicating that the offspring of neo-male tend to sex-reverse to neo-male (Fig. 1-3-37) (Chen et al., 2014; Shao et al., 2014).

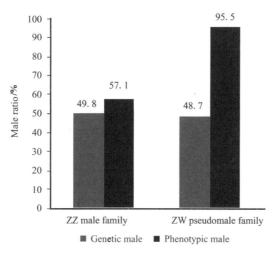

Fig. 1-3-37 Sex ratio of offspring from ZZ male and ZW pseudomale

2.7.4 Epigenetic modulation in sex reversal—Why the offspring of neo-male become male more easily?

Dmrt1 in ZW female showed hypermethylation, no expression, while in ZW neo-male showed demethylation, high expression, and then it became male and the methylation pattern passed to offspring (Fig. 1-3-38) (Chen et al., 2014; Shao et al., 2014). We uncovered the epigenetic mechanism of sex reversal.

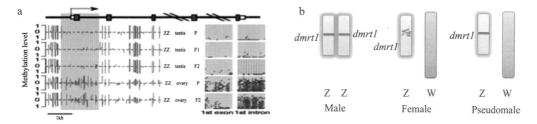

a. Methylation status across the differentially methylated region (DMR) of *dmrt1* in the gonads of an adult WZ female, a ZZ male and a WZ female compared to male sex-reversed fish; b. Schematic model of epigenetic regulation for *dmrt1* in ZZ male, ZW female and ZW pseudomale.

Fig. 1-3-38 *Dmrt1* in ZW female, male and pseudomale

2.7.5 Breakthrough of fish genome and sex control study in China

Data about genome sequencing, evolutionary analysis of sex chromosome and epigenetic mechanism of sex reversal have been published on *Nature Genetics* and *Genome Research* in 2014 (Fig. 1-3-39).

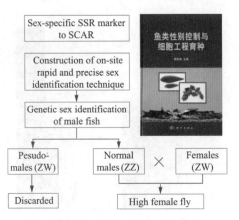

Fig. 1-3-39   Technique for production of high-female ratio fries

2.7.6 Development of technique for fries with high female ratio

Technique for producing Chinese tongue sole fries with higher female ratio was established and applied in many sole farms. The phenotypic female ratio was increased from 20% to 40%. This is a successful example of sex specific markers applied in aquaculture in China (Fig. 1-3-40)

Fig. 1-3-40   Group members in the factory for industrial support

3 Conclusion

A. Sex specific AFLP and SSR markers were discovered for the first time, and genetic sexing techniques were developed.

B. Sex-specific molecular markers-assisted sex control techniques have been developed and widely utilized and promoted the development of the aquaculture of Chinese tongue sole.

C. Chinese tongue sole genome has been sequenced, providing genomic information and powerful tools to perform the genetic dissection of important economic traits.

D. A male sex determining gene, *drmt1*, has been discovered, and sex determination mechanism has been elaborated in Chinese tongue sole.

E. Genome editing technology has been established in Chinese tongue sole, and the genome editing breeding technology will be developed.

4 Acknowledgements

This research was supported by Changwei Shao, Qisheng Tang, Na Wang, Yongsheng Tian, Zhenxia Sha, Xiaolin Liao, Zhongkai Cui, Yang Liu, Yangzhen Li, Siping Deng, Jing Li, Wentao Song, Mingshu Xie, Qiaomu Hu, Xiaoli Dong, Junjie Zhang, Ying Zhu, Shanshan Liu, from YSFBI, Jun Wang, Guojie Zhang, Quanfei Huang, Na An, Qiye Li, Zhiyuan Xie, Bo Li, from BGI at Shenzhen, M. Schart1 from *Wuerzburg University*, CHK Cheng, Yun Liu from *The Chinese University of Hong Kong* (Fig. 1-3-41).

Fig. 1-3-41　Main collaborators

References

Chen S L, et al. Induction of mitogynogenetic diploids and identification of WW super-female using sex-specific SSR markers in half-smooth tongue sole (*Cynoglossus semilaevis*) [J]. Marine Biotechnology, 2012, 14: 120-128.

Chen S L, et al. Isolation of female-specific AFLP markers and molecular identification of genetic sex in half-smooth tongue sole (*Cynoglossus semilaevis*) [J]. Marine Biotechnology, 2007, 9: 273-280.

Chen S L, et al. Whole-genome sequence of a flatfish provides insights into ZW sex chromosome evolution and adaptation to a benthic lifestyle [J]. Nature Genetics, 2014, 46: 253-260.

Cui Z K, et al. Genome editing reveals *dmrt1* as an essential male sex-determining gene in Chinese tongue sole (*Cynoglossus semilaevis*) [J]. Scientific Reports, 2017, 7: 42213.

Deng S P, et al. Molecular cloning, characterization and expression analysis of gonadal P450 aromatase in the half-smooth tonguesole, *Cynoglossus semilaevis* [J]. Aquaculture, 2009, 287: 211-218.

Dong X L, et al. Molecular cloning, characterization and expression analysis of *Sox9a* and *Foxl2* genes in half-smooth tongue sole (*Cynoglossus semilaevis*) [J]. Acta Oceanologica Sinica, 2011, 30(1): 68-77.

Liao X L, et al. Construction of a genetic linkage map and mapping of a female-specific DNA marker in half-smooth tongue sole (*Cynoglossus semilaevis*) [J]. Marine Biotechnology, 2009, 11(6): 699–709.

Shao C W, et al. Epigenetic modification and inheritance in sexual reversal of fish [J]. Genome Research, 2014, 24: 604–615.

Shao S W, et al. Construction of two BAC libraries from half-smooth tongue sole *Cynoglossus semilaevis* and identification of clones containing candidate sex-determination genes [J]. Marine Biotechnology, 2010, 12: 558–568.

Song W T, et al. Construction of a high-density microsatellite genetic linkage map and mapping of sexual and growth-related traits in half-smooth tongue Sole (*Cynoglossus semilaevis*) [J]. PLOS One, 2012, 7(12): e52097.

Zhang J J, et al. A first generation BAC-based physical map of the half-smooth tongue sole (*Cynoglossus semilaevis*) genome [J]. BMC Genomics, 2014, 15: 215.

Zhou L Q, et al. The karyotype of the tonguefish *Cynoglossus semilaevis* [J]. Journal of Fisheries of China, 2005, 29(3): 417–419. (In Chinese with English abstract)

History, status and prospect of shrimp farming in China

Author: Wang Xiuhua
E-mail: wangxh@ysfri.ac.cn

1 History of shrimp cultivation in China

1.1 Development course of *Penaeus chinensis* industry in China

P. chinensis is one of the main shrimp species in the world, which can live in the water with a wide range of temperature and salinity. The development course of *P. chinensis* industry in China showed in Fig. 1-4-1.

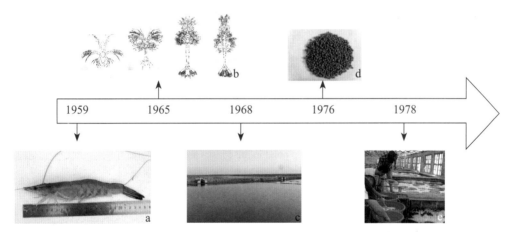

a. Began with artificial breeding; b. Understood the developmental process of larva metamorphosis;
c. Outdoor shrimp farming began in 1968; d. Completed the research of artificial diet; e. Established the technologies of breeding and cultivation of shrimp.

Fig. 1-4-1 Development course of *Penaeus chinensis* industry in China

1.2 Development course of *P. vannamei* industry in China

P. vannamei is an exotic species (Fig. 1-4-2). It was first introduced into China from the United States in 1988 and widely cultivated in the coastal and inland area (Fig. 1-4-3). Now it has been a primary cultured species in China (Fig. 1-4-4).

78 | Mariculture Modes in China

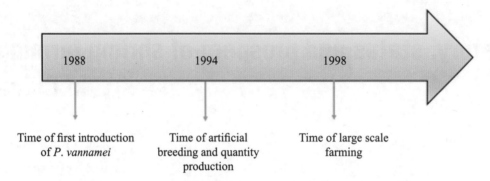

Fig. 1-4-2 Development course of *P. vannamei* industry in China

a. Shrimp pond in coastal area; b. Shrimp in inland area; c. Shrimp pond in saline-alkali land; d. Shrimp pond in green house.

Fig. 1-4-3 Shrimp pond in the coastal and inland area

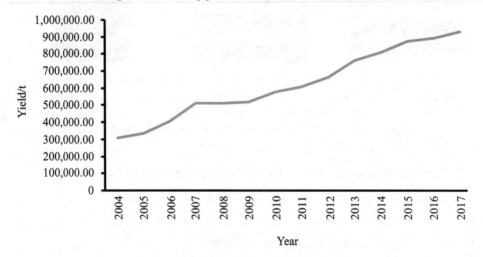

Fig. 1-4-4 Yield of *P. vannamei* in China

Advantages of *P. vannamei* are as following:

A. Can grow in water with wide range of salinity (from 0.2 to 40).

B. Can be cultivated at high density and get a high yield.

C. Has higher disease resistance against virus (WSSV) than other shrimp species.

1.3 Development course of *P. monodon* industry in China

P. monodon is a native species in the South China Sea. Taiwan was the main producing areas in the 1980s in China.

Research on overwintering and artificial breeding of *P. monodon* (Fig. 1-4-5) in cement ponds was carried out early in 1985. *P. monodon* was firstly cultivated in the north coast of China in 1989.

Fig. 1-4-5 *P. monodon*

1.4 Development course of *P. japonicus* industry in China

P. japonicus (Fig. 1-4-6) was firstly cultivated in Fujian Province in 1988. Now it becomes a very popular species in Chinese coastal line.

A. It can live in low temperature, but can not live well in water with low salinity (Chen, 1994).

B. It can not be cultivated with high density like *P. vannamei*.

C. It needs a higher protein level than *P. vannamei*, and live food is often used in grow-out stage.

D. It has a high market price and often be harvested at small body length.

Fig. 1-4-6 *P. japonicus*

1.5 Development course of shrimp culture industry in China

P. chinensis farming in China had experienced 20 years of scientific research and industrial promotion, then entered into a 10-year period of rapid development. In 1993, due to the impact of shrimp white spot syndrome disease, the industry has experienced a slump. Later, with the introduction of *P. vannamei*, shrimp farming industry returned to rapid growth stage. Since 2012, due to the impact of a variety of emerging diseases, the growth rate of shrimp aquaculture industry has slowed down (Fig. 1-4-7).

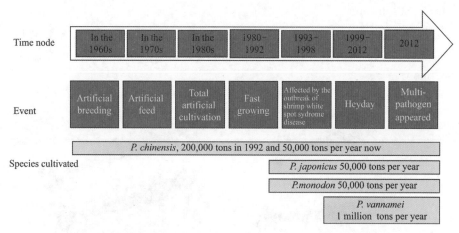

Fig. 1-4-7 Development course of shrimp culture industry in China

1.6 Shrimp yield in China 1978-2018

The statistical results of shrimp annual production including wild and cultivated in sea water in China from 1978 to 2016 are shown in Fig. 1-4-8. It can be seen that the annual output of wild shrimp is relatively stable, while the production of cultured shrimp increased rapidly, and the current annual output exceeds 1.2 million tons.

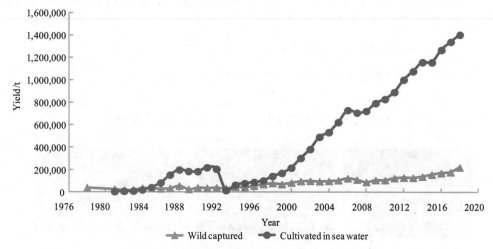

Fig. 1-4-8 Shrimp yield in China 1978-2018
(Data from *China Fishery Statistic Yearbook* 1978-2018)

1.7 Marine culture area of shrimp

The aquaculture area of marine shrimp in China showed a trend of increasing from 1981 to 2016 in general (Fig. 1-4-9). In the early 1990s, the occurrence of shrimp white spot syndrome disease caused a decrease in shrimp aquaculture area and now it maintained at about 250,000 hectares.

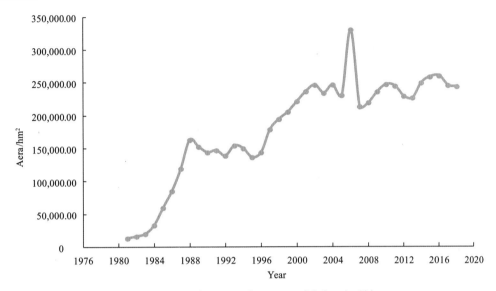

Fig. 1-4-9　Marine aquaculture area of shrimp in China
(Data from *China Fishery Statistic Yearbook* 1982-2019)

1.8 Yield of shrimp cultivated in fresh water

P. vannamei farming in freshwater also developed rapidly in China, the production rising from 300,000 t in 2003 to more than 700,000 t in 2016 (Fig. 1-4-10).

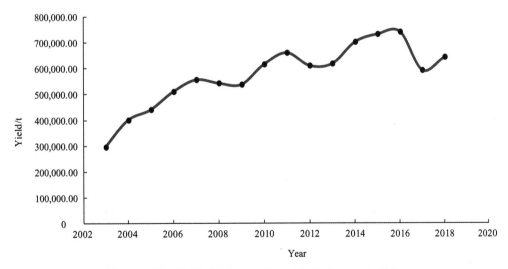

Fig. 1-4-10　Yield of shrimp cultivated in fresh water in China
(Data from *China Fishery Statistic Yearbook* 1982-2019)

1.9 Yield of different shrimp species in China

The main shrimp species cultivated in sea water in China are *P. vannamei*, *P. japonicus*, *P. monodon* and *P. chinensis*. The yields of different cultured shrimp species from 2003 to 2018 are shown in Fig. 1-4-11. Among them, the annual production of *P. japonicus*, *P. monodon* and *P. chinensis* is around 50,000 t each, it show that the production of *P. vannamei* is the highest.

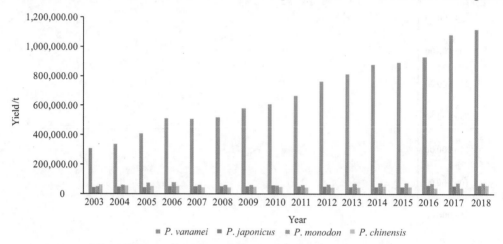

Fig. 1-4-11 Yield of different shrimp species in China
(Data from *China Fishery Statistic Yearbook* 1982-2019)

2 Status of shrimp cultivation in China

2.1 Main shrimp species cultured in China currently

There are six shrimp species cultured currently in China, including four main species *P. vannamei*, *P. japonicus*, *P. monodon* and *P. chinensis* and two niche species *P. penicillatus* and *P. merguiensis* (Fig. 1-4-12).

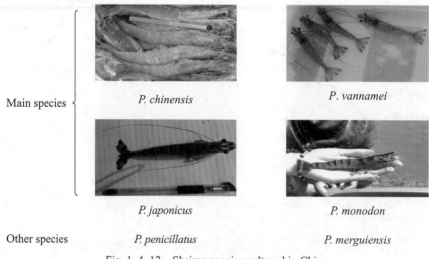

Fig. 1-4-12 Shrimp species cultured in China

2.2 Distribution of shrimp species cultured in China

P. monodon, *P. japonicus* and *P. vannamei* can be cultured in all the coastal line of China, while *P. chinensis* is limited to the coastal line of northern China (from the Bohai Sea to the Yellow Sea).

2.3 Shrimp cultivation mode in China

A. Extensive culture in the large pond.

B. Semi-intensive culture in the small pond.

C. Intensive culture in small the lined pond.

D. Polyculture with fish, crab or sea cucumber.

E. Cultivation in the nursery pond with clean water.

F. Bifloc technology.

G. Cultivation in recirculating aquaculture systems.

2.3.1 Management of extensive culture

A. Kill all the wild fish before stocking shrimp fry.

B. Fill the pond (Fig. 1-4-13) with water 1-2 months before stocking of shrimp to culture the food organisms.

C. Artificial food is not needed at the early stage until the shrimp begin to search for food along the pond side.

D. Shrimp cultivation density is below 100,000 individuals per hectare.

E. Water exchange is not needed at the early stage.

F. No aerator is needed.

G. In order to control and prevent shrimp disease, some fish, jellyfish, crab and sea cucumber can be polycultured.

H. The expected yield of shrimp is 1.5-3.0 t/hm^2.

Fig. 1-4-13　Shrimp ponds for extensive aquaculture

2.3.2 Intensive culture pond in greenhouse

A. The building structures of greenhouses for shrimp cultivation are different in different regions (Fig. 1-4-14), and the area of the pond is usually 100-500 m^2.

B. 2-3 crop cycles one year.

C. Shrimp cultivation density is 200-300 individuals per hectare.

E. Live feed (*Artemia* and Copepoda) is used 1-2 times a day.

F. Daily water exchange is 20%-40% of the whole volume.

G. Use aerator to oxygenate.

H. In order to control and prevent shrimp disease, probiotics are often used.

I. The expected yield is about 4 kg/m^2.

Fig. 1-4-14   Intensive aquaculture ponds in greenhouse

2.3.3 Land-based intensive culture

A. Land-based shrimp intensive ponds are mostly lined with the area of 1,000-2,000 m^2 (Fig. 1-4-15).

B. 1-2 crop cycles one year.

C. Shrimp cultivation density is 150-250 individuals per hectare.

D. Daily water exchange is 20%-40%.

E. Use aerator to oxygenate.

F. In order to control and prevent shrimp disease, probiotics are often used.

G. The expected yield of shrimp is about 2-4 kg/m^2.

Fig. 1-4-15 Land-based intensive aquaculture ponds

2.3.4 Intensive culture in nursery pond with clean water

A. The building structure of the pond for intensive culture with cleans water is similar to the shrimp hatchery pond which area is 20-50 m² (Fig. 1-4-16).

B. 3-5 crop cycles one year.

C. Shrimp cultivation density is 400-500 individuals per hectare.

D. Live food used 1-2 times a day (*Artemia* and Copepoda).

E. Daily water exchange is 100%-200%.

F. Use aerator to oxygenate.

G. Clean the pond bottom 2-3 times a day to remove the carapace, feces and residues of feed.

H. The expected yield of shrimp is around 5-8 kg/m².

Fig. 1-4-16 Nursery ponds for intensive aquaculture

2.3.5 Biofloc technology

A. Use aerator to oxygenate.

B. Keep C/N ratio at 15%–25% to promote the formation of biofloc.

C. Use lime to balance the pH.

D. Add probiotics to increase the function of biofloc (Fig. 1-4-17).

E. This technology can reduce water exchange rate.

F. Special pond structure is needed to keep the floc floating (Fig. 1-4-18).

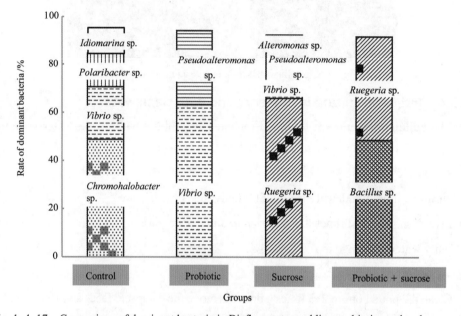

Fig. 1-4-17　Comparison of dominant bacteria in Biofloc system adding probiotics and carbon source (Data from Zhang et al., 2016)

Fig. 1-4-18　Ponds for Biofloc aquaculture

2.3.6 Recirculating aquaculture system (RAS)

RAS (Fig. 1-4-19) requires supporting facilities such as particle separation, protein separation, denitrification (Fig. 1-4-20) and sterilization.

Fig. 1-4-19　Recirculating aquaculture system

Fig. 1-4-20　Equipments for denitrification in RAS

2.4　Problems in the cultivation of shrimp in China

A. The infrastructure and facilities used for shrimp cultivation are updated slowly (Fig. 1-4-21a).

B. Emerging disease and multi-pathogen infection occur frequently (Fig. 1-4-21b).

C. Native genetic improved shrimp varieties develop slowly (Fig. 1-4-21c).

D. The cyclic utilization rate of aquaculture water is low.

E. Risk from disease and larvae quality is increasing (Fig. 1-4-21d).

a. Shrimp pond with low yield; b. Shrimp infected by multi-pathogen; c. Shrimp without genetic improvement grow slowly; d. Disease post-larvae.

Fig. 1-4-21   Diseased shrimp and extensive aquaculture pond

Shrimp disease have caused great economic loss in recent years in China, and the main diseases are listed in Table 1-4-1. Shrimp white spot disease (WSD) caused by white spot syndrome virus (WSSV) can impact on the production of a variety of shrimp (Fig. 1-4-22); AHPND caused by *Vibrio parahaemolyticus* (VP_{AHPND}) occur in *P. vannamei* from post larvae to adult (Fig. 1-4-23); HPM caused by microsporidian *Enterocytozoon hepatopenaei* seriously affects the growth of shrimp (Fig. 1-4-24); Other diseases like white feces symptoms have been prevalent in recent years though the pathogenesis is unclear (Fig. 1-4-25), and harmful filiform algae sometimes cause the death of shrimp (Fig 1-4-26).

Table 1-4-1   Shrimp diseases in China in 2013 and 2019

Disease	Extent of damage	
	2013	2019
White spot disease (WSD)	+++	+
Acute hepatopancreas necrosis disease (AHPND)	+++	++
Hepatopancreatic microsporidiosis (HPM)	++++	++
Infectious hypodermal and hematopoietic necrosis (IHHN)	++	+
Viral covert mortality disease (VCMD)	++	+
Infectious myonecrosis (IMN)	+	-
Yellow head disease (YHD)	+	-
Taura syndrome (TS)	+	-

(to be continued)

Disease	Extent of damage	
	2013	2019
Shrimp iridovirus disease (SHID)	–	?

Notes: +, means the severity of shrimp disease, the more plus signs, the more serious the disease; "-" means the disease has no impact on the shrimp; "/" means the disease has not been founded; "?" means the damage of the disease is unclear.

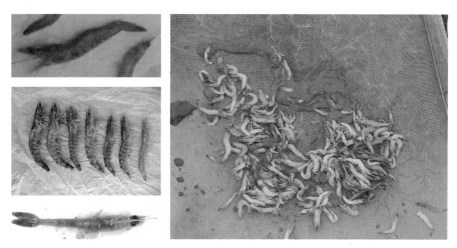

Fig. 1-4-22　Shrimp infected by WSSV

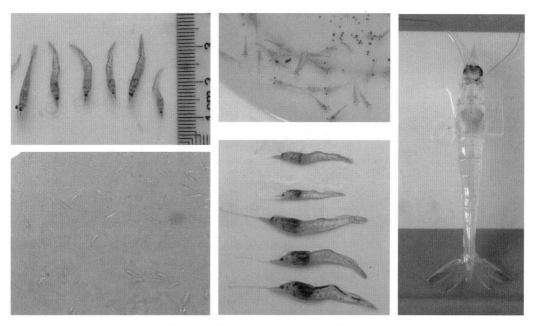

Fig. 1-4-23　Shrimp infected by VP_{AHPND}

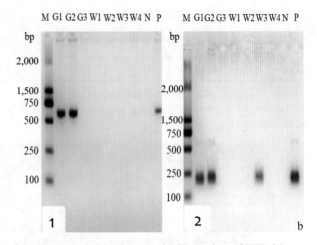

a. Shrimp infected by EHP grow slowly (Data from Jie Huang); b1. First round PCR product of EHP; b2. Second round PCR product of EHP; M, DL2000 DNA marker; G1, G2, G3, W1, W2, W3 and W4, Shrimp sample numbers; N, Negative control; P. Positive control.

Fig. 1-4-24　Symptom of shrimp infected by EHP and the results of pathogen detection by PCR

(Data from Wu et al., 2019)

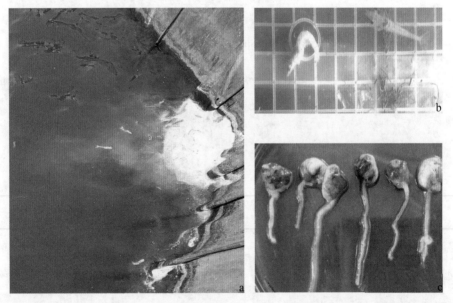

a. Shrimp floating in water; b. Shrimp infected by *Vibrio* showing white feces; c. Hepatopancreas and intestinal tract getting white.

Fig. 1-4-25　White feces symptoms of shrimp

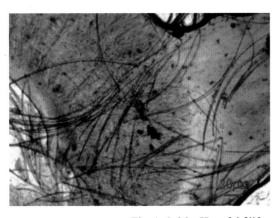 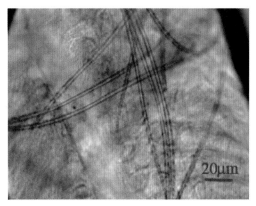

Fig. 1-4-26　Harmful filiform algae wrapping shrimp gill

2.5　Strategies for the prevention and control of shrimp disease

A. Shrimp post larvae is subject to strict origin quarantine and harmless treatment to shrimp infected by pathogens like WSSV, IHHNO (infectious hypodermal and hematopoietic necrosis virus, IHHNV), EHP, VP_{AHPND} are carried out.

B. Pond and water pumped into the pond should be disinfected and purified before using in the whole production process.

C. Install disinfection facilities in farms to prevent the introduction of pathogens by visitor and vehicles.

D. Use pellet diet instead of fresh feed (trash fish/fish side by-products).

E. Construct healthy environmental microflora to inhibit pathogenic bacteria by adding anti-pathogen probiotics.

F. Add functional probiotics (denitrifier, photosynthetic bacteria) and carbon source or water quality improver to improve the water quality.

G. Lower the stocking density and increase DO to reduce the stress on shrimp.

H. Select right aquaculture mode (polyculture, biofloc, intensive culture, etc.) based on the pond condition.

2.5.1　Shrimp pathogen detection methods developed and applied in different periods

Shortly after the onset of the white spot syndrome disease, in order to detect pathogen WSSV, monoclonal antibody detection method was firstly developed in early 1990's (Fig. 1-4-27a), then the nucleic acid hybridization technology was developed in early 2000's (Fig. 1-4-27b). With the popularity of PCR instruments, PCR detection methods for various pathogens of shrimp are widely used in laboratories (Fig. 1-4-27c), and now the detection kit based on the technology of loop-mediated isothermal amplification (LAMP) has been used (Fig. 1-4-27d).

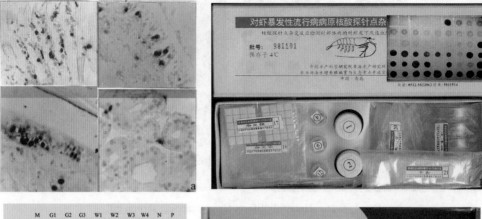
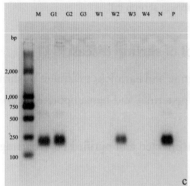
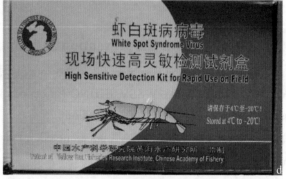

a. Monoclonal antibody detection (Data from Jie Huang); b. Detection kit based on dot blot hybridization technology; c. PCR method; d. Detection kit based on LAMP technology.

Fig. 1-4-27　Development of detection technology for shrimp pathogens

2.5.2　Species and function of probiotics in shrimp aquaculture

A. Probiotic species: A variety of bacillus (*B. subtilis*, *B. licheniformis*, *B. amyloliquefaciens*), lactic acid bacteria, photosynthetic bacteria and so on.

B. Function: Increase shrimp immunity, improve water quality, antagonize pathogens, improve the intestinal function.

C. Application scope: Added to the aquaculture water or shrimp feed directly or scale-up cultivation before use.

A train of *Pseudoalteromonas* isolated from shrimp pond had a broad-spectrum antagonistic activity against *Vibrio* (Fig. 1-4-28).

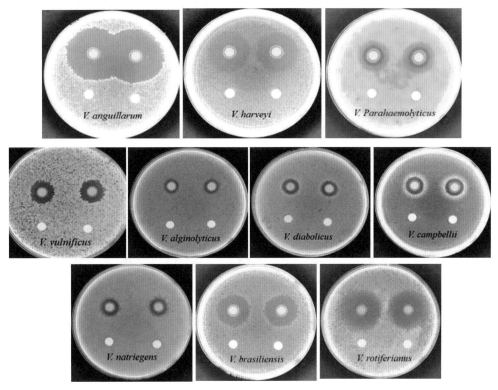

Fig. 1-4-28 Antibacterial effect of *Pseudoalteromonas* sp.
(Data from Zhang et al., 2016)

3 Future prospect of shrimp farming

A. Develop intelligent fishery technologies (such as automatic water exchange system, automatic feeding machine, automatic water quality detecting and control system).

B. Develop disease resistant and fast growing varieties to support the shrimp industry.

C. Develop microbial technologies for shrimp cultivation such as biofloc technology, multi-functional probiotics to improve water quality and reduce water exchange rate.

D. Utilize biosecurity strategy in shrimp farming to prevent and control shrimp diseases.

References

Chen J, Bian P. An influence of the low salinity upon tigershrimp *Penaeus japonicus* [J].Journal of Zhejiang College of Fisheries,1994, 13(4): 289-291.

Wu H Y, Wang X H, Yang B, et al. Tracking of shrimp multiple pathogens in a shrimp farm [J]. Progress in Fishery Sciences, 2019, 40(4): 104-114.

Zhang H H, Wang X H, Li C, et al. Isolation and identification of a *Bacillus* sp. strain and its role in Bioflocs for the shrimp culture system [J]. Progress in Fishery Sciences, 2016, 37(2): 111-118.

Zhang H H, Wang X H, Li C, et al. Isolation and identification of a bacterial strain with *Vibrios*-Antagonism from shrimp ponds [J]. Progress in Fishery Sciences, 2016, 37(3): 85–92.

Quantitative genetic basis of shrimp selective breeding

Author: Luan Sheng
E-mail: luansheng@ysfri.ac.cn

1 What is selective breeding?

Selective breeding (also called artificial selection) is the process by which humans use animal breeding and plant breeding to selectively develop particular phenotypic traits (characteristics), typically by choosing animal or plant males and females which will sexually reproduce and have offspring together (Fig. 1-5-1).

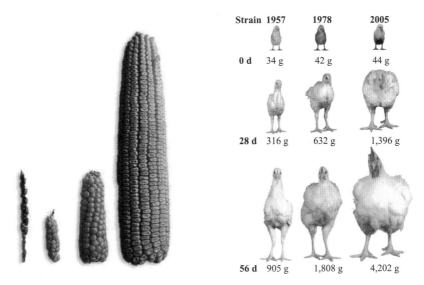

Fig. 1-5-1 Selective breeding in plants and animals
(Data from Zuidhof, 2014)

1.1 Family

Family includes full-sibs and half-sibs (Fig. 1-5-2).

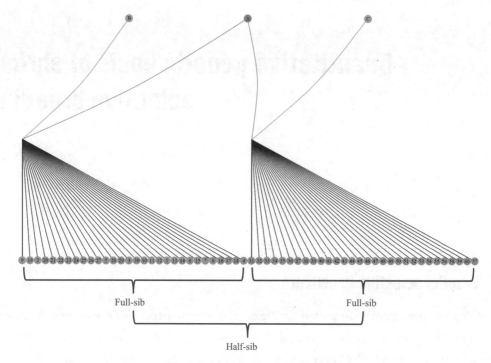

Fig. 1-5-2　Full-sibs and half-sibs

1.2　Family-based selection

Family-based selection process was shown in Fig. 1-5-3.

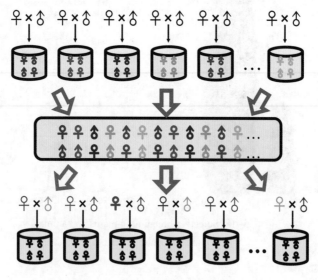

Fig. 1-5-3　Family-based selection

1.3　Genetic gains from artificial selection

Genetic gains from artificial selection were shown in Fig. 1-5-4.

Why is quantitative genetics important for aquaculture?

A. Economically important traits (*i.e.* growth rate, feed efficiency traits, resistance to disease) are complex quantitative traits.

B. Environmental variation reduces efficiency of selection.

C. Quantitative genetics theory improves the selection accuracy and is the basis for selective breeding programs.

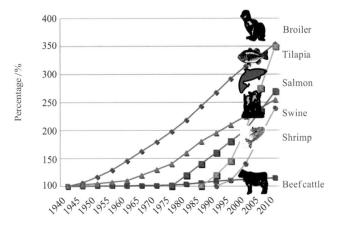

Fig. 1-5-4   Genetic gains from artificial selection
(Data from Chamberlain, 2010)

2   Definition of quantitative genetics

Quantitative genetics, also referred to as the genetics of complex traits, is the study of such characters and is based on a model in which many genes influence the trait and in which non-genetic factors may also be important (Hill, 2010). The analysis of traits shows that the variation is determined by both a number of genes and environmental factors (Figs. 1-5-5 to 1-5-6).

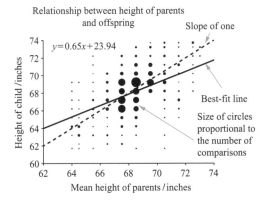

Fig. 1-5-5   Phenotype is highly uninformative as to underlying genotype
(Data from https://o.quizlet.com/0TucI-IuwX2IFN0VD-zCJQ_m.png2018)

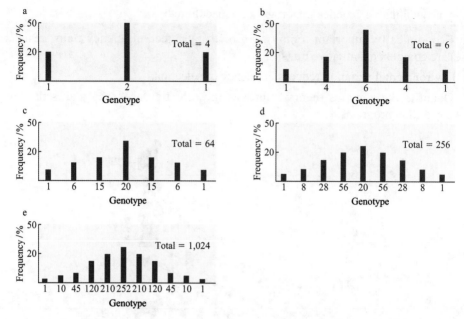

a. 1 gene pair $Aa \times Aa$; b. 2 gene pairs $AaBb \times AaBb$; c. 3 gene pairs $AaBbCc \times AaBbCc$; d. 4 gene pairs $AaBbCcDd \times AaBbCcDd$; e. 5 gene pairs $AaBbCcDdEe \times AaBbCcDdEe$.

Fig. 1-5-6   Effect of genotype

3   Contents of quantitative genetics

3.1   Contents

A. Continuous phenotypic variation within populations.

B. Causes of variation:

a. Genes *vs*. environment;

b. Interactions between genes and environment;

c. Components of genetic variation;

d. Components of environmental variation.

3.2   Goals

A. The goals of quantitative genetics are first to partition total traits variation into genetic *vs*. environmental components.

B. This information (expressed in terms of variance components) allows us to predict resemblance among relatives.

Is variation genetic or non-genetic? (Fig. 1-5-7)

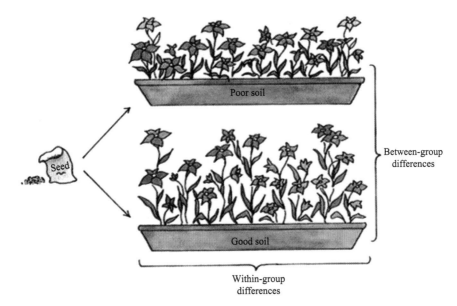

Fig. 1-5-7 Factors influencing variation
(Data from http://slideplayer.com/slide/2329426/8/images/34/Heritability.jpg2018)

Different cultivation sites may lead to genotype by environment interaction (Fig. 1-5-8).

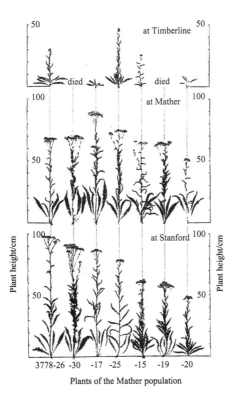

The numbers at the bottom are those of the individual plants.
Fig. 1-5-8 Genotype by environment interaction

Different environment factors (*eg.*, salinity) may lead to genotype by environment interaction (Figs. 1-5-9-10).

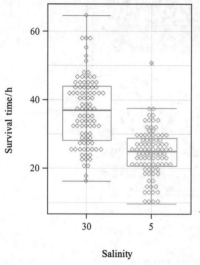

Fig. 1-5-9 Effect of salinity on survival   Fig. 1-5-10 Survival time at different salinity

4 Basic conceptions and methods

The basic formulas of quantitative genetics are as follows:

$$P = G + E, \qquad \text{(Equation 1)}$$

$$V_P = V_G + V_E. \qquad \text{(Equation 2)}$$

P, phenotypic value;

V_P, phenotypic variance;

G, genotypic value;

V_G, genotypic variance;

E, environmental deviation;

V_E, residual variance.

4.1 Decomposition of genotypic value

The formula of decomposition of genotypic value is as follows:

$$G = A + D + I \qquad \text{(Equation 3)}$$

G, genetic effect;

A, additive effect (also called breeding value): Effect of each allele being added together; heritable from parents to offspring;

D, dominant effect (not heritable): Effect of interaction between alleles of the same locus from parents within the individual;

I, epistatic effect (not heritable): Effect of interaction between alleles of different loci from parents within the individual.

4.2 Heritability of quantitative traits

Narrow sense heritability: straightforward to estimate ratio of V_A to V_P.

$$h^2 = \frac{V_A}{V_P} \times 100\%. \qquad \text{(Equation 4)}$$

V_A, additive genetic variance;

V_P, phenotypic variance;

h^2: The degree to which offspring resemble their parents.

Simple example is as Fig. 1-5-11.

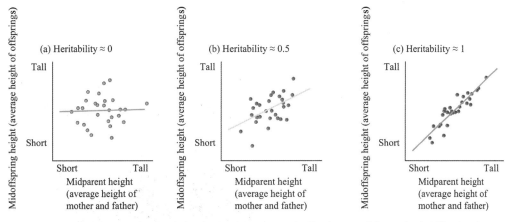

Fig. 1-5-11 Phenotypic correlation between parents and offsprings at different heritability levels
(Data from http://slideplayer.com/slide/7635146/25/images/24/Heritability+Determination.jpg)

Genetic parameters for traits in *Penaeus vannamei* are shown in Table 1-5-1.

Table 1-5-1　Genetic parameters for traits in *P. vannamei*

Traits	$h^2 \pm$ SE	$c^2 \pm$ SE
Body weight	0.335 ± 0.087	0.084 ± 0.031
Feed efficiency (FER)	0.577 ± 0.232	–
Residual feed intake (RFI)	0.641 ± 0.237	–
Survival rate	0.051 ± 0.034	0.093 ± 0.018
Resistance to WSSV	0.145 ± 0.073	0.015 ± 0.031
Tolerance to low temperature (CDH)	0.026 ± 0.021	–
Tolerance to high ammonia (postlarva, survival time)	0.154 ± 0.045	–

(*to be continued*)

Traits	$h^2 \pm SE$	$c^2 \pm SE$
Tolerance to high ammonia (adult shrimp, survival rate)	0.078 ± 0.033	-
Fillet yield	0.120 ± 0.040	-
Reproductive traits (egg number, spawning frequency, spawning success)	0.12, 0.06, 0.19	-

Notes: h^2, heritability; c^2, common environment effect; SE, standard error.

4.3 Breeding value (BV)

Breeders focus on breeding value (BV) because it is heritable.

The expected value of an offspring is the average breeding value of its parents:

$$BV_{Offspring} = \frac{BV_{Sire}}{2} + \frac{BV_{Dam}}{2} \quad \text{(Equation 5)}$$

$BV_{Offspring}$, breeding value of offspring;

BV_{Sire}, breeding value of sire;

BV_{Dam}, breeding value of dam.

High throughput phenotype is shown as Fig. 1-5-12.

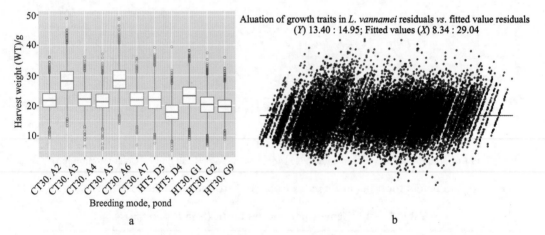

a. Harvest weight in different culture environments; b. High throughput phenotype analysis.

Fig. 1-5-12 Records per generation over 40,000 individuals

4.4 Genetic gain

The formula of genetic gain is as the follows:

$$\Delta G = ih\sigma_A \quad \text{(Equation 6)}$$

i, selection intensity;

h, selection accuracy;

σ_A, additive genetic standard deviation.

Genetic gain of body weight is shown in Fig. 1-5-13.

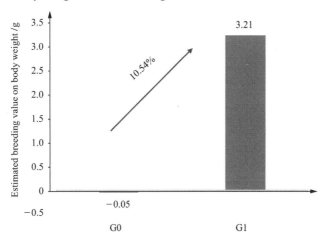

Fig. 1-5-13 Genetic gain of body weight

Genetic gain of body weight in P. vannamei

Genetic gain of body weight response to selection is 2.16%-3.89% (Fig. 1-5-14) per generation.

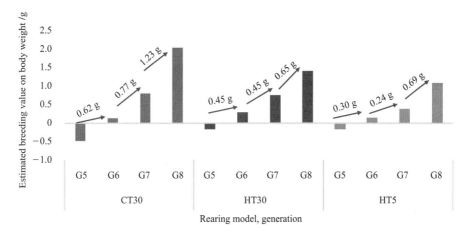

Fig. 1-5-14 Genetic gain of body weight in *P. vannamei*

A. Tank-salinity 30: the total genetic gain is 2.50 g; the body weight increases by 11.66%; the total genetic gain is 0.83 g per generation on average; the body weight increases by 3.89% per generation on average.

B. Pond-salinity 30: the total genetic gain is 1.55 g; the body weight increases by 7.51%; the total genetic gain is 0.52 g per generation on average; the body weight increases by 2.50% per generation on average.

C. Upland pond-salinity 5: the total genetic gain is 1.23 g; the body weight increases by

6.49%; the total genetic gain is 0.41 g per generation on average; the body weight increases by 2.16% per generation on average.

4.5 Inbreeding coefficient

The coefficient of inbreeding is the probability of inheriting two copies of the same allele from an ancestor that occurs on both sides of the pedigree (Fig. 1-5-15).

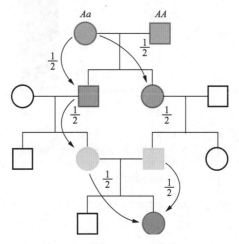

Fig. 1-5-15　Inbreeding coefficient
(Data from http://bio3400.nicerweb.net/Locked/media/ch25/25_18-coefficient_of_inbreeding.jpg2018)

A sample of pedigree is shown in Fig. 1-5-16.

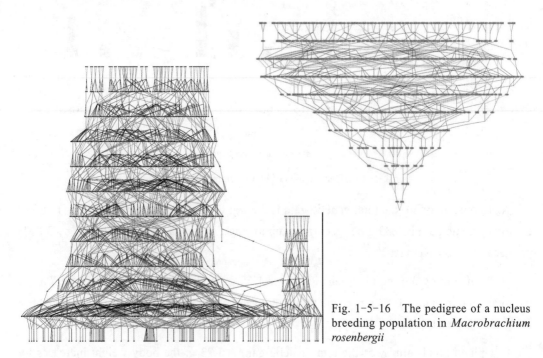

Fig. 1-5-16　The pedigree of a nucleus breeding population in *Macrobrachium rosenbergii*

Inbreeding coefficient of the nucleus breeding population is shown in Fig. 1-5-17 and Fig. 1-5-18.

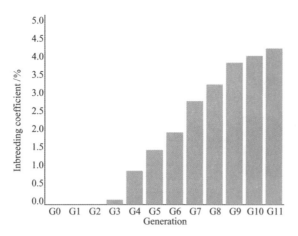

Fig. 1-5-17　Inbreeding coefficient of 12 generations in *M. rosenbergii*

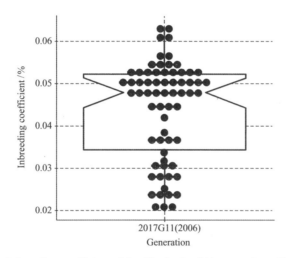

Fig. 1-5-18　Inbreeding coefficient of families in the G11 generation of *M. rosenbergii*

5　How to construct a selective breeding population

Selective breeding process is shown in Fig. 1-5-19.

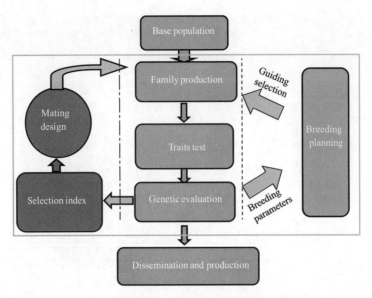

Fig. 1-5-19　Selective breeding process

5.1　Objective traits

The objective traits need to be determined. They usually include one or more of the following:

A. Growth and survival.

B. Stress resistance.

C. Disease resistance.

D. Feed efficiency.

E. Group uniformity.

F. Meat yield.

G. Fecundity or quality traits.

5.2　Population collection

Genetic background analysis is in Figs. 1-5-20 to 1-5-22.

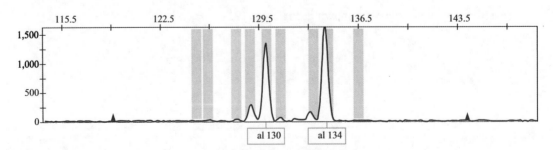

Fig. 1-5-20　Microsatellite typing

	CP-1	SIS-1	CP-2	PRI-1	PRI-2	CP-3	PRI-3	CP-4	CP-5	PRI-4	PRI-5	TR
CP-1	1											
SIS-1	0.6147	1										
CP-2	0.4288	0.4305	1									
PRI-1	0.3454	0.5548	0.3546	1								
PRI-2	0.2665	0.3711	0.4875	0.5771	1							
CP-3	0.8144	0.5637	0.4805	0.4934	0.3844	1						
PRI-3	0.3945	0.5093	0.3795	0.8365	0.6536	0.4854	1					
CP-4	0.7257	0.5884	0.5151	0.3355	0.4101	0.6109	0.35	1				
CP-5	0.7349	0.6412	0.6395	0.5243	0.5357	0.7377	0.4436	0.7967	1			
PRI-4	0.3247	0.48	0.3185	0.7823	0.6845	0.3919	0.8933	0.3865	0.4006	1		
PRI-5	0.4417	0.4594	0.5353	0.7004	0.6555	0.4725	0.7328	0.5168	0.5657	0.7433	1	
TR	0.2754	0.2992	0.3454	0.2896	0.222	0.2709	0.3197	0.2834	0.3062	0.2774	0.2574	1

Fig. 1-5-21 Genetic similarity analysis

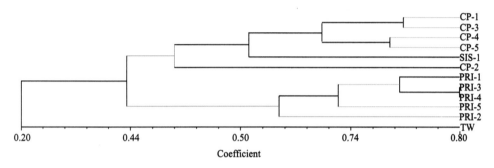

Fig. 1-5-22 Population cluster analysis

Founder pedigree reconstruction is in Fig. 1-5-23 and Fig. 1-5-24.

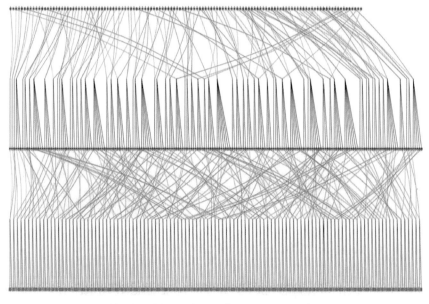

Fig. 1-5-23 Before pedigree reconstruction

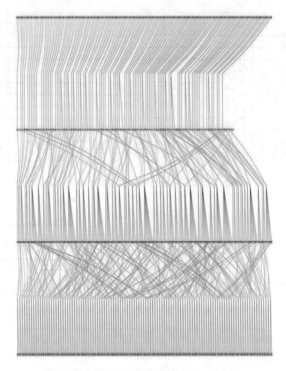

Fig. 1-5-24　After pedigree reconstruction

5.3　Individual tag

Family-based selective breeding program for shrimp is in Fig. 1-5-25.

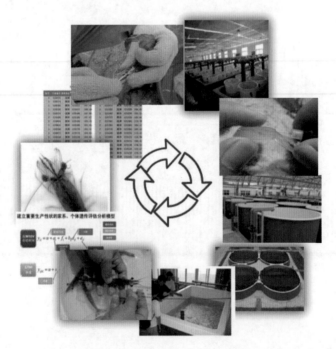

Fig. 1-5-25　The process of family-based selection breeding program for shrimp

Key technology contains the following aspects:

A. Family production.

B. Family/individual tagging.

C. Trait test and genetic evaluation.

D. Optimum mating scheme.

...

6 Acknowledgements

The above research is supported by the following members:

A. The members of Dr. Jie Kong's team from YSFRI.

B. Dr. Yutao Li: Introduction to quantitative genetics, genome wide association studies and genomic selection.

C. Dr. Ponzoni: Estimation of phenotypic and genetic parameters.

D. Prof. Qin Zhang: The basic properties of quantitative traits.

References

Zuidhof M J, Schneider B L, Carney V L, et al. Growth, efficiency, and yield of commercial broilers from 1957, 1978, and 2005 [J]. Poult Sci, 2014, 93(12): 2970-2982.

https://www.aquaculturealliance.org/wp-content/uploads/2015/04/goal10-Chamberlain.pdf.

Breeding program for Pacific white shrimp, *Penaeus vannamei*

Author: Sui Juan

E-mail: suijuan@ysfri.ac.cn

1 Current situation of Pacific white shrimp seed Industry

The advantages of Pacific white shrimp (Fig. 1-6-1): rapid growth, good survival in high-density culture, disease tolerance.

Fig. 1-6-1　Pacific white shrimp

1.1　Cultivation in China

Annual output exceeds 1.6 million tons, with 90% yield of farmed shrimp (Figs. 1-6-2 to 1-6-3) (MOA, 2017).

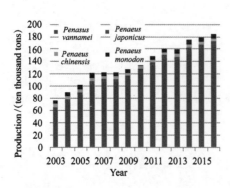

Fig. 1-6-2　Shrimp production in China

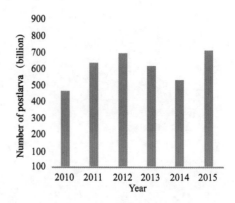

Fig. 1-6-3　Seedling demand of *Penaeus vannamei* in China

1.2 Problems

1.2.1 Rely on foreign companies, high cost, uncontrollable

Major international shrimp companies include SIS, Kong Bay, HHA, University of Guam, Charoen Pokphand Group, Molokai, Primo, Molokai, Global GEN, Blue-Genetics, SyAqua, etc (Figs. 1-6-4 to 1-6-5).

Fig. 1-6-4 Parent shrimp demand in China   Fig. 1-6-5 International shrimp companies

Fig. 1-6-6 shows the statistics on China's import of *P. vannamei* in 2013-2017.

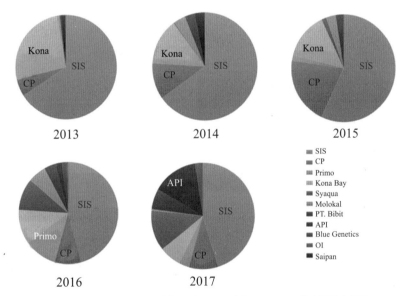

Fig. 1-6-6 Statistics of China's import of *P. vannamei* in 2013-2017

New varieties of *P. vannamei* selected independently in China is shown in Table 1-6-1.

Table 1-6-1 Nine new varieties of *P. vannamei* selected independently in China

New varieties	Breeding technique	Growth	Tolerance	Egg amount	Others
Zhongxing No.1	FS:30	−	WSSV tolerance	−	−
Guihai No.1	FS:60	+	−	−	Survival rate

(*to be continued*)

New varieties	Breeding technique	Growth	Tolerance	Egg amount	Others
Zhongke No.1	MS + FS: 20-40	+	-	-	Freshwater stress
Kehai No.1	FS: >200	+	-	-	Survival rate
Renhai No.1	FS + CB: >100	+	-	-	Survival rate
Haixingnong No.2	FS: >100	+	-	-	Factory + high level pool
Guangtai No.1	FS: 320	+	-	+	Survival rate
Xinghai No.1	FS + CB: 60	+	Low temperature and salt tolerance at seedling stage	-	Survival rate
Zhengjinyang No.1	FS: >100	+	Ammonia nitrogen tolerance	-	-

Notes: MS, Mass Breeding; FS, Family Breeding; CB, Cross Breeding. +, Different breeding techniques were used at the same time.

1.2.2 Introduced species can not meet the requirement of diverse breeding modes and broad areas in China

Fig. 1-6-7 shows different breeding modes.

a. Indoor tank; b. High pond; c. Indoor tank; d. High pond.

Fig. 1-6-7 Different breeding modes

2 Selection and breeding technology system of Pacific white shrimp

2.1 Population collection

A. Foundation for improved shrimp cultivation: abundant genetic variation, high genetic diversity.

B. Shrimp resources come from seven different regions of Singapore and United States, Which were named as SIN, UA1, UA2, UA3, UA4, UA5, UA6.

2.2 Breeding Objective

The breeding program was carried out in marine genetic breeding center of agriculture ministry.

2.2.1 Requirements

A. Economic or critical traits.

B. Hereditable variation.

C. Accurately test at low cost.

2.2.2 Objective traits

A. Growth and survival.

B. Stress resistance.

C. Disease resistance.

D. Feed efficiency.

E. Group uniformity.

F. Meat yield.

G. Fecundity or quality traits.

H. Shrimps should adapt to different culture environment.

2.3 Virus detection and genetic background analysis

Candidate parent should go through virus detection and genetic background analysis (Figs. 1-6-8 to 1-6-9).

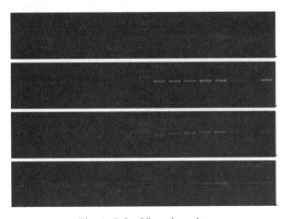

Fig. 1-6-8 Virus detection

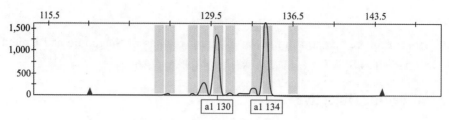

Fig. 1-6-9　Microsatellite typing

The relationship between individuals can be improved by molecular markers such as microsatellites (Figs. 1-6-10 to 1-6-11).

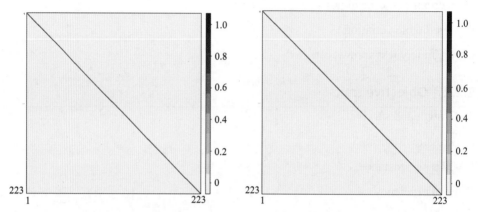

Fig. 1-6-10　Before pedigree reconstruction　　　Fig. 1-6-11　After pedigree reconstruction

2.4　Family construction

Important operation techniques for family construction include directional mating technique, seedling breeding technique and family and individual marker technique (Figs. 1-6-12 to 1-6-15).

Fig. 1-6-12　250 m^2 tank　　　　　　　　　Fig. 1-6-13　Parent shrimp

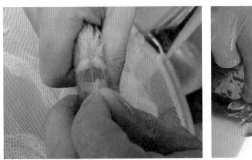
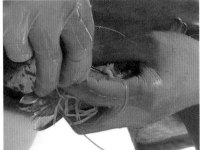

Fig. 1-6-14　Male　　　　Fig. 1-6-15　Female

2.5　Family cultivation

Standard breeding rules for seedling are shown as Figs. 1-6-16 to 1-6-18.

Fig. 1-6-16　The first standardization nauplii: 5,000　　Fig. 1-6-17　The second standardization post-larvae: 1,500　　Fig. 1-6-18　The third standardization post-larvae: 500

2.6　Family tag

VIE tagging process is shown as Fig. 1-6-19.

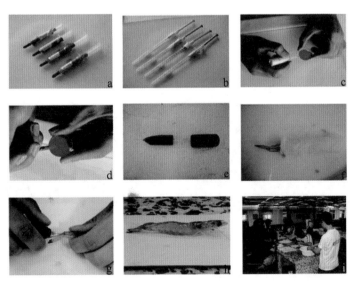

a. Pigment; b. Coagulant; c. The coagulant is added to the pigment in a certain proportion and mix evenly; d. Inhale the mixture into the syringe; e. Put the syringe in the booster; f. Wrap the shrimp with wet gauze; g. Pour the mixture into the tail of the shrimp; h. Marked shrimp; i. The scene of shrimp tagging.

Fig. 1-6-19　VIE tagging process

2.7 Growth and survival test

The marked shrimps were cultured in different sites (Figs. 1-6-20 to 1-6-21).

a. Communally rearing; b. Family conservation.
Fig. 1-6-20　Industries in Huanghua city, Hebei Province

Fig. 1-6-21　Industries in Qingdao city, Shandong Province

Eye ring is used to tagging shrimp individuals (Fig. 1-6-22).

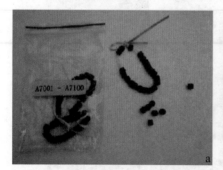

a. Eye ring; b. Shrimp after putting on the eye ring.
Fig. 1-6-22　Individual tagging

2.8 Disease resistance testing

Shrimps need to be tested for infection (Fig. 1-6-23).

a. Disease resistance testing site; b. Indoor culture tank; c, d, e. Shrimps after eating poison bait.

Fig. 1-6-23 Disease resistance testing

2.8.1 Size of breeding population

The more families and individuals there are, the more favorable it is for selection and inbreeding control (Figs. 1-6-24 to 1-6-25).

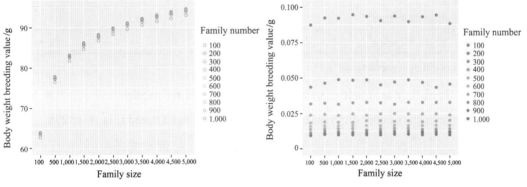

Fig. 1-6-24 Individual number in the family   Fig. 1-6-25 Individual number in the family

The genealogical map of single tail individuals needs to be tested (Fig. 1-6-26).

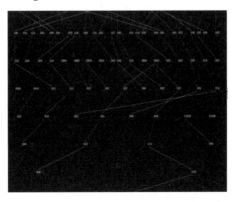

Fig. 1-6-26 Genealogical map of single tail individuals in G6 generation of *P. vannamei* breeding population

2.8.2 High throughput phenotypic data

A. A large number of individuals need to be weighed.

B. A lot of weight data need to be analyzed.

2.8.3 Heritability of different traits

Heritability of different traits (Table 1-6-2) are calculated using Equation 1.

$$h^2 = \frac{a}{a+c+e} \qquad \text{(Equation 1)}$$

h^2, heritability;

a, additive genetic variance;

c, common environment variance;

e, residual error.

Table 1-6-2 Heritability of different traits

Traits	$h^2 \pm SE$	$c^2 \pm SE$
Harvest body weight	0.335 ± 0.087	0.084 ± 0.031
Survival rate	0.051 ± 0.034	0.093 ± 0.018
WSSV resistance	0.145 ± 0.073	0.015 ± 0.031
Cold temperature tolerance	0.026 ± 0.021	-
Ammonia-nitrogen tolerance*	0.154 ± 0.045	-
Ammonia-nitrogen tolerance*	0.078 ± 0.033	-

Note: *, Result in different populations.

2.8.4 Genetic parameters under different breeding modes

Genetic parameters under different breeding modes need to be tested (Fig. 1-6-27).

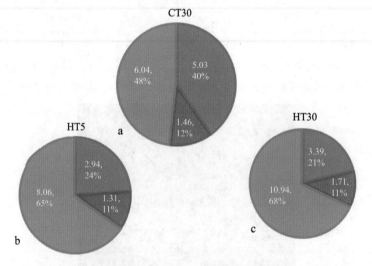

■ Additive genetic variance   ■ Common environment variance   ■ Residual variance

a. Cement tank, salinity 30; b. High pond, salinity 5; c. High pond, salinity 30.

Fig. 1-6-27 Genetic parameters under different breeding modes

2.8.5 Genetic correlation among important economic traits

Table 1-6-3 shows the genetic correlation among important economic traits.

Table 1-6-3 Genetic correlation among important economic traits

Traits	Correlation coefficient
Body weight with low temparature	-0.77 ± 0.46
Body weight with ammonia tolerance	0.83 ± 0.12
Body with meat yield	0.32 ± 0.28

2.8.6 Genetic and environment interaction

Table 1-6-4 shows the genetic and environment interaction.

Table 1-6-4 Genetic and environment interaction

Generation	Site	HBHH
G0	QDAS	0.813 ± 0.122
G1	QDAS	0.976 ± 0.106
Across	QDAS	0.943 ± 0.066

Notes: QDAS, Aoshan Town, Qingdao; HBHH, Huanghua Ctly, Hebei.

2.8.7 Predicted genetic gain based on estimated breeding value

The selection response is different under different cultivation modes (Fig. 1-6-28).

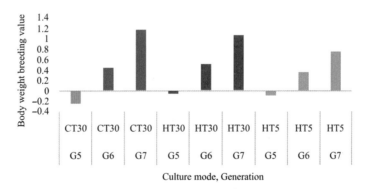

CT30: 1.42 g;   HT30: 1.14 g;   HT5: 0.86 g.

Fig. 1-6-28 Selection response per generation 4.39%-6.31%

2.8.8 Comparative analysis of new varieties and common commercial varieties

New varieties need to be compared with commercial seedlings (Fig. 1-6-29).

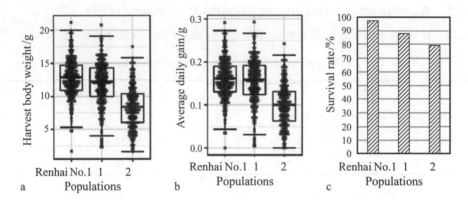

a. Harvest body weight: 7.38%–53.23%; b. Average daily gain : 6.67%–60%;
c. Survival rate: 10.33%–22.05%.

Fig. 1-6-29　Comparative analysis of new varieties and common commercial varieties

Application extension and benefits

In 2014–2016, the accumulative extension area amounts to more than 2,666.6 hm^2, the new output value is 940 million *yuan*, and new profit is 280 million *yuan* (Fig. 1-6-30).

Fig. 1-6-30　Harvest of new varieties

3　Acknowledgements

The above research is supported by members of Dr. Jie Kong's team from YSFRI.

References

Lu X, Luan S, Cao B X, et al. Estimation of genetic parameters and genotype-by-environment interactions related to acute ammonia stress in Pacific white shrimp (*Litopenaeus vannamei*) juveniles at two different salinity levels [J]. PLoS ONE, 12(3): e0173835.

MOA (Ministry of Agriculture of the People's Republic of China). China fishery statistical yearbook 2017[M]. Beijing: China Agricultural Press, 2017.

Ruan X H, Luo K, Luan S, et al. Evaluation of growth performance in *Litopenaeus vannamei* populations introduced from other nations [J]. Journal of fisheries of China. 2013, 37(1): 34-42. (In Chinese).

Sui J, Luan S, Luo K, et al. Genetic parameters and response to selection for harvest body weight of Pacific white shrimp, *Litopenaeus vannamei* [J]. Aquaculture Research, 2016, 47, 2795-2803.

Technical system of ecological culture of shrimp in China

Authors: Li Jitao, Li Jian
E-mail: lijt@ysfiac.cn; lijian@ysti.ac.cn

1　Background

1.1　Main way of mariculture in China

From Fig. 1-7-1 we can see, raft culture and bottom sowing culture occupy absolute proportion, and pond culture has developed rapidly recently, accounting for 12.06% of the total.

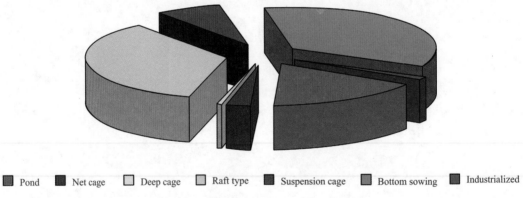

■ Pond　■ Net cage　□ Deep cage　□ Raft type　■ Suspension cage　□ Bottom sowing　■ Industrialized

Fig. 1-7-1　Main way of mariculture in China

1.2　The distribution of marine pond

Guangdong Province has the maximum pond culture output in China, followed by Jiangsu Province, Shandong Province, Fujian Province, Zhejiang Province, Liaoning Province and Guangxi Zhuang Autonomous Region (Fig. 1-7-2).

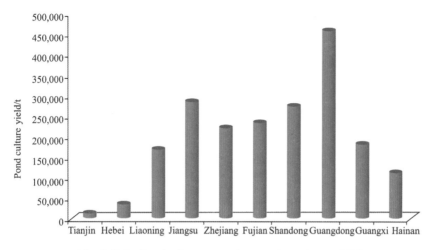

Fig. 1-7-2　Pond culture output in coastal provinces of China

1.3　The mariculture area and production of shrimp in China

The mariculture area and production of shrimp in China is shown in Fig. 1-7-3.

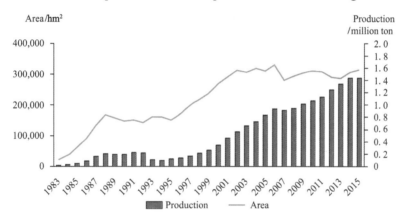

Fig. 1-7-3　The mariculture area and production of shrimp in China

1.4　Main mariculture shrimps in China

White-leg shrimp *Penaeus vannamei* (Fig. 1-7-4) is the main culture shrimp in China. Chinese shrimp *Penaeus chinensis* (Fig. 1-7-5) is the main culture shrimp in North China. Giant tiger prawn *Penaeus monodon* (Fig. 1-7-6) is the main culture shrimp in South China. Kuruma prawn *Penaeus japonicus* (Fig. 1-7-7) is the most economic culture shrimp. Ridgetail white prawn *Exopalaemon carinicauda* (Fig. 1-7-8) is the characteristic culture shrimp.

Fig. 1-7-4　White-leg shrimp
P. vannamei

Fig. 1-7-5　Chinese shrimp
P. chinensis

  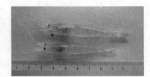

Fig. 1-7-6　Giant tiger prawn　　Fig. 1-7-7　Kuruma prawn　　Fig. 1-7-8　Ridgetail white prawn
　　　P. monodon　　　　　　　　　*P. japonicus*　　　　　　　　　*E. carinicauda*

2　Main culture mode

2.1　Extensive culture

Extensive culture was used in the early stage of industrial development in the 20th century, using the bay to build large ponds, realizing extensive culture by tidal inflow. In the southern area it is called "fish culture", and in the northern area it is called "harbor culture". The growth of shrimp depends entirely on the natural baits and in the water area.

The features of extensive culture are: Firstly, low cost, low yield and poor effect; Secondly, it can not drain off the pond water, resulting in sediment accumulation causing shrimp pond aging, and pathogenic microorganisms proliferation causing shrimp disease outbreak.

2.2　Semi-intensive culture

Semi-intensive culture was used widely in the 1980s–1990s of the 20th century.

The culture ponds are built on the intertidal flat, widely used in northern China, and the pond area is 15–30 m^2 with water depth over 1 m. The mode is tide-dependent, artificial bait, lack of water exchange or aerator and low artificial regulation. The culture density is 15,000–30,000 individuals per square meter and the output is 50–200 kg.

The shortcomings of this mode were:

A. The water is eutrophic;

B. The artificial regulation and control is low;

C. Disease spread is serious, and the output is unstable.

2.3　Intensive culture

2.3.1　High level pond culture mode

The mode used plastic bottom and cement wall. The pond area is 0.13–0.6 hm^2 and the depth is 2.0–2.3 m (Fig. 1-7-9). The mode is mainly distributed in coastal inner bay area (Fig. 1-7-10).

Intensive culture has high intensive degree, equipped with sand filter well, water storage disinfection tank, water intake and drainage system, oxygen system, central sewage system, etc.

The stocking density is 150×10^4 tails per hectare, and the yield is generally 20–30 t/hm^2, which is 6–8 times of that in the traditional shrimp culture model.

Fig. 1-7-9 High level pond   Fig. 1-7-10 Cultivation area

2.3.2 Greenhouse cultivation mode

The mode includes mainly fresh pond greenhouse, also the high level greenhouse and the cement greenhouse (Fig. 1-7-11).

The mode is mainly built in advance, extending the cultivation time, staggering the harvest season, and avoiding the disease peak period. The mode can achieve 2-3 harvest in one year, and the annual yield can reach 30-45 t/hm^2.

Fig. 1-7-11 Greenhouse culture

Features:

A. The mode is intensive, staggering peak with good economic benefits.

B. The mode is enclosed with less air convection and harmful substances.

C. The mode uses different temperature and it is difficult to change water quality.

D. The mode has insufficient photosynthesis of algae.

2.3.3 Industrialized culture

The mode constructs high standard pond and cultures shrimps using filtered seawater. Special equipment is adopted to complete work of aeration, heat preservation and sewage discharge, keeping the water quality stable through controlling water factors (Fig. 1-7-12).

Fig. 1-7-12 Industrialized culture

The stocking density is 400-500 tails per square meter, and the culture period is 100-120 d. The yield is 3-5 kg/m² and the survival rate is 60%-70%.

Features:

A. The mode is strong artificial controlled, high yield and stable income, but need high investment and technique.

B. The mode is a high-density culture using 6%-21% nitrogen. 50%-60% nitrogen converting into soluble ammonia nitrogen can cause serious pollution.

Recirculating aquaculture mode can achieve the recycling of aquaculture water through a series of water treatment links, and greatly improve the utilization of water resources (Fig. 1-7-13).

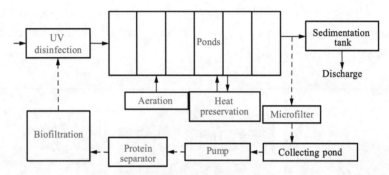

Fig. 1-7-13 Recirculating aquaculture mode

The water treatment facilities are matched with microfilter, protein separator, biological filter, ozone disinfection device (Figs. 1-7-14 to 1-7-15). The wastewater of this mode is filtered by microfilter, protein separator and other mechanical filters, then filtered and disinfected by UV light and recycling (Fig. 1-7-16). The mode needs 7 times of the daily circulation of aquaculture water, and the exchange rate of the daily water is 5% (Fig. 1-7-17).

Fig. 1-7-14 Microfilter   Fig. 1-7-15 Protein separator   Fig. 1-7-16 Biofiltration   Fig. 1-7-17 Industrial pond

3 Main problems in aquaculture industry

Challenges in aquaculture industry development mode are as follows:

A. Production safety: Culture space compression and lack of support for technological innovation.

B. Ecological safety: Terrigenous pollution is a major threat, and pollution caused by aquaculture production is serious.

C. Food safety: Quality safety and market supervision are worrying.

3.1 Development direction: ecological culture and industrial culture

3.1.1 Culture environment worsens, causing stress effects

Human activities, such as industrial pollution, agricultural pollution, domestic waste, cause eutrophication of water, which also result in environmental pollution (Fig. 1-7-18) and frequent diseases.

Fig. 1-7-18   Environmental pollutions

3.1.2 Self problem of aquaculture

The aquaculture ecosystem is characterized by simple structure, low material recycling efficiency, high yield and vulnerability of ecosystem and lack of regulation for culture ecosystems.

Only 75% of the bait is fed, and the rest is wasted in pond. 85% of feed nitrogen is assimilated by shrimp, and 15% of nitrogen was discharged into water and sediment (Fig. 1-7-19) (Simon and Matthew, 1998).

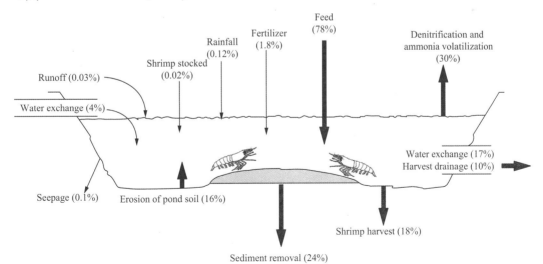

Fig. 1-7-19   Nutrient transport efficiency

About 0.27 kg (dry weight) of feces, 0.25 kg of suspended particulate matter and 6.12 g of ammonia nitrogen are produced by 1 kg feed of shrimp. Organic pollution deposites at the bottom, and aquaculture wastewater pollutes offshore and destructs coastal biodiversity, causing red tides (Fig. 1-7-20).

Fig. 1-7-20　Pond bottom pollution

3.2　Germplasm degradation and quality instability

The germplasm of culture shrimp (Fig. 1-7-21) is deteriorated, which causes lack of good varieties with outstanding economic traits and resistance to ecology. The germplasm resources of industrial culture (*P. vannamei*) are totally imported from abroad, and germplasm deteriorates seriously (Fig. 1-7-22). Meanwhile, the quality of shrimp seedling is not perfect, and specific pathogen free (SPF) control technology is weak (Fig. 1-7-23).

Fig. 1-7-21　Commericial shrimp seeds　　Fig. 1-7-22　Various brands of seeds　　Fig. 1-7-23　Size difference of seeds

3.3　Weak technological innovation and infrastructure

The increase of mariculture yield is mainly achieved by feeding high-density concentrated feed and high-intensity feed. The genetic improvement rate of varieties is about 16%, and the coverage rate of improved varieties is less than 50%. The main problem is the lack of ecological management and control measures. Fish meal is used in large quantities. Nutrient budgets of semi-intensive ponds in Honduras indicated that 72% of nitrogen was discharged into the external environment with exchanged water. As a result, the output of pollutants in the aquaculture process may lead to self deterioration and eutrophication of coastal waters. Energy consumption is extremely high.

3.4 Safety of aquatic products restrict industrial development

The event of chloramphenicol residues exceeding the standard occurred in 2002 (Fig. 1-7-24). The event of enrofloxacin in roasted eel occurred in 2003, and the maleic green event happened in 2005 (Fig. 1-7-25). Mean while, illegal drugs in turbot exceeded the standard in 2006, and melamine replacing protein was added to feed in 2008 (Fig. 1-7-26). Therefore EU banned the importation of shellfish products from China for 19 years.

Fig. 1-7-24　Chloramphenicol incident

Fig. 1-7-25　Enrofloxacin incident

Fig. 1-7-26　Turbot

4　Ecological shrimp culture in pond

Integrated multi-trophic pond aquaculture, IMTPA

Fig. 1-7-27 shows the pattern diagram of IMTPA.

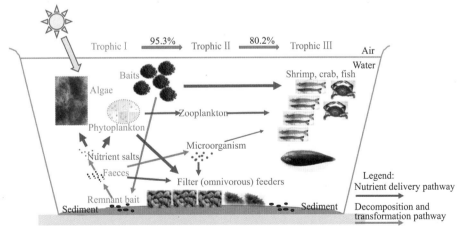
Fig. 1-7-27　The pattern diagram of IMTPA

The nutrient delivery pathway, and the decomposition and transformation pathway were shown in Fig. 1-7-28.

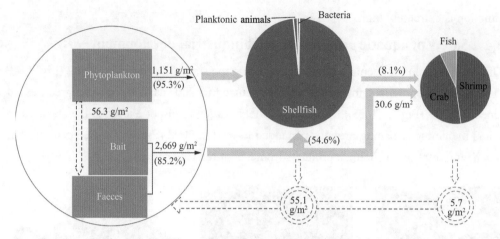

Fig. 1-7-28　The nutrient transport pathway and efficiency in IMTPA

Features:

A. The IMTPA mode improves nutrient utilization and water quality.

B. The IMTPA mode increases economic values of mixed culture species.

C. The IMTPA mode improves disease resistance of cultured species.

The principle of multi-trophic aquaculture construction:

A. The mode must be consistent with environmental requirements and adaptability.

B. The mode need focus on specification and density of polyculture species. Over-densities of cultured species in combination will reduce the growth rate of the host species.

C. The mode must be with no antagonism among polyculture species.

The freshwater Nile tilapia can be mixed in seawater pond after domestication. No one-way or two-way predation occurs among mixed species. The survival rate of crab and shrimp is 100% in mixed culture of shrimp and river crab.

D. Some successful cases are introduced, such as blue shrimp-sea cucumber, shrimp-rock oyster, white prawn-lobster, shrimp-grouper, etc.

4.1　Development of multi-trophic aquaculture in China

4.1.1　IMTA in shallow sea

A. *Laminaria japonica*-mussels inter culture existed in Shandong Province in 1975.

B. *L. japonica*-mussels inter culture existed in Fujian Province in 1979.

C. Three-dimensional cultivation of kelp, *Undaria pinnatifida*, scallop, mussel, sea cucumber, abalone and oyster existed in Changdao, Shandong Province in 1984.

D. The bottom sowing culture of macrophyllous algae and sea treasures existed in Chudao, fish-shellfish (abalone) -kelp (asparagus) existed in Sanggou Bay, and abalone-kelp-sea cucumber culture existed in Lidao Bay in Shandong Province in 2000.

4.1.2 IMTA in marine pond

A. Shrimp-barracuda existed in Ganyu, Jiangsu Province in 1979.

B. Shrimp-*Meretrix* existed in Qidong, Jiangsu Province in 1980.

C. At present, shrimp polyculture with sea cucumber, shellfish, crabs, fish and algae are mixed in pond.

4.2 IMTA pond modes in China

4.2.1 Shrimp-fish mode

Kuruma prawn-fish mode

A. Northern area: Kuruma prawn mixed with puffer fishes (Zhang, 2012).

B. Southern area: Kuruma prawn mixed with grouper (Lei and Lin, 2014).

Whiteleg shrimp-fish mode

A. Southern area: Whiteleg shrimp mixed with green spot, tilapia and moustache catfish (Qiu et al., 2012).

B. Northern area: Whiteleg shrimp mixed with *Cyprinus carpio*, grass carp, tilapia (Zhang et al., 2011).

Chinese shrimp-fish mode

A. Chinese shrimp mixed with puffer fish, gentian grouper, tongue sole, black bream; culture success rate can reach 90%–100%; shrimp culture production increased by 10% (Liu et al., 2007).

B. Chinese shrimp mixed with culture perch: Perch (body length > 15 cm) preys on 4 cm long healthy shrimp and weakly diseased shrimp, but does not feed on dead shrimp. It is not suitable for multi-nutrient culture (Wang and Li, 1999).

4.2.2 Shrimp-crab-shellfish-sea cucumber mode

There were *P. monodon-Scylla serrate* mode (Dong et al., 2012), shrimp-razor clam mode (Wang et al., 1999), shrimp-clam-gracilaria mode (Wang et al., 2006), shrimp-razor clam-tilapia mode (Li et al., 1994), shrimp-crab-clam mode (Li, 2010) and shrimp-fish-shellfish-seaweed mode (Fig. 1-7-29) (Shpied et al., 1992; Shen et al., 2007).

Fig. 1-7-29 The shrimp-crab-shellfish-sea cucumber mode

4.3 A successful example mode

We established the shrimp-crab-shellfish-fish mode (Fig. 1-7-30), which is characterized by ecological, efficient, safe, emission reduction and energy saving, named as IMTPA.

Fig. 1-7-30 The shrimp-crab-shellfish-fish mode

There are six ecological approaches in shrimp culture, including selection of new variety with good traits, application of bottom micro-pore aeration, microbial regulation approach, immune enhancement approach, application of living creature diets, and establishment of ecological culture model.

4.3.1 Selection of new variety with good traits

We established a sustainable shrimp and crab aquaculture breeding technology system.

Five new varieties with good traits were obtained, including *P. chinensis* "Huanghai No. 1" (GS01001-2003), *P. chinensis* "Huanghai No. 3" (GS-01-002-2013), *E. carinicauda* "Huangyu No. 1" (GS-01-005-2017), *Portunus trituberculatus* "Huangxuan No.1" (GS-01-002-2012) and *P. trituberculatus* "Huangxuan No.2" (GS-01-006-2018).

4.3.2 Application of bottom micropore aeration

The advantages of micropore aeration are less resistance, small bubbles, complete dissolved oxygen, and high DO values in the middle of pond bottom as 4.5 or so (Figs. 1-7-31 to 1-7-33).

Fig. 1-7-31 Micro-pore aeration setup 　　Fig. 1-7-32 Application of bottom micro-pore aeration in aquaculture

Fig. 1-7-33 Application of micro-pore aeration in culture pond

The effect on the DO

After 60 d, dissolved oxygen content of micro-pore aeration (6.2–8.1 mg/L) was higher than stone aeration (4.4–6.5 mg/L) significantly (Han et al., 2012) (Fig. 1-7-34). The content of DO of micropore aeration was higher than that of waterwheel (Fig. 1-7-35). There was little difference in DO between the upper, middle and lower layers of the pond.

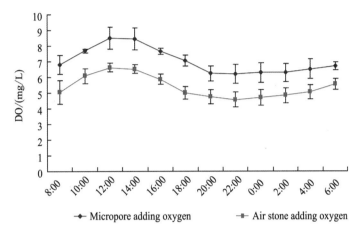

Fig. 1-7-34   Comparison of microporous and aerolite oxygenation

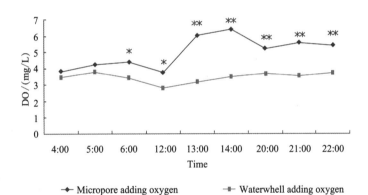

Fig. 1-7-35   Comparison of microporous and waterwheel oxygenation

The effect on the immune enzymes

The technique of micro-hole culture could improve the serum SOD, LYZ, antibacterial enzyme activity, and enhance the immune defense ability of shrimp ($P < 0.05$).

4.3.3 Microbial regulation approach

We screened three probiotics, including *Bdellovbrio bacteriovorus* (Bb), *Rhodopseudomlonas sphaeroides* (Rs) and *Rhodotorula glutinis* (Rg) (Fig. 1-7-36).

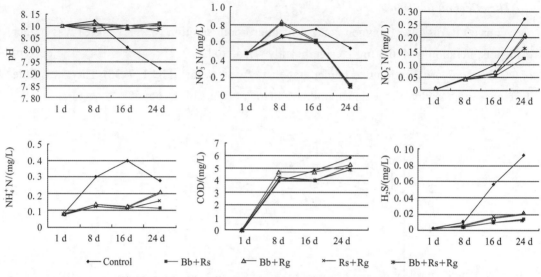

Fig. 1-7-36 The effect of microbial probiotics on water quality
(Data from Xu et al., 2009)

Probiotics (photosynthetic bacteria, *Bacillus*, *Lactobacillus*, yeast, *Bdellovibrio*, etc.) can significantly reduce ammonia nitrogen, nitrite and other stress on the shrimp, and reduce eutrophication.

The probiotics can regulate planktonic unicellular algae populations, and promote growth and reproduction of good algae. The probiotics can form beneficial biological organic particles as supplementary feed, and promote metabolite recycling. Mixing *Bdellovibrio phage* (Bp) and *Rhodobacter sphaeroides* (Rh) can significantly increase the activity of nonspecific immune enzyme (Fig. 1-7-37).

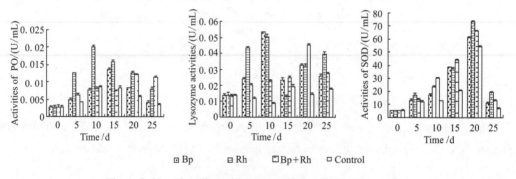

Fig. 1-7-37 The effect of mixing *B. phage* and *R. sphaeroides*
(Data from Xu et al., 2009)

4.3.4 Immune enhancement approach

Immunostimulants can improve the animal's resistance to disease by acting on the non-specific immune factors (Fig. 1-7-38).

Fig. 1-7-38　The products of immune enhancement

The immunostimulants includes anti-stress vitamins (Vitamin C, Vitamin E), Chinese herbs (Baicalin, *Astragalus*, *Rheum officinal*, *Glycyrrhiza uralensis*, *Astragalus* polysaccharides), and alga extract (astaxanthin and *Schizochytrum* sp.).

Vitamin C could enhance the activities of iNOS, CAT and LZM and 1% addition showed the better immune effects in *P. chinensis* (Fig. 1-7-39).

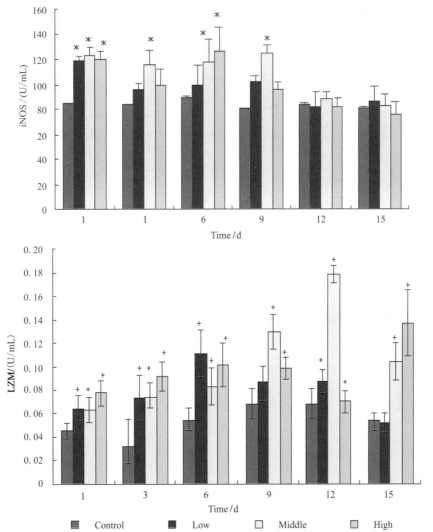

Fig. 1-7-39　The effects of Vitamin C on the enzyme activities of *P. chinensis*

Vitamin E can up-regulatory mRNA expression of TLR and NF-κB in *P. chinensis* (Fig. 1-7-40).

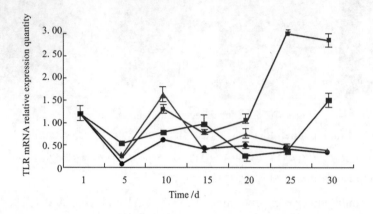

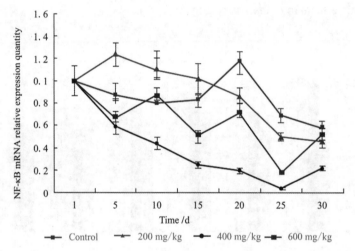

Fig. 1-7-40 The effects of Vitamin E on TLR and NF-κB gene expression in *P. chinensis*

Chinese herbs

Vibrio parahaemolyticus challenge test indicated that the immune protective rate of the two groups was 68.3% and 27.2% at the 14th day and was 91.6% and 45.8% at the 28th day, respectively (Fig. 1-7-41).

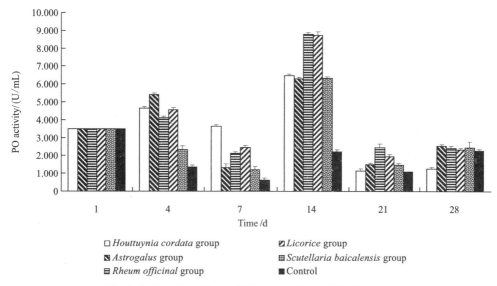

Fig. 1-7-41　The effects of Chinese herbs on PO activity

PO and LSZ activities of *P. vannamei* in *Astragalus*, *R. officinal* and *G. uralensis* groups were higher than the control (Figs. 1-7-42 to 1-7-43).

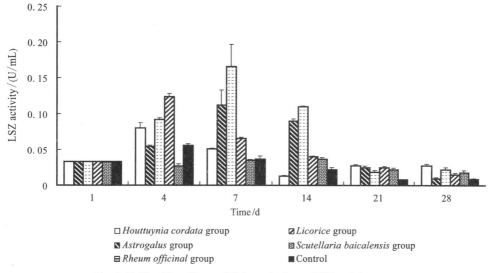

Fig. 1-7-42　The effects of Chinese herbs on LSZ activity

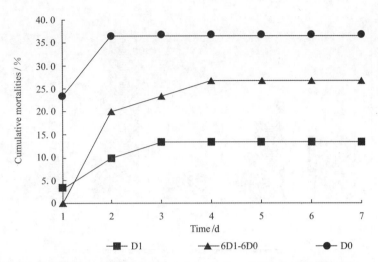

Fig. 1-7-43　Cumulative mortalities of *P. vannamei* after challenge with *V. parahaemolyticus*

Astaxanthin and Schizochytrium sp.

Fig. 1-7-44 shows the effects of astaxanthin (Asta) and *Schizochytrium* sp. (Schi) on T-SOD, CAT and MDA activities.

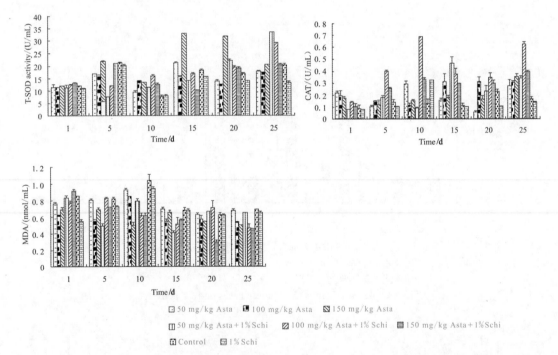

Fig. 1-7-44　The effects of astaxanthin (Asta) and *Schizochytrium* sp. (Schi) on T-SOD, CAT and MDA activities

100 mg/kg astaxanthin and 1% crack algae powder could improve the immunity function of *P. chinensis*.

4.3.5 Application of living creature diets

Rotifers and Artemia strengthen bait

Larvae survival of shrimp had little difference in the metamorphosis rate of nauplii and zoea I among three groups (Table 1-7-1, Figs. 1-7-45 to 1-7-46). From zoea II, the metamorphosis rate of artificial diet group reduced gradually to 61.68% at postlarvae, while the other groups remained about 80% ($P < 0.05$).

Table 1-7-1 The effects of different diets on larvae cultivation

Treat	Average metamorphosis/%				
	N_1-N_6	Z_1	Z_2-Z_3	M_1-M_3	P_1-P_{10}
Artificial diet	82.56	80.45	71.35	65.52	61.68
Live diet	82.03	80.67	78.23	80.12*	79.37*
Strengthen diet	82.10	81.23	82.05	83.40*	83.33*

Note: * indicated different significantly.

Fig. 1-7-45　Low value clam

Fig. 1-7-46　Live diet

Gammarid Ampithoe valida

Ampithoe valida could significantly improve the growth rate of 2.07 g Chinese shrimp, and improve nonspecific immunity of 0.25 g shrimp (Fig. 1-7-47). *A. valida* could improve anti-diease ability of *P. japonicus* after challenging WSSV compared with formulated feed group (Fig. 1-7-48). *A. valida* should be used at the early stage of culturing shrimps.

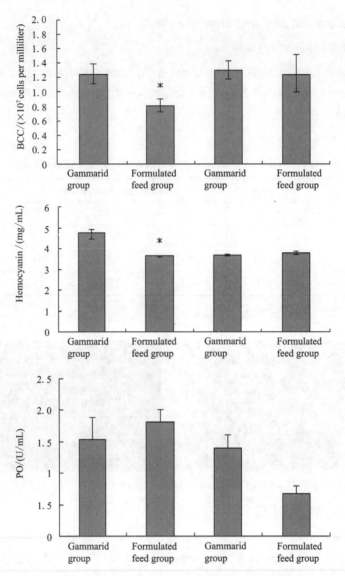

Fig. 1-7-47　The effects of Gammarid on growth and immunity

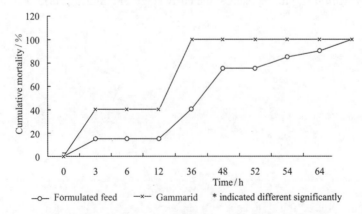

Fig. 1-7-48　The effects of Gammarid on cumulative mortality

4.3.6 Establishment of ecological culture mode

Establishment of IMTA in pond

The main species was Chinese shrimp *P. chinensis* (Fig. 1-7-49). The mix aquaculture species were swimming crab, shellfish and tongue sole (Fig. 1-7-50).

Fig. 1-7-49　*P. chinensis*

Portunus trituberculatus　　　　Philippines clam　　　　Tongue sole

Fig. 1-7-50　The mix culture species

After exposure, the IMTA pond was conducted by excavator (Fig. 1-7-51). The nets were covered in the bottom to protect the shellfish.

Fig. 1-7-51　The IMTA pond bottom covered with net for shellfish culture

In the middle of April, The shrimp seeds (length: 1 cm of juvenile) were put into the pond. The stocking density was 6,000 individuals per mu (Fig. 1-7-52).

Fig. 1-7-52　Stocking time and density of shrimp seeds

In early May, crab seeds (period Ⅱ of juvenile) were put into the pond. The stocking density was 3,000 individuals per mu (Fig. 1-7-53).

Fig. 1-7-53　Stocking time and density of crab seeds

In early June, tongue sole (weight: 100 g, juvenile) were put into the pond. The stocking density was 20-30 individuals per mu.

To block WSSV transmission, and to improve shrimp survival rate, tongue sole were put into the pond as biological control (Fig. 1-7-54).

Fig. 1-7-54　Stocking time and density of fish

Achievements of the farming

The shrimp and crab were harvested by different nets (Fig. 1-7-55).

Fig. 1-7-55 Harvest shrimp and crab

The tongue sole were harvested used nets (Fig. 1-7-56).

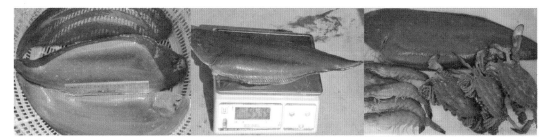

Fig. 1-7-56 Harvest fish

The shellfish were harvested after draining the pond lastly (Fig. 1-7-57).

Fig. 1-7-57 Harvest shellfish

Fig. 1-7-58 was the harvest scene.

Fig. 1-7-58　Harvest scene

Economic statistics of shrimp IMTA in pond

The economic statistic of shrimp IMTA in pond was shown in Table 1-7-2.

Table 1-7-2　Economic statistics of IMTA in pond

Species	Seeds quantity/ (individuals per mu)	Size/ (individuals per kilogram)	Production /(kg/mu)	Survival rate/%	Output/ (*yuan*/mu)
Chinese shrimp	6,000	28-35	75	28	6,000
Shellfish	50,000-60,000	120-160	280-380	84.0	2,000
Swimming crab	3,000	♂: 5-7	30	11.4	1,200
		♀: 4-7	32		1,500
Half-smooth tongue sole	40	1-1.5	28	84	2,780
Total	-	-	-	-	13,480

5　Future research and prospects

The future development of aquaculture in China should follow the concept of green, low carbon and environmental friendliness. To achieve the sustainable development goal of "high efficiency, high quality, ecology, health and safety", the development of ecological aquaculture in seawater ponds should focus on the following research directions:

A. Strengthen the basic theoretical research on the ecological culture of seawater ponds to promote the efficient utilization of material and energy in the pond IMTA system.

B. Establish standardized and meticulous aquaculture technical specifications to improve the success rate and benefit of breeding.

C. Study the new technology of water eutrophication and explore a new environment-friendly marine ecological culture mode.

References

Chang Z Q, Amir N, He Y Y, Li J T, Qiao L, Steven I P, Liu P, Li J. Development and current state of seawater shrimp farming, with an emphasis on integrated multi-trophic pond aquaculture farms, in China-a review. Reviews in Aquaculture, 2020, DOI: 10.1111/raq.12457.

Lei C G, Lin H C. Experiment on mixed culture of *Penaeus japonicus* and green spot [J]. Journal of Aquaculture, 2014, 8: 1-3. (In Chinese)

Liu G, Ding Z M, Li C B. Experiment on polyculture of *Penaeus chinensis* and Fugu rubripes in ponds [J]. Journal of Aquaculture, 2007, 28(2): 2-3. (In Chinese)

Qiu Y, Yan F, Luo J, et al. Study on economic benefits of polyculture of *Litopenaeus vannamei* and *Epinephelus punctatus* [J]. Guangdong Agricultural Science, 2012, 18: 148-154 (In Chinese).

Simon J F-S, Matthew R P B. Nutrient budgets in intensive shrimp ponds: implications for sustainability [J]. Aquaculture, 1998, 164: 117-133.

Wang J Q, Li D S. Comparative study on efficiency and benefit of different integrated culture systems in shrimp ponds [J]. Acta fisheries Sinica, 1999, 23(1): 45-52. (In Chinese)

Zhang Z D, Wang F, Dong S L, et al. Study on the structure optimization of polyculture system of grass carp, silver carp and *Litopenaeus vannamei* [J]. Journal of Ocean University of China (Natural Science edition), 2011, 41(7): 60-66. (In Chinese)

Zhang Z F. High yield demonstration of polyculture of *Fugu rubripes* and *Penaeus japonicas* [J]. Hebei Fisheries, 2012, 5: 19-20. (In Chinese)

Selective breeding of swimming crab *Portunus trituberculatus* in China

Author: Lv Jianjian
E-mail: jianjian997@163.com

1 Introduction

1.1 Introduction of *Portunus trituberculatus*

1.1.1 The taxonomic status

P. trituberculatus (Fig. 1-8-1) (Crutacea: Decapoda: Brachyura) is commonly known as the swimming crab.

Fig. 1-8-1 *P. trituberculatus*

1.1.2 Biological characteristics

A. Fast growth.

B. High yield.

C. Euryhaline.

1.1.3 Distribution and culture

P. trituberculatus is widely distributed in the coastal waters of China, Korea and Japan. It has been farmed for more than 30 years in Asian countries.

1.2 Swimming crab culture in China

Commercial farming commenced in China in the 1970s. Crab farming is mainly distributed in coastal provinces and cities, including Liaoning, Hebei, Shandong, Jiangsu, Zhejiang Provinces. The production is about 100 thousand tons in China each year.

1.3 Problems in the crab farming

Wild seedlings are widely used in aquaculture. The problems in the crab farming are as follows:

A. Overfishing leading to resource exhaustion.

B. Instability of economic traits.

C. Germplasm degradation.

D. Uncontrollable pathogen (Fig. 1-8-2).

Fig. 1-8-2 *P. trituberculatus* with uncontrollable pathogen

1.4 The new variety—"Huangxuan No. 1"

In 2005, four wild geographic populations distributed in the Yalu River Estuary, Laizhou Bay, Zhoushan Bay and Haizhou Bay were collected individually (Li et al., 2013).

Population selection were carried out in order to breed new strains.

The characteristics of the new variety are as follows:

A. The average body weight increases 20.12%.

B. The survival increases 32.00%.

C. Full carapace width coefficient of variability is less than 5%.

2 The breeding of the new variety —"Huangxuan No. 1"

2.1 Breeding goal and strategy

A. The goal is to develop a more bigger and faster-growing cultured crab.

B. Breeding character: Growth.

C. Breeding strategy: Population selection strategy.

2.2 Technology roadmap of breeding process

Fig. 1-8-3 shows the technology roadmap of breeding process.

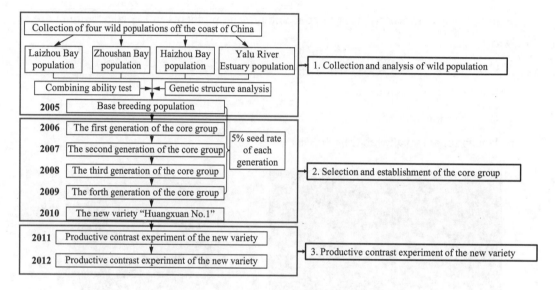

Fig. 1-8-3 Technology roadmap of breeding process

2.3 The collection and evaluation of wild geographical groups

Fig. 1-8-4 4 wild geographical groups
(Data from Gao et al., 2015)

1,261 crabs were collected from 4 wild geographical groups (Fig. 1-8-4). The ratio of male to female was 1 : 1.

Cluster analysis and discriminant show that the carapace color and spots were significantly different among four populations. Genetic variation analysis of 4 wild groups based on isozyme, 16S, COI and AFLP showed that genetic diversity of each population is higher.

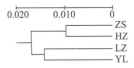

Fig. 1-8-5 Cluster analysis of 4 wild geographical groups

The cluster tree showed that crabs from Haizhou Bay and Zhoushan Bay were pooled into one group and crabs from Yalu River Estuary, and that Laizhou Bay were pooled into another group (Fig. 1-8-5).

2.4 Selection and establishment of the core group

In 2006, 50 female crabs (with large body size, strong vitality and healthy mating) were selected from each basic group (Fig. 1-8-6), and the growth rate was used as an indicator for population selection.

 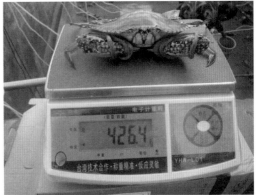

Fig. 1-8-6 Female crabs selection for basic group

2.5 Parent selection and breeding

During 2006-2009, selective intensity of each generation was around 5% for the selection of the bigger ones. After mating, the females migrate to the pond indoor from outdoor for overwintering.

When the fertilized eggs were about to hatch, crabs were transferred to the hatching pond. The hatched larvae, termed zoea, went through four zoea-stage larvae, one megalopa and juvenile I stage in the hatching pond.

At juvenile II stage, the crabs were transferred to outdoor pond to culture.

Abdominal morphology of mating, mated and non-mated female crab was shown in Fig. 1-8-7.

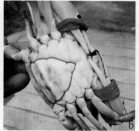

Fig. 1-8-7 Abdominal morphology of mating (a), mated (b) and non-mated (c) female crab

2.6 Formation of new variety

The new variety "Huangxuan No. 1" for fast-growthing of the swimming crab was selected successfully in 2010 after five generation selections.

The survival rate and growth rate were significantly improved as shown in Table 1-8-1. The average body weight increased 20.12%, and the survival increased 32.00%.

Table 1-8-1 Character statistics of new varieties during breeding

Year	Body weight/g			Survival rate/%		
	"Huangxuan No. 1"	Control	Improved/%	"Huangxuan No. 1"	Control	Improved
2006	194.91	189.40	2.91	11.42	10.51	8.65
2007	206.35	192.10	7.42	11.76	10.13	16.09
2008	217.58	191.04	13.89	12.86	10.49	22.55
2009	227.64	192.00	18.56	13.21	10.43	26.65
2010	229.43	191.00	20.12	13.58	10.40	32.00

2.7 Productive contrast culture experiment

In 2011, the growth rate of the new variety "Huangxuan No.1" was 20.79% higher than the wild control group, and the survival rate was 32.07% higher as shown in Table 1-8-2.

In 2012, the growth rate of the new variety "Huangxuan No.1" was 20.91% higher than that of the wild control group, and its survival rate was 32.73% higher as shown in Table 1-8-2.

Table 1-8-2 Information of productive contrast culture experiment

Growth trait	2011			2012		
	"Huangxuan No.1"	Control	Improved /%	"Huangxuan No.1"	Control	Improved /%
Body weight/g	226.00	187.10	20.79	228.12	188.67	20.91
Survival rate/%	14.00	10.60	32.07	14.03	10.57	32.73
Carapace width variability/%	4.79	9.47	4.68	4.71	9.49	4.78

2.8 Harvest and extension of new variety

From 2010 to 2016, we had breded 250 million second-stage crabs. The area of "Huangxuan No.1" radiating culture had reached 7,000 hm^2, including Qingdao, Weifang, Rizhao, and Yantai of Shandong Province, Lianyungang of Jiangsu Province, Ningbo of Zhejiang Province and Cangzhou of Hebei Province.

The result of culture is very excellent, such as better survival and standard. A remarkable effect and a considerable economy benefit were obtained (Fig. 1-8-8).

Fig. 1-8-8   Harvest and promotion of "Huangxuan No.1"

3 Marker-assisted selection

3.1 Development of molecular markers

A large number of molecular markers have been obtained by high-throughput sequencing technology (Figs. 1-8-9 to 1-8-10).

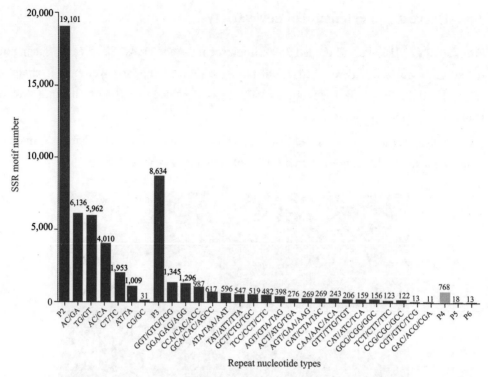

Fig. 1-8-9 Distribution of simple sequence repeat (SSR) nucleotide classes among different nucleotide types
(Data from Meng et al., 2015)

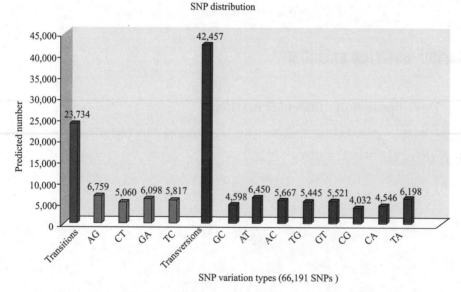

Fig. 1-8-10 Distribution of putative single nucleotide polymorphisms (SNP)
(Data from Meng, et al., 2015)

The microsatellite genotypes and abundance differences between the two groups can be used as for the identification of "Huangxuan No.1".

Amplification map of microsatellite Pot09, Pot42 and PTR70 are shown in a, b and c respectively of Fig. 1-8-11. 1-10 were the result of amplification of the new species, and the 1′-10′ were the result of amplification of the wild *P. trituberculatus*. M is the DNA Maker Ⅱ.

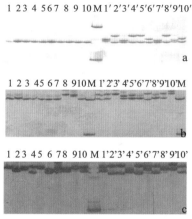

a. Pot09; b. Pot42; c. PTR70.

Fig. 1-8-11　Amplification map of microsatellite Pot09, Pot42 and PTR70

3.2　High-resolution genetic linkage map

A high-resolution genetic linkage map is an essential tool for decoding genetics and genomics (Fig. 1-8-12).

Fig. 1-8-12　The first generation genetic linkage map of *P. trituberculatus*
(Data from Liu et al., 2012)

3.2.1　Materials and methods

A. Mapping population: F1 full-sib family.

B. Marker discovery and genotyping: SLAF (Fig. 1-8-13).

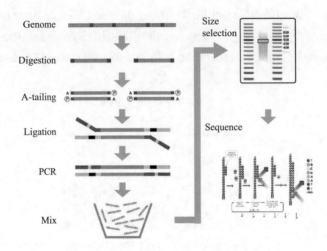

Fig. 1-8-13　SLAF experiment process

C. Genetic linkage map construction: Highmap strategy (Fig. 1-8-14, Table 1-8-3).

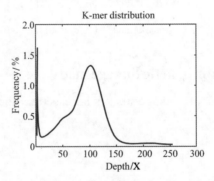

Fig. 1-8-14　K-mer frequency distribution of sequencing reads

Table 1-8-3　Statistics of the genome survey and assembly

Category	Data
Genome size/Mb	805.92
Heterozygosity	0.96%
Repeat sequences	39.42%
Data/Gb	113.83
Depth/X	141.24
Number of contigs	1,268,724
Total length of contigs/bp	833,944,844
Contig N50/bp	756
Number of scaffolds	898,300
Total length of scaffolds/bp	842,129,340
Scaffold N50/bp	1,154

3.2.2 Results

After SLAF sequencing, 317,918,396 paired-end reads were generated for the mapping family (two parents and 116 progenies). A total of 60,319 polymorphic SLAF markers were identified from 152,449 SLAF markers, of which 11,068 could be successfully genotyped in both parents and offspring (Table 1-8-4). SLAF markers with the five segregation patterns (ab × cd, ef × eg, hk × hk, lm × ll, nn × np) could be used in linkage map construction for the full-sib family, of which nn × np was the major pattern (41.0%), followed by lm × ll (28.0%) (Fig. 1-8-15). The average read depth of genotyped markers were 25.10, 74.71 and 100.23 in the offspring, male and female parents, respectively.

Table 1-8-4 Statistics of the SLAF

Category	Data
Number of reads	317,918,396
Number of high quality SLAF	152,449
Polymorphic SLAF	60,319
Average depth in male parent/X	74.71
Average depth in female parent/X	100.23
Average depth in offspring/X	25.10

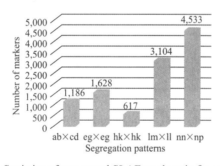

Fig. 1-8-15 Statistics of genotyped SLAF markers in five segregation patterns

Table 1-8-5 Summary of *P. trituberculatus* linkage map data

Category	Sex-averaged map	Male map	Female map
Number of markers mapped	10,963	7,001	7,875
Number of linkage group	53	53	53
Minimum length of linkage group/cM	15.74	5.25	12.21
Maximum length of linkage group/cM	188.47	192.48	225.12
Minimum markers number per linkage group	21	16	17
Maximum markers number per linkage group	473	322	394
Average marker interval/cM	0.51	0.64	0.74
Observed genome length/cM	5,557.85	4,454.68	5,759.27
Estimated genome length/cM	5,622.790526	4,528.713	5,854.632
Genome coverage/%	0.988450481	0.983653	0.983712

A pseudo-testcross strategy was used to construct linkage maps in this study. Finally, we obtained a linkage map containing 53 linkage groups with 10,963 markers. The total map distance of the sex-averaged map (Fig. 1-8-16) was 5,557.85 cM, which covered 98.85% of the genome based on the estimated total length of the genome map. The distribution of markers among linkage groups was evaluated by statistics of the marker interval. On average, 96% of the sex-averaged map were covered by markers with an interval distance of less than 5 cM and the mean distance between two markers was 0.51 cM (Table 1-8-5).

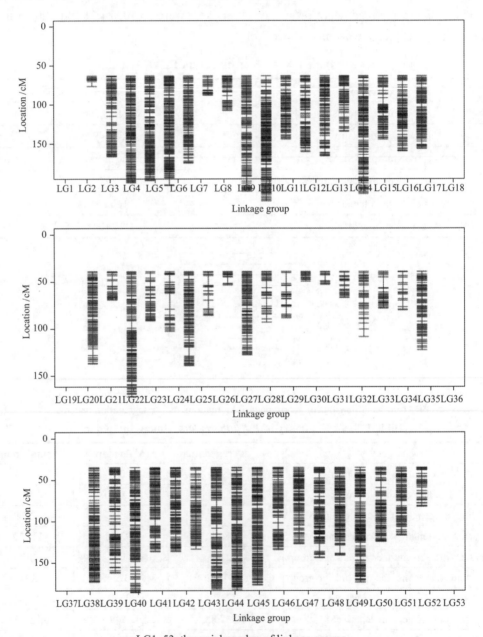

LG1-53, the serial number of linkage groups.

Fig. 1-8-16　The sex-averaged map of *P. trituberculatus*

(Data from Lv et al., 2017)

Based on information from blast, the markers of linkage maps, genomic scaffolds/contigs and unigenes of transcriptome could be integrated (Fig. 1-8-17). Finally, a total of 7,487 markers could be aligned to the genomic scaffolds/contigs or transcriptome unigenes after integrating, of which 2,378 markers could be explicitly annotated via blastX with the public Nr and Swissprot databases.

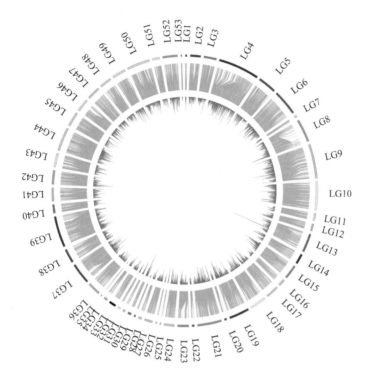

Fig. 1-8-17 Integration of linkage map, genomic scaffolds and transcripts

3.3 QTL mapping

3.3.1 Methods

A. Interval mapping method and multiple-QTL model mapping.

B. Methods (MQM) in the program Map QTL 4.0 software (Tables 1-8-6 to 1-8-7).

Table 1-8-6 Characteristics of growth related QTLs

Trait	QTL	LOD threshold	Linkage group	Start	End	Marker number	Max LOD	Max PVE
BW	qBW-1	3	7	7.082	7.082	1	3.11	40.9
	qBW-2	3	8	0	0	5	3.15	12.3
	qBW-3	3	10	125.941	125.941	1	3.12	61.5
	qBW-4	3	34	1.755	2.624	3	3.22	16.4

(to be continued)

Trait	QTL	LOD threshold	Linkage group	Start	End	Marker number	Max LOD	Max PVE
FCW	qFCW-1	3	24	0.870	16.015	17	3.75	20.3
	qFCW-2	3	24	49.437	49.437	1	3.03	12.0
	qFCW-3	3	34	0	7.034	9	3.81	21.2
CW	qCW	3	10	10.944	10.944	1	3.1	12.2
CL	qCL-1	3	24	0.87	19.597	26	4.14	22.0
	qCL-2	3	24	49.002	49.872	7	3.27	12.8
	qCL-3	3	34	1.755	2.624	3	3.27	17.2
BH	qBH	5.5	24	0.87	16.015	17	6.91	35.9

Table 1-8-7 Statistics of growth-related genes located in QTL

Marker ID	Linkage group	Position/cM	Annotation	Marker location
Marker 26271	LG24	2.609	Xylulokinase	5'UTR
Marker 46391	LG24	9.431	Glycoprotein 6-alpha-L-fucosyltransferase	Intron
Marker 6733	LG24	10.301	Mitogen-activated protein kinase	Intron
Marker 25749	LG24	13.810	REI-silencing Transcription factor	ORF (nonsynonym.ous)
Marker 10494	LG24	13.810	Poly[ADP-ribose] polymerase	3'UTR
Marker 26391	LG34	0	RNA-directed DNA polymerase	ORF (nonsynonym.ous)
Marker 25832	LG34	1.755	Integrin alpha 8	Intron
Marker 25048	LG34	2.624	Monocarboxylate transporter	5'UTR

3.3.2 Results

QTLs of different traits is represented by different colors (Fig. 1-8-18).

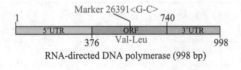
RNA-directed DNA polymerase (998 bp)

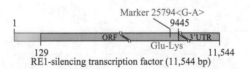
RE1-silencing transcription factor (11,544 bp)

Fig. 1-8-18 Structural features and nonsynonymous sites of REST and RNA-directed DNA polymerase genes

The cDNA sequences of two genes were assembled based on transcriptome unigene data. Marker 26391 with a SNP (G–C) on RNA-directed DNA polymerase led to the nonsynonymous change of Val to Leu; Marker 25794 with a SNP (G–A) on REST led to the nonsynonymous change of Glu to Lys (Fig. 1-8-18).

4 Summary

The new variety "Huangxuan No. 1" which is the fast-growthing of the swimming crab (Fig. 1-8-19) was selected successfully in 2010 after five generation selections.

Fig. 1-8-19 *P. trituberculatus*

A smaller genome with a lower ratio of repetitive and heterozygous sequences was observed in this species. A genetic linkage map with 10,963 markers was constructed via SLAF-seq. An integrated map was building, on which 2,378 markers could be explicitly annotated. Ten growth-related QTLs and eight growth-related genes were identified.

Acknowledgements

PI: Liu Ping; Staff: Gao Baoquan, Meng Xianliang, Ren Xianyun (Fig. 1-8-20).

Funding financial support for this study was provided by the National High Technology Research and Development Program of China (Project 2012AA10A409), the National Natural Science Foundation of China (Nos. 41776160 and 41576147), the China Agriculture Research System (No. CARS-48). This research was supported by the National Natural Science Foundation of China (Nos. 41576147 and 41306177), the Efficient Eco-Agriculture Innovation Project of Taishan Leading Talent Project (No. LJNY2015002), and the Scientific and Technological Innovation Project financially supported by Qingdao National Laboratory for Marine Science and Technology (No. 2015ASKJ02).

Fig. 1-8-20　Photo of research team

References

Gao B Q, Liu P, Li J, Liu L. Analysis of body color variations among four wild populations of *Portunus trituberculatus* [J]. Progress in fishery sciences, 2015, 36(3): 79-84. (In Chinese)

Li J, Liu P, Gao B Q, Chen P. The new variety of *Portuns trituberculatus* "Huangxuan No.1" [J]. Progress in Fishery Sciences, 2013, 34(5): 51-57. (In Chinese)

Liu L, Li J, et al. A genetic linkage map of swimming crab (*Portunus trituberculatus*) based on SSR and AFLP markers [J]. Aquaculture, 2012: 66-81.

Lv J, Gao B, Liu P, et al. Linkage mapping aided by de novo genome and transcriptome assembly in *Portunus trituberculatus*: applications in growth-related QTL and gene identification [J]. Scientific Reports, 2017, 7(1): 7874.

Meng X, Liu P, Jia F, et al. De novo transcriptome analysis of *Portunus trituberculatus* ovary and testis by RNA-Seq: Identification of genes involved in gonadal development [J]. Plos One, 2015, 10(7): e0133659.

Studies on techniques of the artificial propagation of *Portunus trituberculatus*

Author: Gao Baoquan
E-mail: gaobq@ysfri.ac.cn

1 Overview of marine crab culture

1.1 Overview of international marine crab culture

There are abundant species of crabs in the ocean, and there are more than 2,000 species of crabs in the Indian-Pacific region (Fig. 1-9-1). There are some species that have large scales of aquaculture: *Portunus trituberculatus*, *Scylla* spp., etc. Among them, the aquaculture and stock enhancement of *P. trituberculatus* is very representative in marine crabs.

a. *Callinectes sapidus*; b. *Portunus pelagicus*; c. *Charybdis feriatus*.
Fig. 1-9-1   Marine crabs

1.1.1 Overview of international swimming crab culture

The swimming crab, *P. trituberculatus*, is widely distributed throughout the coastal waters of Asian-Pacific nations, and it is an important economic species in this region (Fig. 1-9-2). This species is dominant in portunid crabs fisheries around the world and supports a large aquaculture industry in China. Annual farming output exceeds 100,000 t.

Fig. 1-9-2　Swimming crabs

1.1.2　Overview of international mud crab culture

The main farming countries and regions of mud crab are China, Southeast Asia and India (Fig. 1-9-3). In addition to China, the farming of mud crabs in other countries is very small in scale. According to the data released by the FAO, the top five countries and regions in the world in 2010 are Chinese mainland, the Philippines, Indonesia, Malaysia and Taiwan China. The production of mud crabs in Chinese mainland exceeds 100,000 t, and that of Philippines is only 10,000 t. The output of other countries and regions is less than 10,000 t.

Fig. 1-9-3　Mud crabs

1.2　Overview of marine crab culture in China

1.2.1　Overview of swimming carb culture in China

There are only two marine crabs that have a larger scale of aquaculture in China: swimming crab and mud crab.

In the late 1980s, small scale artificial farming of swimming crab began. The farming methods mainly were shrimp-crab polyculture. In 2016, the swimming crab culture area has reached 26,224 hm^2, with a production of 125,300 t and annual value of nearly 10 billion *yuan* in China. It has developed into one of the leading species of marine aquaculture in China (Fig. 1-9-4).

a. Culture pond; b. Putting seedling; c. Farming harvest.
Fig. 1-9-4 The culture of swimming crab

1.2.2 Overview of mud crab culture in China

As early as the 1890s, people began to raise and fatten the mud crabs in Guangdong Province, China. Before the 1980s, the farming of mud crab was mainly fattening and promoting ovarian maturation. Since the 1990s, it is mainly based on pond culture. According to the farming site, the farming mode includes pond culture, shoal fence net farming, mangrove farming and facility farming. In China, the species of mud crab culture is mainly *Scylla paramamosain*, and its aquaculture production accounts for more than 90% of the total production of mud crab farming.

2 Propagation technology for swimming crab

2.1 Cultivation and selection of parent crab

2.1.1 Wintering facilities

The bottom area of the wintering pond should be 30-40 m^2 and the water depth is 0.8-1.2 m (Fig. 1-9-5). It can also be adjusted according to actual conditions. The bottom of the pond is sanded with a depth of 10-15 cm, and 20%-30% of the bottom is blank used as the bait table.

Fig. 1-9-5 Wintering pond

2.1.2 Selection of parent crab

The quality of the parent crabs should meet the following conditions: mating females, strong and lively, with good appendages, no trauma, no adhesion or parasitic harmful organisms, body weight 250 g or more. The eggs need to meet the following conditions: contour is complete, and egg is compact and full. Meanwhile, the egg surface is smooth and bright in color. The crab should not carry specific pathogens after regular sampling. So before the crab is placed in the pond, a 1–2 g/m^3 povidone-iodine bath for 15 min is needed.

2.1.3 Stocking density

The wintering pond, sand, tools, etc. should be strictly disinfected before the crabs enter the pond. In the wintering pond, the best density of no egg-bearing crab is 3–5 individuals per square meter, and the egg-bearing crab is 2–3 individuals per square meter.

2.1.4 Water quality regulation

After the crabs enter the wintering pond, the temperature of pond water is controlled when the natural water temperature drops to 9 ℃ (Table 1-9-1). During the overwinter, the water temperature is maintained at about 9 ℃, and the daily temperature difference does not exceed ± 0.5 ℃. If the natural water temperature is stable above 9 ℃, the natural water temperature is maintained over the winter.

Table 1-9-1 The water quality index requirements of wintering pond

Parameter	Index	Parameter	Index
Salinity	20–33	Dissolved oxygen	≥ 5.0 mg/L
Ammonia nitrogen	≤ 0.5 mg/L	Nitrite	≤ 0.1 mg/L
pH	7.8–8.6		

2.1.5 Feeding of bait

Live silkworms or shellfish should be fed during winter to the crabs. The daily feeding amount is 3% to 5% of the weight of the parent crab, and it is increased or decreased according to the specific feeding situation, and the bait is taken at 4 o'clock in the afternoon.

2.1.6 Light intensity

The light intensity of the wintering pond is not more than 500 lx. The light intensity and time are appropriately increased or decreased according to the requirement of reproducing.

2.1.7 Strengthening cultivation

During the intensive cultivation of the crab, the water temperature is increased with 0.5 ℃ each day, until reaching to 21 ℃.

If the natural water temperature was higher than 21 ℃, the crabs should be bred by the natural water temperature (Fig. 1-9-6).

The daily feed amount increases from 10% to 15% of the body weight.

Fig. 1-9-6 Heating facilities

2.2 Seed cultivation and daily management

2.2.1 Facilities

The nursery pond is an indoor cement pond with the water volume of 30–40 m^3. The depth of water is 1.5–1.8 m. First, it should be soaked with 6.0 mg/L bleaching powder for 24 h, then rinsed, and dried for 1–2 day. At last the pond should be rinsed with clean sea water before use (Fig. 1-9-7).

Fig. 1-9-7 Nursery pond

2.2.2 Water for nursery

Water for nursery should be precipitated, filtered, disinfected, in which filtering takes sand and nets. Add 120–150 g/m^3 chloro to the nursery water, among which 8%–10% of effective chlorine is contained. Sodium hyposulfite is used for eliminating residual chlorine in the water after 12 h. After dechlorination, the water should be inflated.

2.2.3 Ovulation of the crab

When the abdominal egg mass is black-gray, the heartbeat of the larvae reaches about 170 times per minutes. The parent crab should be promptly removed. After being disinfected and rinsed, the parent crab should be placed in a 10-mesh cage and the cage should be hoisted into the cultivation pond. According to the volume of full pond water, the density of larvae is controlled with in $3 \times 10^4 - 5 \times 10^4$ individuals per cubic meter.

2.2.4 Feeding of bait

The bait at the first zoeal stage is rotifers, and the density of rotifers in seedling water is not less than 10 individuals per milliliter. The bait at the second and third zoeal stages are fed with is *Artemia* nauplii, and the feeding times is four to six one day. The feeding amount is 1.5–3 individuals per milliliter at every turn. The fourth zoeal stage larvae is also fed with *Artemia* nauplii, and the feeding times is six to eight one day. The feeding amount is 2–3 individuals per milliliter at every turn. At the megalopa stage, I juvenile crabs and II juvenile crabs are mainly fed with fresh live *Artemia* adults, and the daily feeding amount is 10 individuals per milliliter, 20 individuals per milliliter, 30 individuals per milliliter (Fig. 1-9-8).

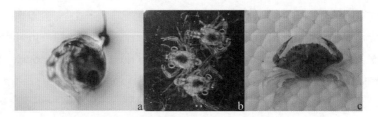

a. Zoeal; b. Megalopa; c. Juvenile crab; d. Feeding of bait.

Fig. 1-9-8   The larva of swimming crab

2.2.5 Daily management

At the beginning, the height of the nursery pond is generally 50–60 cm. In the early stage, the water is added, with 10 cm/d. The pond is gradually filled, and then the water exchange rate is 10% one day (Fig. 1-9-9). When the parent crabs discharge the larvae, the water temperature of the nursery pond should be 21.0 ℃. Then the water temperature is increased by 0.5 ℃ each day gradually, until to 26 ℃ ± 0.5 ℃. By adding *Chlorella* ($2 \times 10^6 - 3 \times 10^6$ individuals per milliliter) or *diatom* ($2 \times 10^4 - 3 \times 10^4$ individuals per milliliter), the pond water becomes yellow-green.

Fig. 1-9-9   Exchange water

2.2.6 Seed harvest and transportation

When the larvae develop to I juvenile crabs or II juvenile crabs, they will be transfered to outdoor pond. Seedling yield can reach 10,000 individuals per cubic meter. The juvenile crabs are collected through the seedling trough.

The transportation method often adopts the new rice bran as a shelter. Generally, every

250 g seeding is put into to a 30 cm × 50 cm polyethylene double-layer plastic bag, which is mixed evenly with the same amount of rice bran. Then the plastic bag was put into the foam box after oxygenation (Fig. 1-9-10).

a. Net seeding; b. Put in rice bran; c. Mix well; d. Oxygen filling.
Fig. 1-9-10 Seedling packaging

3 The culture technology of *P. trituberculatus*

3.1 Breeding facilities

3.1.1 Breeding pond

A. The bottom of the culture pond should be sandy or sandy mud.

B. The area of pond is about 1-2 hm^2.

C. The depth of the pond is about 2.0-3.0 m.

D. The water inlet gate and drainage gate should be installed at end of the breeding pond (Fig. 1-9-11).

a. Breeding pond; b. Sandy bottom; c. Slope protection ; d. Inlet channel.
Fig. 1-9-11 Breeding facilities

3.1.2 Access and drainage channels

A. The water inlet and outlet should be kept as far as possible.

B. The width of the drainage channel should be greater than that of the inlet channel.

C. The height of the bottom of the drainage channel should be lower than the bottom of each crab pond drainage gate by more than 30 cm.

3.1.3 Aeration equipment

The breeding ponds should be equipped with aeration equipment, and the power of the microporous aerator should be 1.5-4.5 kW per hm^2 (Fig. 1-9-12).

a. Inlet channel; b. Microporous aerator; c. Biofan.

Fig. 1-9-12 Aeration equipment

3.1.4 Analysis and testing

The farm should be equipped with an environmental testing and analysis room, equipped with a biological microscope, a salinity meter, a water temperature meter, a dissolved oxygen meter, a pH meter, a transparency plate, etc (Fig. 1-9-13).

a. Microscope; b. Water quality monitoring equipment.

Fig. 1-9-13 Environmental testing and analysis room

3.2 Preparing for seedling delivery

3.2.1 Decontaminate and arrange pond

Before the breeding, the water in the pond, the reservoir, the inlet and outlet channels, etc. should be drained. The sludge and debris should be removed, and the bottom of the pond should be insolated repeatedly (Fig. 1-9-14).

Fig. 1-9-14　Turn over the soil of pond

3.2.2　Disinfection

About 20 d before seedling delivery, water should be added to the culture pond to 20 cm. The original pests pathogens and sleeping eggs are killed in the pond, by splashing mixture in the whole pond. The mixture is mixing quicklime with water. The dose of quicklime is 1,200-1,500 kg/hm^2 (Fig. 1-9-15).

Fig. 1-9-15　Disinfection of pond

3.2.3　Basic bait culture

After disinfection for 10-15 d, we add 60-80 cm of water to the pond. Then algae and basic biological bait are cultivated, by adding 15-37.5 kg nitrogen fertilizer and 0.1-0.5 kg phosphate per hm^2. The chicken manure can also be used with 150-225 kg/hm^2 (Fig. 1-9-16). Tea seed cake with 10 g/m^3 was used to completely remove wild fish and fish eggs, and it also played a role of fertilizer. When the pond water is brown or yellowish green, we can seed.

Fig. 1-9-16　Basic bait culture

3.3 Putting seedlings

3.3.1 Water quality requirements

Aquaculture water should meet the following conditions: salinity 20-32, pH 7.8-8.6, transparency 30-40 cm, dissolved oxygen > 4.5 mg/L, ammonia nitrogen < 1.0 mg/L, hydrogen sulfide < 0.01 mg/L.

3.3.2 Seed selection

We should choose a healthy seed with strong vitality, strong constitution, transparent body, full stomach with no damage, no deformity and uniform specifications.

3.3.3 Seedling time

The swimming crab seedlings were stocked at a water temperature of 18 ℃ or higher in mid-May.

3.3.4 Seedling specifications

The size of second-stage crab of the swimming crab is 16,000-24,000 individuals per kilogram.

3.3.5 Seedling density

The density in the polyculture is generally controlled at 15,000-30,000 individuals per hectare, and in the monoculture is generally controlled at 30,000-45,000 individuals per hectare.

3.4 Growing period management

3.4.1 Aquaculture water environment management

Maintain water level and change water

Before the first 10 d of July, the water depth is more than 1.5 m.

In the middle of the culture, the water exchange rate should be increased to once every 3-5 d and 30%-40% each time. In the later stage of culture, the water exchange amount and frequency should be reduced.

Aeration

The start-up time is 1-2 hours in the morning and at noon within 30 d after the seedling is released. The start-up time can be prolonged after 30 d of culture, so that the dissolved oxygen content in the water can always be maintained above 5 mg/L (Fig. 1-9-17).

The start-up time should be increased appropriately in cloudy and rainy days, and the machine should be stopped for 0.5 hour when feeding.

Fig. 1-9-17 Aeration

Water quality protection agent

Each half of the month, zeolite powder and calcium peroxide is added as the main component of the water quality protection agent. The method of use is that every 15-20 d, the dosage is 300-450 kg/hm^2. The total alkalinity of the pond water is required to be 80-120 mg/L.

Beneficial bacteria

Beneficial microbial agents include photosynthetic bacteria and heterotrophic bacteria. In the prophase of culture, once every 10-15 d and once every 3-5 d in the anaphase of culture, they can not be used with disinfectant and antibiotic simultaneously.

3.4.2 Bait management

Feed types

Fresh feed should be selected for breeding of the crab. The early stages (stage II to V) are dominated by *Artemia* adults.

After V childish crab, the feeds are dominated by *Potamocorbuila laevis* and fresh miscellaneous fish (Fig. 1-9-18).

a. *Potamocorbula laevis*; b. Feeding clam.

Fig. 1-9-18 Bait

Bait quantity

The feeding amount is 8%-10% of the crab's body weight, twice a day in the early stage.

And the feeding amount is 10%-15% of the crab's body weight, twice a day in the middle stage. Then the feeding amount is 8% of the crab's body weight, twice a day in the later stage.

Feeding method

The feeding time is 3:00-4:00 and 16:00-17:00. The evening feeding amount accounted for 70% of the total daily feeding. Before feeding, the bait must be sterilized and bait should be thrown into the shallow water of the pond.

3.4.3 Daily management

Water quality determination

Dissolved oxygen and water temperature are measured in the morning every day, and the water quality indexes such as salinity, transparency, pH value and ammonia nitrogen are measured regularly.

Microbial detection in aquaculture pond

The species and quantity of plankton in the pond are constantly observed (Fig. 1-9-19).

a. Water quality determination; b. Microbial detection.

Fig. 1-9-19　Water quality and microbial detection

3.4.4 Set up hidden objects

The shelter (tile, mesh) is put into the culture pond to prevent massacre, especially to protect the newly molted (soft shell) crab (Fig. 1-9-20).

a. Net; b. Tile; c. Adding seawater.

Fig. 1-9-20　Hidden objects

3.5 Disease prevention

3.5.1 Touring pond

The farm should tour the pond at least once every morning, afternoon and evening in order to observe the activity, distribution and feeding of the crab. Attention should be paid to the disease, dead crab, and checking the cause of death and dealing with it as well.

3.5.2 Cut off the pathogen

It should be prohibited that the inlet water is discharged from morbidity shrimps and crab ponds. Crabs should not be fed with bait with pathogens and should be the spread of pathogens cut off in time.

3.5.3 Pathogen detection

The pathogenic organisms such as *Vibrio alginolyticus*, ciliates, *Vibrio anguillarum* and white spot virus were detected regularly (Fig. 1-9-21).

a. Disinfection of bait; b. Pathogen detection.
Fig. 1-9-21 Disease prevention

3.5.4 Drug use

Drug selection

The use of drug should meet the requirements of NY 5071-2001. The drug of high efficiency, low toxicity and low residue should be used, and ecological agents should be advocated.

Medication prescription

The prescription of the drug should be prescribed by qualified personnel and the farm should use drug according to the prescription strictly.

Medication record

During the breeding process, the medication records should be made and executed according to the prescribed drug withdrawal period to ensure that the drug residues meet the requirements of NY 5070-2002.

Production record

The breeding production records, medication records, water quality monitoring records, etc. should be carefully done.

3.6 Farming harvest

3.6.1 Quality and safety inspection

The swimming crab grows fast by shelling. It can be harvested after 5 months farming. Before harvest, it should be ensured that the crab meet the requirements of the drug withdrawal period, and the quality, safety and hygiene indicators are checked according to the provisions of NY 5163-2002.

3.6.2 Harvesting equipment

Tools such as nets, containers, and packaging should be clean and hygienic.

3.6.3 Harvesting method

A. Netting: Catch crab with net (Fig. 1-9-22).

B. Trap: Catch crab with chicken head or miscellaneous fish. When fish is clamped by the crab, pull it up and catch the crab (Fig. 1-9-22).

a,b. Netting; c. Trap.

Fig. 1-9-22　Harvesting method of crab

4　Integrated multi-trophic aquaculture (IMTA) of swimming crab

4.1　Monoculture

There are some disadvantages in monoculture of swimming crab, such as inefficient use of feed resources, low water quality and low profit (Fig. 1-9-23).

Fig. 1-9-23 Monoculture of swimming crab

4.2 IMTA

The IMTA for swimming crab includes shrimp-crab, shrimp-crab-shellfish, shrimp-crab-shellfish-fish, and so on (Fig. 1-9-24). The IMTA is increasingly popular in China for some advantages, *e.g*, improving the utilization of nutrients and water quality, reducing the occurrence of disease and increasing the survival rate of culture species.

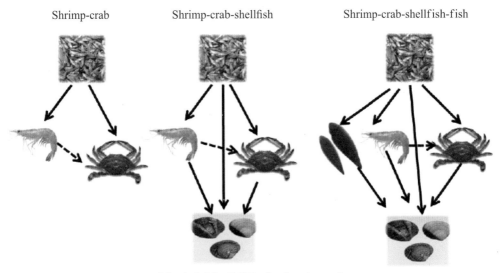

Fig. 1-9-24 IMTA of swimming crab

4.3 IMTA for the crab

An IMTA of shrimp-carb-shellfish-fish is shown in Table 1-9-2.

Table 1-9-2   IMTA of shrimp-crab-shellfish-fish

Species	Seedling quantity/individuals per hectare	Production/(kg/hm^2)	Output/($yuan$/hm^2)
Chinese shrimp	90,000	1,200	108,000
Shellfish	750,000-900,000	4,200-5,700	30,000
Swimming crab	30,000	1,200	82,500
Half smooth tongue sole	600	417	41,700
Total	-	-	262,200

Fig. 1-9-25   Harvest of IMTA

References

Dai A Y, Feng Z Q, Song Y Z. Preliminary investigation on fishery biological resources of *Portunus trituberculatus* [J]. Journal of Zoology, 1977(2): 30-33.

Fishery Bureau of the Ministry of Agriculture. China fisheries yearbook [M]. Beijing: Chinese Agriculture Express, 2017.

Wang G Z, Li S J, Chen Z G. Status of mud crabs (*Scylla* spp.) farming and studies on the population biology of *S. paramamosain* [J]. Journal of Xiamen University (Natural Science), 2016, 55(5): 617-623.

An overview of marine shellfish industry in China

Author: Wu Biao

E-mail: wubiao@ysfri.ac.cn

1 Shellfish has a long historic standing in China

1.1 Origin of the Chinese character "贝"

Today, the shellfish is denoted by the Chinese character "贝" which first appeared in oracle bones and eventually evolved into the present form (Fig. 1-10-1). The word in oracle bone inscriptions looks like a clam with opened shell.

Shellfish was used as money for trade in ancient time (Fig. 1-10-2). Thus, "贝" is a component of many wealth-related Chinese characters.

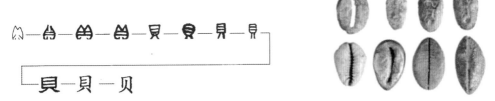

Fig. 1-10-1 The evolution of the glyph of "贝"
(Data from https://baike.baidu.com/item/%E8%B4%9D/84241?fr=aladdin#reference-[7]-11142905-wrap)

Fig. 1-10-2 The sea shells used as money in ancient time
(Data from https://baike.baidu.com/item/%E8%B4%9D%E5%B8%81/1973599?fr=aladdin)

1.2 Evidences about shellfish utilization in ancient time

In China, shellfish utilization has a tradition of about 50 thousand years inferred from the shell fossils of the paleolithic period unearthed near Beijing area. About 3,000 years ago, clam was found to be used as money for exchange. Oysters culture can be traced back to 2,400 years ago. In Ming Dynasty, Chinese people could produce pearl by freshwater mussel and utilize the medicinal function of shellfish (Wang and Wang, 2008; Smaal et al., 2019).

Since the 1950s, the large-scale mariculture of bivalves has been extensively practiced. In the following 70 years, the shellfish farming industry has been developed rapidly due to the persistent exploration of hatchery and wild seed collection techniques. Nowadays, the bivalves cultured in China has rose from around 10 species to approximately 70 since the 1960s.

1.3 Advantage for marine shellfish culture in China

Fig. 1-10-3 shows favorable conditions for the development of marine shellfish culture in China (Wang and Wang, 2008).

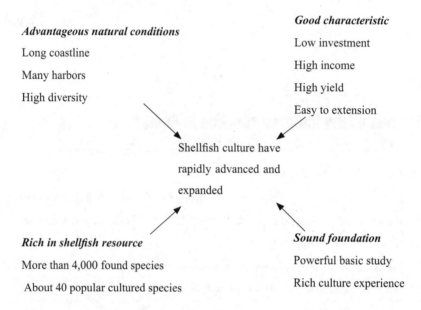

Fig. 1-10-3　Favorable conditions for the development of marine shellfish culture in China

1.4 Shellfish production in China

In China, shellfish is the largest sector of mariculture, accounting for more than 70% of the total production in weight. Thus, it supports a major aquaculture industry which is important to the economy of China's coastal regions.

In 2017, China produced about 14.4 million tons of shellfish. Oysters topped the species list, with production of 4.9 million tons. Clams production ranked the second, with 4.2 million tons (Figs. 1-10-4 to 1-10-5).

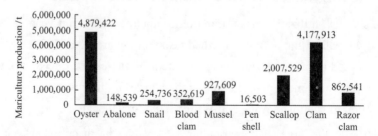

Fig. 1-10-4　Mariculture production of China in 2017

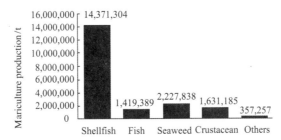

Fig. 1-10-5 Mariculture production of major shellfish in 2017
(Data from *Fisheries administration of the ministry of agriculture of China*, 2018)

2 Major cultured species

2.1 Major cultured species group

China is rich in marine shellfish resources. Before the 1970s, traditional marine shellfish culture was largely limited to four species in China: oyster, razor clam, cockle and manila clam. Until now, about 40 popular species have been cultured commercially in coastal area, intertidal zone and offshore ponds.

China's different climatic zones and eco-environment coastlines including the tropics, subtropical and temperate zones provide varieties of survival and reproduction condition for various shellfish species. The well-known and high-yield representative species mainly belong to bivalvia and gastropoda, including oyster, clam, scallop, mussel, abalone, blood clam, razor clam, and so on.

2.2 Major cultured areas

Table 1-10-1 shows the major cutured areas in China.

Table 1-10-1 Major cultured shellfish species in coastal provinces in China

Province or region	Cultured species group
Liaoning	Clam, scallop, oyster
Hebei	Scallop, clam, cockle
Shandong	Clam, scallop, oyster, mussel
Jiangsu	Clam, scallop, razor clam, oyster
Zhejiang	Razor clam, mussel, oyster, cockle, abalone
Fujian and Guangdong	Oyster, abalone, pen shell
Guangxi	Oyster, clam, pearl oyster
Hainan	Clam, scallop, sea snail, oyster

2.3 Major species

2.3.1 Oysters

China has over 20 recorded species of oysters occurring along its coast, and the classification is sometimes controversial. Generally, oysters are the largest cultured molluscan group in China (Fig. 1-10-6).

  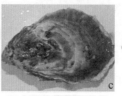 

a. *Crassostrea gigas*; b. *Crassostrea ariakensis*; c. *Crassostrea angulata*; d. *Ostrea denselamellosa*.
Fig. 1-10-6 Major cultured oysters in China

2.3.2 Clams

There are more than 15 clams cultured in China. The major species include Manila clam (*Ruditapes philippinarum*), Chinese venus (*Cyclina sinensis*) and hard clam (*Meretrix meretrix*) (Fig. 1-10-7).

a. *Ruditapes philippinarum*; b. *Meretrix meretrix*; c. *Cyclina sinensis*; d. *Mactra veneriformis*; e. *Paphia undulata*.
Fig. 1-10-7 Major cultured clams in China

2.3.3 Scallops

There are 4 major scallops cultured in China, with approximate annual production of 1.3

million tons in recent years. The noble scallop is mainly cultured in southern China, and the other three are farmed in northern China (Fig. 1-10-8).

a. *Chlamys farreri*; b. *Patinopecten yessoensis*; c. *Argopecten irradias*; d. *Chlamys nobilis*.
Fig. 1-10-8 Major cultured scallops in China

2.3.4 Blood clams

There are four major blood clams cultured in China (Fig. 1-10-9).

a. *Scapharca broughtonii*; b. *Scapharca kagoshimensis*; c. *Tegillarca granosa*; d. *Barbatia virescens*.
Fig. 1-10-9 Major blood clams cultured in China

2.3.5 Other cultured species

Other cultured shellfish in China are shown as Fig. 1-10-10.

a. *Sinonovacula constricta*; b. *Haliotis discus hannai*; c. *Moerella iridescens*; d. *Rapana venosa*;
e. *Pinctada martensi*; f. *Babylonia areolata*; g. *Mytilus coruscus*; h. *Atrina pectinata*;
i. *Saxidomrs purpuratus*; j. *Hyriopsis cumingii*.
Fig. 1-10-10 Other cultured shellfish in China

3 Bivalve hatchery culture

3.1 Development of seed breeding in shellfish

3.1.1 Major methods of seed production

A. Natural seed collection: Collecting the seed in the mudflat or sea.

B. Semi-artificial production: Stimulating spawning of broodstock in a pond and then harvesting the seed.

C. Artificial propagation: Intensive artificial production in hatchery.

3.1.2 Development process

A. Before the 1970s, most farmed seeds of shellfish relied on wild resource. Farmers collected seed from natural inhabit.

B. From 1970s to 1980s, shellfish industry has developed rapidly in China, prompted by the demand from domestic and abroad, and some artificial propagation technologies were developed.

C. In order to meet the increasing demand of seed, the artificial propagation of shellfish has been developed rapidly after 1990. UP to date, over 40 cultured shellfish seeds are artificial-produced.

3.2 Hatchery site and infrastructure

For a hatchery site selection, many key factors should be considered, especially convenient transportation, high-quality seawater, steady electricity, qualified labor force, etc. (Fig. 1-10-11).

a. Reservoir: to store the seawater, sediment; b. Sand filter: to filter out most particulate material for improving good quality seawater; c. Boiler: to increase the normal seawater temperature to produce breeding seed earlier than natural; d. Broodstock holding area, spawning area, larval culture area.

Fig. 1-10-11 Basic facilities for shellfish hatchery

3.3 Algae culture

Unicellular algae is the main food source of bivalve, and is the material basis for the development of gonads and larvae. Several species and strains known to provide good nutrition for developing shellfish should be cultured at the hatchery.

3.3.1 Common species of algae

Common species of alge are *Isochrysis galbana*, *I. zhanjiangensis*, *Phaeodactylum tricornutum*, *Nitzschia closterium*, *Platymonas halgolandica*.

3.3.2 Algae culture process

There are three stages for algae culture. The first stage is to maintain stock, and the second phase is to expand the stock in bigger carboys or tanks. After that, the algae will be transferred to pond for the third-stage culture (Fig. 1-10-12).

a. First-stage culture; b. Second-stage culture; c. Third-stage culture.

Fig. 1-10-12 Algae culture

3.4 Life history of bivalve

3.4.1 Gonadal development

Under the induction of external environment factors, the gonad begins to develop.

Gonadal development contains six phases: proliferating stage, growing stage, maturing stage, spawning stage and resting stage (Fig. 1-10-13).

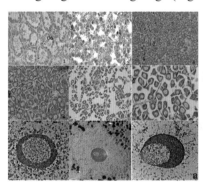 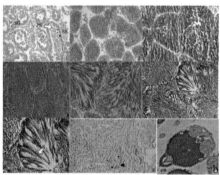

a. Ovary; b. Testis.

Fig. 1-10-13 Section of the overy and testis in different development stages of *A. pectinata*

(Data from Zheng, 2015)

3.4.2 Embryonic and larval development

A. Stages (Figs. 1-10-14 to 1-10-15): Egg-fertilized egg-multicellular-blastocysts-gastrula-trochophore-D-shape larvae-umbones-creeping larvae-early juvenile-adult.

B. The larval stage is divided into two stages: Trochophore and veliger.

C. Metamorphosis: It is an important developmental stage of bivalve from larva to adult. After metamorphosis, the larvae will be benthic sedentary instead of swimming and floating, and then the larvae develop into early juvenile.

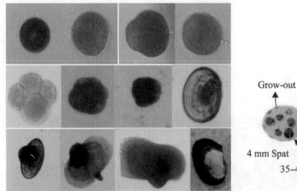

Fig. 1-10-14 Life history of *Scapharca broughtonii*
(Data from Xu *at el*., 2014)

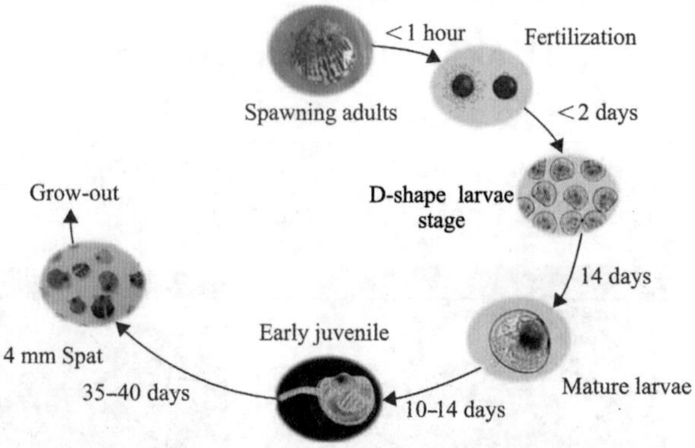

Fig. 1-10-15 Life history of scallop
(Data from Helm and Brourne, 2006)

3.5 Production process

The basic methods for seed production are similar for all bivalves. The hatchery operation mainly includes broodstock conditioning, spawning and fertilization.

3.5.1 Broodstock conditioning

A. Rinse: Before brought into the hatchery, they should be scrubbed and rinsed to remove fouling organisms and sediments.

B. Selection: Strong vitality, complete in shell, uniform size, good gonadal development, larger than the biological minimum (Fig. 1-10-16).

C. Culture density: Cultured in tanks with the density of about 2 kg/t seawater.

D. Promote gonad development: Temperature and food are two most important factors for the gonad development of adult shellfish. To provide conditions of high food availability and to raise temperature gradually are applicable to promoting gonad development.

Fig. 1-10-16　Broodstock selection

3.5.2　Spawning and fertilization

Spawning

With growth, the gonad becomes more and more mature. For some species, fully mature gametes can be "stripped", such as oyster.

Major methods are to induce spawning: physical stimulation and chemical stimulation (Fig.1-10-17). Physical stimulation methods were commonly used, such as temperature changing stimulation, flowing water simulation, and drying in the shade stimulation.

In general, the gametes are released by male and female individuals appear differently in water.

A. Male: Continuous stream of milky fluid.

B. Female: Granular appearance or clumps of eggs shed.

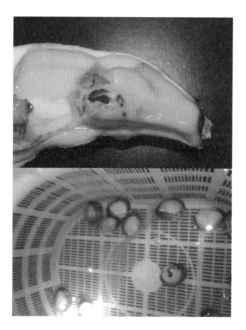

Fig. 1-10-17　Induce spawning

Fertilization procedures

Normally, after induced, females may start spawning as much as 30-60 minutes after the first male begins to liberate sperm. When the density reaches about 20-30 eggs per milliliter in the seawater, keeping 4-8 sperm around each egg, the parents should be transferred out of the spawning rank. Too much sperm can lead to embryo deformities.

After spawning finishing, the seawater should be stirred to make the sperm and eggs suspended and mixed. Fertilization rates almost exceed 90% assuming that the eggs are fully mature. The fertilized eggs will begin to divide at the appropriate temperature for the species.

3.5.3　Larvae rearing

Selection of D-larvae

At proper temperature and good oxygen condition, most embryos will develop into the D-shape larvae stage after 1-2 d.

Once most embryos develop into D-shape larvae stage, the variety selection must be carried out to pick out the imperfectly formed D-shape larvae, and larvae with incomplete or deformed shells, which can rarely develop into next stage.

Daily management of larvae

A. Exchanging water: 12-hour intervals with approximately 1/2 in volume.

B. Density: Less than 10 individuals per milliliter.

C. Feed: Mixed feed, mainly chryso-phyceae.

D. Others: Keep proper temperature and continuous aeration.

Settlement and metamorphosis

A. Settlement: Larvae begin to drop out of the water column onto a substrate, crawl around on the substrate using their feet with the shell upright, and search on the surface for a suitable place to settle.

B. Metamorphosis: The swimming larva develop to a spat during this stage, and it is a irreversible process. It is a critical stage from larva to the juvenile form, and extensive mortalities can occur in this stage (Fig. 1-10-18).

C. Symbol: A pair of darkly pigmented eye-spot, one on either side between the surface of the digestive gland and the shells valves.

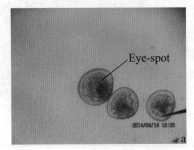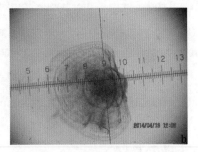

a. Eye-spot appear; b. Eye-spot larvae.

Fig. 1-10-18 Metamorphosis of bivalve

Settlement and metamorphosis

When about 30% larvae appear the eye-spots, substrates should be put into the rearing pond for settlement (Figs. 1-10-19 to 1-10-20).

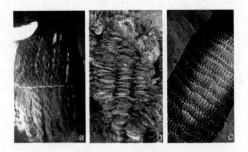

a. Coir rope; b. Scallop shell; c. Mesh netting.

Fig. 1-10-19 Commonly used substrates for bivalve larvae

a. Oysters; b. Blood clams.

Fig. 1-10-20　Larvae on the substrates

3.6　Nursery culture

A. Nursery culture: 1-2 mm in shell length.

B. Sites: Ponds or sea (Fig. 1-10-21).

Fig. 1-10-21　Two nursery culture methods for bivalve

Nursery serve is an interface between hatcheries and the grow-out phase. The purpose is to make small-size seed grow rapidly at low cost to a size, which is suitable for transfering to grow-out trays, bags, or nets with mesh apertures of 7-12 mm.

4　Shellfish culture

Two major ways of shellfish culture are hanging culture and bottom culture.

4.1　Hanging culture

Hanging culture (Fig. 1-10-22) relies on either a raft or longline system that floats on the sea surface from which the cultured shellfish are suspended.

Longline culture is the most popular way of hanging culture. Many bivalves and gastropods species are cultured using longline mode, such as scallops, oysters, mussels and abalone. Area for longline culture should be calm with water depth at low tide greater than 4 m, flow rate about 0.3-0.5 m/s and abundant phytoplankton. The raft rope length for cultivation is usually 80-100 m. The space between two adjacent longlines is about 10-15 m. Polyethylene ropes or net cages are suspended from the raft at intervals of 1 m for cultivation. Floats are used to increase buoyancy. 4-7 long lines form a breeding unit.

Fig. 1-10-22　Hanging culture of bivalve

4.2　Bottom culture

Bottom culture can be applied in conjunction with or as an alternative to hanging culture. The main advantage of using the method of bottom culture is to reduce the cost of buoyancy needs as equipment is supported by the seabed.

Bottom culture includes bottom sow culture, mud flat culture, industry culture and pond culture (Fig. 1-10-23).

a. Bottom sow culture; b. Mud flat culture; c. Industry culture; d. Pond culture.

Fig. 1-10-23　Different methods of bottom culture

4.3 Culture mode

In recent years, poly-culture has become a popular eco-farming mode which can bring double-win with ecological and economic benefits.

A. Integrated culture: Cultivate multiple species in one sea area or pond.

① Shellfish-seaweed;

② Fish-shellfish-seaweed;

③ Shellfish-sea cucumber-seaweed;

④ Shellfish-crab-shrimp;

⑤ Shellfish-fish;

⑥ Shellfish-shrimp.

B. Recirculating culture: Water recycle between different culture units, such as shrimp and mollusk pond.

Potential human power is an advantage for shellfish farming and harvesting in China. Women are the main labor sources for harvest (Fig. 1-10-24).

a, b. Harvesting; c. Harvested siphon-pump; d. Packing and transportation.

Fig. 1-10-24　Harvesting of shellfish

5　Status of genetic breeding of shellfish

5.1　Genetic breeding is very important for shellfish industry development

A. Purpose: Provide high yield and good quality species to support the sustainable development of shellfish culture.

B. Target traits: Growth, resistance, shell color, quality, and so on.

C. Methods: Group selection, family selection, hybridization, GWS, and so on.

5.2 Major task in this field

A. Identification of germplasm resources.

B. Utilization of germplasm resources.

C. Breeding frontier genetic technology innovation.

D. Genetic analysis on important economic traits.

E. New variety breeding.

F. New variety propagation and culture technology.

G. Until 2018, 37 new varieties of shellfish had been certificated by the National New Variety Committee in China.

H. New varieties including oysters, clams, pearl oyster and scallop (Fig. 1-10-25).

a. Jinli No.1; b. Bo hai hong; c. Nanke No.1; d. Wan li hong; e. Bang mage; f. Haixuan No.1; g. Leqingwan No.1.

Fig. 1-10-25 Partial new shellfish in China

References

Helm M M, Bourne N. Hatchery culture of bivalves [M]. Rome: FAO, Fisheries Technical Paper 471. 2006.

National Fisheries Technology Extension Center. 2015 guidelines for the promotion of new aquatic varieties [M]. Beijing: China Agricultural Press, 2015.

National Fisheries Technology Extension Center. 2016 guidelines for the promotion of new aquatic varieties [M]. Beijing: China Agricultural Press, 2016.

National Fisheries Technology Extension Center. 2017 guidelines for the promotion of new aquatic varieties [M]. Beijing: Ocean Press, 2017.

Smaal A C, Ferreira J G, Grant J, et al. Goods and services of marine bivalves [M]. Switzerland: Springer, 2019.

Wang R C, Wang Z P. Science of marine shellfish culture [M]. Qingdao: China Ocean University Press, 2008.

Xu W D, Wu B, Yu X Q, et al. Key technology of large-scale breeding and observation on ontogenetic process of Korea Scapharca broughtonii [J]. Shandong Fisheries, 2014, 31(10): 14-15.

Zheng Y X. Studies on reproductive cycle and larval-rearing technology of *Atrina pectinata* [D]. Shanghai: Shanghai Ocean University, 2015.

Seedling of gastropods with high economic value in China—taking the Pacific abalone as an example

Author: Li Jiaqi
E-mail: lijq@ysfri.ac.cn

1 Introduction

The Pacific abalone (*Haliotis discus hannai*) is one of the most important and favored mollusk in China (Figs. 1-11-1 to 1-11-2; Table 1-11-1).

Table 1-11-1   Taxonomic status

Kingdom	Phylum	Class	Subclass	Superfamily	Family	Genus
Animalia	Mollusca	Gastropoda	Vetigastropoda	Haliotoidea	Haliotidae	*Haliotis*

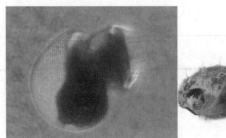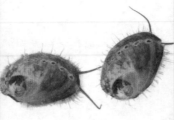

Fig. 1-11-1   *Haliotis discus hannai*

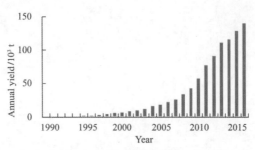

Fig. 1-11-2   Annual yield of abalone in China

2 Artificial breeding

2.1 Artificial maturation

Abalone is fed with fresh kelp (*Laminaria japonica*) and cultured in 20 ℃ seawater for about 3 months. Effective accumulated temperature is 1,300 ℃.

The male and female abalone can be easily distinguished (Fig. 1-11-3) by the appearance of the gonad once they reach maturation. The color of the gonads of male and female are white and dark blue (or dark green), respectively.

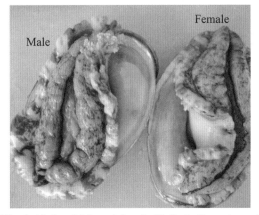

Fig. 1-11-3 Male and female *Haliatis discus hannai*

2.2 Larvae production

The process of abalone larvae production is shown in Fig. 1-11-4.

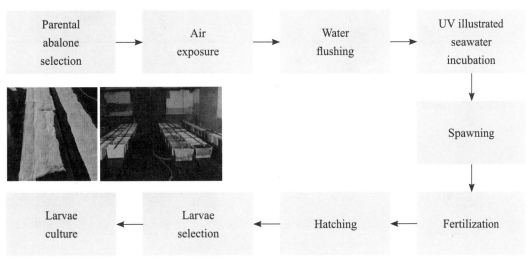

Fig. 1-11-4 Larvae production

2.3 Induced spawning

Elevating seawater temperature to 23.5 ℃ and flushing the parents with seawater can induce them spawning (Fig. 1-11-5).

Fig. 1-11-5 UV light radiated seawater

2.4 Fertilization and hatching

The process of fertilization and hatching of abalone is shown in Fig. 1-11-6.

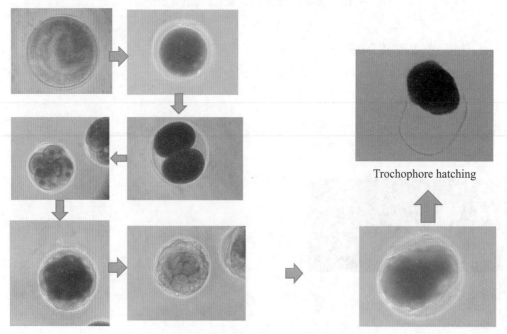

Fig. 1-11-6 Fertilization and hatching

2.5 Larvae culturing

Fig. 1-11-7 shows the larva culturing process of abalone.

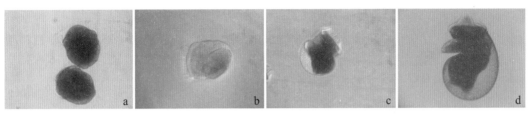

a. Trochophore; b. Initiation of calcification; c. Partial calcified veliger; d. Fully calcified veliger.

Fig. 1-11-7 Larvae culturing

Larvae culturing is satisfied with the following conditions:

A. Culturing density: 10 larvae per milliliter.

B. Culturing seawater temperature is from 20.5 ℃ to 22.5 ℃.

C. DO>6 mg/L.

In about 4 d, the abalone embryo develops from fertilized eggs to larvae that are ready to metamorphosis. Larvae do not feed in these days and begin to feed on micro-algae after metamorphosis.

3 Intermediate culture

3.1 Benthic diatom feeding stage

The benthic diatom feeding stage of juvenile abalone is shown in Fig. 1-11-8.

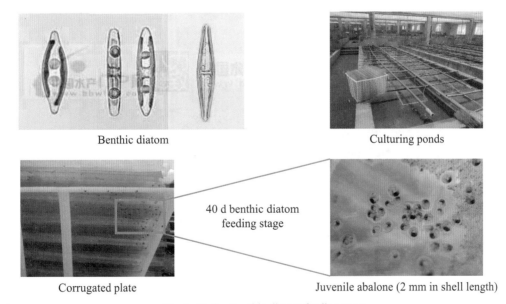

Fig. 1-11-8 Benthic diatom feeding stage

3.2 Benthic diatom feeding to artificial feeding stage

The 4 mm (shell length) juveniles were transferred from corrugated plate to culturing ponds and fed with artificial feed (Fig. 1-11-9).

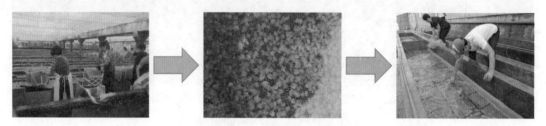

Fig. 1-11-9 Benthic diatom feeding to artificial feeding stage

3.3 Artificial feeding stage

Artificial feeding stage (Fig. 1-11-10) is 140 d. The abalone shell is from 4 mm to 2.5 cm in length.

Fig. 1-11-10 Artificial feeding stage

3.4 Marketing and transportation

Fig. 1-11-11 shows the marketing and transportation of abalone.

Fig. 1-11-11　Marketing and transportation

3.5 Over-wintering

The over-wintering of abalone is shown in Fig. 1-11-12.

Fig. 1-11-12　Over-wintering abalone

4 Grow-out

There are several abalone mariculture models for abalone in China.

4.1 Land-based farming

Fig. 1-11-13 shows land-based farming of abalone.

Fig. 1-11-13　Fresh kelp and salted kelp (*Gracilaria*)

4.2 Intertidal farming

The intertidal farming of abalone is shown in Fig. 1-11-14.

Fig. 1-11-14　Intertidal farming

4.3 Long-line farming

The long-line farming mode is shown in Fig. 1-11-15.

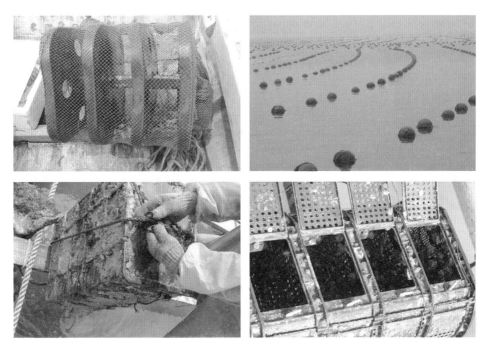

Fig. 1-11-15　Long-line farming

4.4 North-south-relay farming

Fig. 1-11-16 shows the north-south-relay farming of abalone.

Fig. 1-11-16　North-south-relay farming

5 Genetic breeding

5.1 Advancement in the production capacity

Advancement in the production capacity of the world's main livestock and the Pacific abalone is shown in Fig. 1-11-17 and Table 1-11-2.

Fig. 1-11-17 The world's main livestock and the Pacific abalone

Table 1-11-2 Advancement in the production capacity of the world's main livestock and the Pacific abalone

Animal	Traits	Years	Improvement	Reference
Broiler chicken	Slaughter weight	45	400%	Havenstein, 2003
Turkey	Slaughter weight	37	100%	Havenstein, 2004; 2007
Layer chicken	Daily egg production	43	43%	Anderson, 1996
Swine	Slaughter weight	43	12,126 g	Chen, 2002
Cattle	Slaughter weight	50	62%	USDA
The Pacific abalone	One year shell length	20	About 100%	Zhang et al., 2004; Li et al., (unpublished)

5.2 Application of strain-cross and genetic improvement

Great contributions of genetic improvement have been achieved in the advancement of abalone breeding since 1997, such as strain-cross in commercial breeding of the Pacific abalone (Fig. 1-11-18).

Fig. 1-11-18 Application of strain-cross and genetic improvement

5.3 Parental strain

Parental strain has different genetic backgrounds and complementary production traits (Table 1-11-3).

Table 1-11-3 Parental strain

Strain	Selection age	Genetic background	Advantages	Disadvantages
Sire	One year old	Descendants of Chinese and Japanese wild population hybrids	One year/generation; fast growing; high survival rate	Low-fecundity of female abalone; low success rate of embryonic development
Dam	Three years old	Chinese wild population	High-fecundity of female abalone; high success rate of embryonic development	Three years/generation

5.4 Embryonic development

The one year old and stock seed of abalone are shown in Fig. 1-11-19.

a. One year old; b. Stock seed.

Fig. 1-11-19 Embryonic development

The different letters above the dark grey bars indicate significant differences ($P < 0.05$) in average yolk diameters (Fig. 1-11-20). The different letters above the dark grey (capital letters) and grey bars (lower case letters) indicate significant differences ($P < 0.05$) in the fertilization and metamorphosis rates, respectively (Fig. 1-11-21).

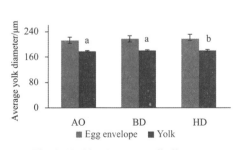 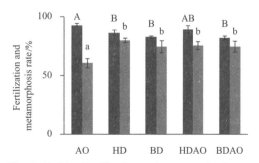

Fig. 1-11-20 Average yolk diameter   Fig. 1-11-21 Fertilization and metamorphosis rate

Table 1-11-4 shows AO, HD, BD strain and its advantages.

Table 1-11-4 AO, HD, BD strain and its advantages

Strain	Advantages
AO strain	Lower fecundity
	Lower yolk volume
	Lower metamorphosis rate
HD and BD strain	Higher fecundity
	Higher yolk volume
	Higher metamorphosis rate

5.5 Juvenile growth and survival

At 15 d, the average shell length (ASL) of juvenile abalone from the AO group was significantly smaller than that of individuals from the other four groups (Table 1-11-5). The ASLs of HD and HDAO individuals were significantly larger than that of AO individuals at the age of 40 d, while no significant difference was seen among the AO, BD and BDAO groups. The ASL of AO individuals was significantly larger than that of individuals from the other four groups at 120 d ($P < 0.05$). At 150 d, individuals from the AO group had the largest ASL, which was significantly larger than those of individuals from the HD and HDAO groups ($P < 0.05$) and slightly larger than those of individuals from the BD and BDAO groups ($P > 0.05$).

Table 1-11-5 The average shell length of five groups of abalone measured post fertilization

Age/d	Shell length of experimental group/mm				
	AO	HD	BD	HDAO	BDAO
15	$0.47 \pm 0.04^{a*}$	0.52 ± 0.05^b	0.51 ± 0.06^b	0.53 ± 0.06^b	0.51 ± 0.06^b
40	2.71 ± 0.37^a	3.09 ± 0.44^b	2.78 ± 0.44^a	3.00 ± 0.43^b	2.90 ± 0.39^{ab}
120	18.25 ± 2.87^a	17.51 ± 2.52^{bc}	17.30 ± 2.53^{bc}	16.99 ± 2.46^b	17.76 ± 2.29^{ab}
150	24.23 ± 2.98^a	23.10 ± 3.24^b	23.42 ± 3.16^{ab}	22.98 ± 3.34^b	23.78 ± 3.47^{ab}

Note: The different superscript letters indicate significant differences ($P<0.05$) between the groups at the same age.

AO group has slow early stage shell growth rate, fast shell growth and high survival rate during artificial feeding stage. Crossbred abalone has faster shell growth and higher survival rate than purebred dam strains (Figs. 1-11-22 to 1-11-23).

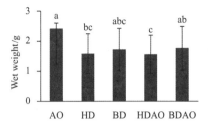
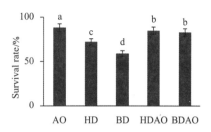

The different letters above the dark grey bars indicate significant differences ($P<0.05$) in average wet weight between the groups.

The different letters above the dark grey bars indicate significant differences ($P<0.05$) in survival rates between the groups.

Fig. 1-11-22　Average wet weight of 150 d

Fig. 1-11-23　Survival rate

The mid-parent heterosis for shell length and body weight was not obvious, and the survival rate of the single parent (relative to the female parent) was obvious (Table 1-11-6).

Table 1-11-6　Mid-parent (H_M) and single-parent heterosis (H_D, relative to the dam line) of two hybrid groups for ASL, AWW and survival rate

Production traits	Mid-parent heterosis (H_M)		Single-parent heterosis (H_D)	
	HDAO	BDAO	HDAO	BDAO
15 d ASL	7.2	4.3	2.1	0.4
40 d ASL	3.3	5.4	−3	4.1
120 d ASL	−5	−0.1	−3	2.6
150 d ASL	−2.9	−0.2	−0.5	1.5
150 d ASL	−11	−3.2	−1	2.3
Survival rate	5.9	12.9	17.8	41.4

Note: Positive and negative values reflect favorable and unfavorable effects, respectively.

5.6　Commercial application

Strain-cross has been applied in commercial abalone breeding of the dominant hatcheries in northern China and the annual production grows year by year (Fig. 1-11-24).

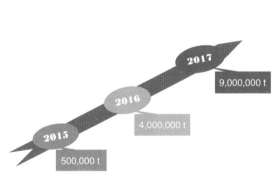

Fig. 1-11-24　The annual production of strain-cross juvenile abalone in Xunshan Group from 2015 to 2017

5.6.1 Production traits of one year old juveniles

Both genetic breeding programs and improvements in cultivation techniques have contributed to this advancement (Fig. 1-11-25).

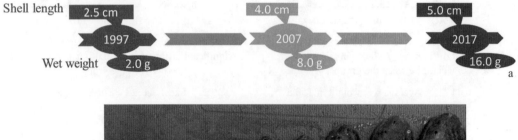

a. The shell length of one year old abalone grew from 2.5 cm to 5.0 cm within 20 years since 1997; b. Shells indicating the improvement in shell length since 1997.

Fig. 1-11-25 Shell length and wet weight

5.6.2 Comparison between selected and wild strains

The metamorphosis rate of wild population was significantly lower than that of selected strain. The metamorphosis rate of hybrids was significantly improved (Fig. 1-11-26).

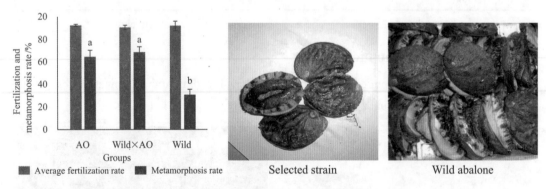

Average fertilization (light grey bars) and metamorphosis rates (dark grey bars) for the experimental groups AO, Wild×AO and Wild. Different letters above the dark grey bars indicate significant differences ($P < 0.05$) in the metamorphosis rate.

Fig. 1-11-26 Photos show representative

5.6.3 Growth and survival rate

At 15 d, ASL of juvenile abalone from the AO group was significantly shorter than that in individuals from the other two groups ($P < 0.05$) (Table 1-11-7). At 40 d, no significant difference was observed among the ASL of all the three groups. The ASL of AO individuals was significantly longer than that of individuals from the Wild group at 90 d and 130 d ($P < 0.05$), but no significant difference in ASL was shown between the Wild×AO and the Wild groups or between the Wild×AO and the AO groups. At 190 d, individuals from the AO group had the longest ASL, which was significantly longer than that of individuals from the other two groups ($P < 0.05$).

Table 1-11-7 The average shell length of the experimental groups post fertilization

Age/d	Shell length of experimental groups		
	AO	Wild	Wild × AO
15	$0.49 \pm 0.03^{a*}$ μm	0.52 ± 0.06^b μm	0.52 ± 0.05^b μm
40	2.73 ± 0.32 mm	2.80 ± 0.40 mm	2.81 ± 0.42 mm
90	9.70 ± 1.17^a mm	8.17 ± 1.16^b mm	9.22 ± 1.22^{ab} mm
130	17.85 ± 1.75^a mm	15.63 ± 1.50^b mm	17.44 ± 1.94^{ab} mm
190	29.64 ± 1.95^a mm	25.17 ± 2.24^b mm	26.69 ± 2.05^b mm

Note: The different letters indicate significant differences ($P < 0.05$) in shell length among the groups at the same age.

The offspring growth and survival rate of the wild strain of the Pacific abalone were inferior to that of the selected strain (Figs. 1-11-27 to 1-11-28).

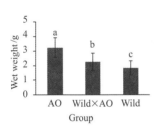

The different letters above the bars indicate significant differences ($P < 0.05$) in AWW between the experimental groups.

Fig. 1-11-27 The average wet weight (AWW) of abalone measured at 190 d

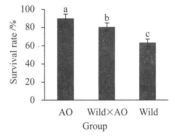

The different letters above the bars indicate significant differences ($P < 0.05$) in survival rates between the groups.

Fig. 1-11-28 The survival rates of the experimental groups during the artificial diet feeding period (40–190 d post fertilization)

5.6.4 Over-wintering production traits

The growth and survival rate of wild strain were lower than that of selected and hybrid groups (Table 1-11-8).

Table 1-11-8 The shell length, wet weight and overwintering survival rate of the three experimental groups

Group	Shell length/mm		Wet weight/g		Survival rate/%
	Before winter	After winter	Before winter	After winter	
AO	29.64 ± 1.95^{a*}	55.07 ± 5.21^a	3.24 ± 0.66^a	19.29 ± 5.42^a	84.96 ± 4.73^a
Wild	27.64 ± 2.11^b	47.44 ± 6.23^c	2.77 ± 0.68^b	13.41 ± 5.67^c	70.65 ± 5.22^c
Wild×AO	27.95 ± 1.85^b	50.62 ± 5.28^b	2.76 ± 0.60^b	16.03 ± 5.49^b	80.68 ±6 .58^b

Notes: Shell length and wet weight were measured both before and after overwintering. The different letters indicate significant differences ($P < 0.05$) in averages among the groups within the same column.

ASL and AWW of the AO group were significantly ($P < 0.05$) larger than those of the other groups both before and after over-wintering. There were no significant differences between ASL and AWW of the Wild group and the Wild×AO group before over-wintering. However, significant differences ($P < 0.05$) between ASL and AWW of the Wild group and the Wild×AO group were observed after over-wintering. A significant difference ($P < 0.05$) was observed in the over-wintering survival rates of the experimental groups. The highest and lowest over-wintering survival rates were found in the AO and Wild groups, respectively (Table 1-11-8).

5.6.5 Contribution of genetic improvement

According to the calculated superiority based on the observed data, in the artificial diet feeding and the over-wintering phase, the genetic improvement in shell growth rate were over 17% and 28%, respectively. However, if we consider the survival rate and normalize the shell growth rate to the wet weight yield per pond or cage, increases in the shell growth rate of over 44% and 51% were observed (Table 1-11-9).

Table 1-11-9 The observed and normalized (to wet weight yield per pond or cage) performance superiority of the selected strain over the wild strain of the Pacific abalone at three different phases

Stages	Metamorphosis	Artificial diets feeding		Overwintering		
Traits	Metamorphosis rate	Shell growth rate	Survival rate	Shell growth rate	Wet weight growth rate	Survival rate
Observed superiority/%	109	17.8	30.0	28.4	50.8	35.1
Normalized superiority/%	–	44.5	–	51.5	92.2	–

Production traits of dam and sire strain can be complementary. Strain-cross is useful in improving production traits. In the last 20 years, the contribution of genetic improvement is over 40%.

References

Havenstein G B, Ferket P R, Grimes J L, et al. Comparison of the performance of 1966-versus 2003-type turkeys when fed representative 1966 and 2003 turkey diets: Growth rate, livability, and feed conversion [J]. Poultry Science, 2007, 86: 232-240.

Havenstein G B, Ferket P R, Qureshi M A. Growth, livability, and feed conversion of 1957 versus 2001 broilers when fed representative 1957 and 2001 broiler diets [J]. Poultry Science, 2003, 82: 1500-1508.

Li J, Wang M, Fang J, et al. Reproductive performance of one-year-old Pacific abalone (*Haliotis discus hannai*) and its crossbreeding effect on offspring growth and survival [J]. Aquaculture, 2017, 473: 110-114.

Li J, Wang M, Fang J, et al. A comparison of offspring growth and survival among a wild and a selected strain of the Pacific abalone (*Haliotis discus hannai*) and their hybrids [J]. Aquaculture, 2018, 495: 721-725.

Zhang GF, Que HY, Liu X, et al. Abalone mariculture in China [J]. Journal of Shellfish Research, 2004, 23: 947-950.

Breeding and culture of sea cucumber (*Apostichopus japonicus*) in China

Author: Liao Meijie
E-mail: liaomj@ysfri.ac.cn

1 Biology of *Apostichopus japonicus*

1.1 Taxonomy

Sea cucumber (Echinodermata: Holothuroidea), also commonly called holothurians or holothuroids, has traditionally been classified through their morphological phenotype (Fig. 1-12-1). It is one of the most important marine resources supporting coastal livelihoods around the world.

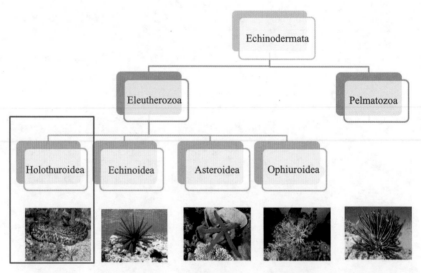

Fig. 1-12-1　Taxonomy of sea cucumbers

1.2 Species in Holothuroidea

Totally there are about 1,400 sea cucumber species in the world, and all of the species live in the ocean (Fig. 1-12-2). In China, 147 species have been identified, 20 species are edible and 10 species have commercial value. But only two species have been successfully artificially bred, and only *A. japonicus* can be largely farmed until now.

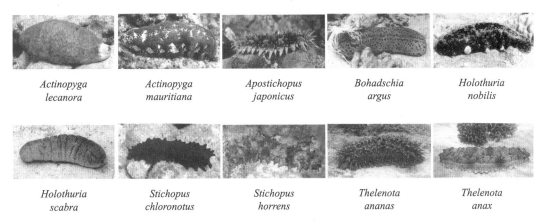

| Actinopyga lecanora | Actinopyga mauritiana | Apostichopus japonicus | Bohadschia argus | Holothuria nobilis |

| Holothuria scabra | Stichopus chloronotus | Stichopus horrens | Thelenota ananas | Thelenota anax |

Fig. 1-12-2 Commercially important sea cucumbers of the world

1.3 Natural distribution of *A. japonicus*

A. japonicus is distributed mainly in the Western Pacific Ocean, such as the Yellow Sea, the Sea of Japan, the Sea of Okhotsk. The northern limits of its geographic distribution are the coasts of the Sakhalin Island, Russian Federation and Alaska (USA). The southern limit is Tanega-Shima in Japan. In China, it is commonly distributed along the coast of Liaoning Province, Hebei Province and Shandong Province. Its southern limit in China is the coast of Lianyungang, Jiangsu Province.

1.4 Appearance of *A. japonicus*

Dorsal surface of *A. japonicus* is variable in color from brown to grey or olive green, while ventral surface is variable from brown to grey. Body length and body width of adult individual are 15-35 cm and 4-6 cm, respectively. Body is squarish in cross-section and tapered somewhat at anterior and posterior ends. Large conical papillae are present in 4-6 loose rows on the dorsal surface (Figs. 1-12-3a, b). Ventral podia are lined in three irregular longitudinal rows. Mouth is ventral with 20 tentacles (Fig. 1-12-3c). Anus is terminal with no teeth.

  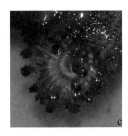

a. The dorsal appearance; b. The ventral appearance; c. Top of the tentacle.
Fig. 1-12-3 Appearance of *A. japonicus*

The sea cucumber has endoskeleton, which is called ossicles. Ossicles are calcareous elements and buried in the connective tissue of the body wall. The shape of ossicles is a key feature in the taxonomic classification of holothuroids. The ossicles in *A. japonicus* include table-shaped ossicles (Figs. 1-12-4a, b, c), rod-shaped ossicles in tube feet (Fig. 1-12-4e) and tentacles (Fig. 1-12-4f), complex ossicles around the anus (Fig. 1-12-4g), and large ossicles (Fig. 1-12-4h) in the tube feet. The shape of the table-shaped ossicles changes with the age of the sea cucumber (Fig. 1-12-5).

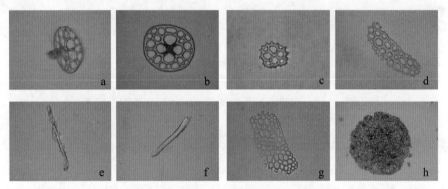

a. Tables of body wall; b. Disc of the table; c. Reduced tables; d. Perforated plate; e. Rod-shaped ossicles in tube feet; f. Rod-shaped ossicles of tentacles; g. Complex ossicles around the anus; h. Large ossicles in the tube feet.

Fig. 1-12-4 Ossicles in *A. japonicus*

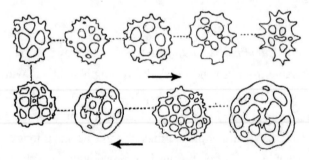

Fig. 1-12-5 Morphological changes of table-shaped ossicles with age

1.5 Anatomy of *A. japonicus*

The body of the sea cucumber is surrounded by a relatively thick and often leather-like body wall, which is the main edible organ. The digestive system is composed of the mouth, pharynx, esophagus, stomach, intestine (anterior intestine, middle intestine, posterior intestine), cloaca and anus (Fig. 1-12-6). The respiratory tree is composed of a trunk that branches off the cloaca and extends into the body cavity to the left and right sides of the digestive tract.

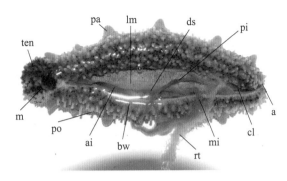

a. Anus; ai. Anterior intestine;
bw. Body wall; cl. Cloaca;
ds. Dorsal sinusoid;
lm. Longitudinal muscle; m. Mouth;
mi. Middle intestine; pa. Papillae;
po. Podia; pi. Posterior intestine;
rt. Respiratory tree; ten. Tentacle.

Fig. 1-12-6　Internal anatomy of *A. japonicus*

A. japonicus is dioecious. It is difficult to distinguish male from female based on external morphology. It can be distinguished by the color of gonad in the reproductive season. The gonad of female in reproductive stage is apricot yellow or orange red, and that of male is yellow or white (Fig. 1-12-7). In the non-reproductive season, the gonad is tiny and it is generally difficult to distinguish genders by its color.

The gonad is a dendritic tubule attached to the dorsal mesentery. It consists of 11–13 main branches with several secondary branches. These main branches converge into a gonoduct that opens into a gonopore through the body wall. The gonopore is located in the interambulacrum on the dorsal side of the anterior end (Fig. 1-12-8). It is circular and has a diameter of 4–5 mm.

The reproductive cycle of *A. japonicus* is annual and has been divided into five major developmental stages according to the gonad index (GI), resting stage, early growth stage, growth stage, mature stage and spent stage.

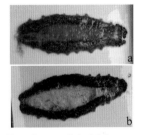

a. Female gonad; b. Male gonad.
Fig. 1-12-7　Mature gonad of *A.japonicus*

Fig. 1-12-8　Gonopore of *A. japonicus* during reproductive season

1.6　Ecological characteristics of *A. japonicus*

The most important elements determining the survival, abundance, and distribution of *A. japonicus* are water temperature, salinity, and bottom type.

A. Temperature: 0–30 ℃, which is the most direct factor causing the sea cucumber to enter aestivation.

B. Salinity: The optimum salinity to *A. japonicus* is between 24.2 and 34.6.

C. pH: 7.9-8.4.

D. Depth: Occurs from the shallows of the intertidal zone at a depth of 3-35 m.

E. Bottom type: Generally, pebbles, gravel, sand and muddy-sand areas close to the coastline, inner bay reefs, and macrophyte habitats unaffected by freshwater runoff constitute good habitats for *A. japonicus* (Fig. 1-12-9).

Fig. 1-12-9　Living sea cucumbers on the bottom of the sea

1.7　Special physiological characteristics of *A. japonicus*

1.7.1　Aestivation

When the seawater temperature rises to a certain level during summer, most individuals of *A. japonicus* migrate to deeper environments and stop moving and feeding, entering a dormant state dubbed aestivation, which lasts up to 100 d.

During this period, the sea cucumber experiences organ atrophy and major weight loss (Fig. 1-12-10). The period of aestivation has positive correlation with the body weight. Juveniles do not aestivate.

Fig. 1-12-10　*A. japonicus* before aestivation (a) and in (b) aestivation

1.7.2　Evisceration and organ regeneration

A. japonicus eviscerate its intestine, respiratory tree and gonad through the cloaca when it is exposed to stressful circumstances. It possesses a striking capacity to regenerate most of its lost tissues or organs, such as the intestine, respiratory tree, gonad and body wall (Fig. 1-12-11a). The time needed for viscera to regenerate is about 30 d.

1.7.3　Autolysis

A. japonicus is a very fragile marine organism. Under the stimulation of the external environment and chemical factors, it can be completely degraded through the process of skin destruction, evisceration and dissolving into water (undergoing autolysis and liquefying) (Fig. 1-12-11b).

a. Evisceration; b. Autolysis.

Fig. 1-12-11 Evisceration and autolysis of *A. japonicus*

1.8 Development and growth of *A. japonicus*

A. japonicus is a dioecious broadcast spawner, *i.e*, the mature individuals shed gametes into seawater where fertilization occurs. Its life cycle can be divided into eight major stages: fertilization, blastula, gastrula, auricularia, doliolaria, pentactula, juvenile and adult (Fig. 1-12-12). The fertilized, mainly demersal eggs measure 140-170 μm in diameter, and develop into freely rotating blastulae in 12-14 h at 20-22 ℃. The embryos then develop into auricularia larvae over the following 31 h, at which time they gain the ability to feed. When the auricularia grow to their maximum size of 800-1,000 μm, they shrink and transform into doliolaria, and subsequently into pentactula after completion of metamorphosis. The development of buccal and ambulacral podia indicates the onset of the juvenile stage. It takes almost two years for *A. japonicus* to reach sexual maturity, and the life span of this species is commonly estimated to be at least five years.

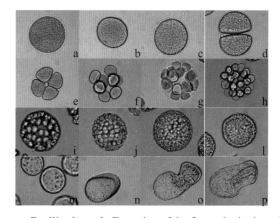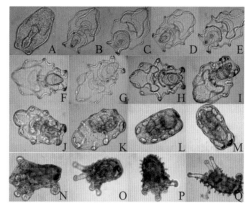

a. Fertilized egg; b. Extrusion of the first polar body; c. Extrusion of the second polar body; d. 2-cell stage; e. 4-cell stage; f. 8-cell stage; g. 16-cell stage; h. 32-cell stage; i. 64-cell stage; j. 128-cell stage; k. 256-cell stage; l. Blastocyst stage; m. Rotational membrane blastocyst; n. Ring blastocyst; o-p. Gastrula stage; A. Early auricularia; B-D. Middle auricularia; E-I. Late auricularia; J-M. Doliolaria; N-Q. Pentactula.

Fig. 1-12-12 Embryonic and larval development of *A. japonicus*

1.9 Nutritional and medicinal function

Fig. 1-12-13 shows the nutritional and medicinal function of *A. japonicus*.

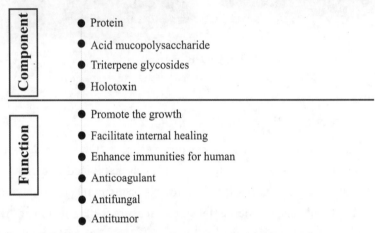

Fig. 1-12-13 Nutritional and medicinal function of *A. japonicus*

2 Historical and present situation of the sea cucumber culture in China

2.1 History of the sea cucumber culture in China

The breeding research began about 70 years ago. The first research project on the breeding of *A. japonicus* was conducted and obtained the first artificially bred juveniles of the species. In the mid-1980s, artificial breeding techniques of *A. japonicus* were successfully established. Sea ranching was then carried out from the 1980's to the late 1990's. Sea cucumber farming has developed rapidly since 2000.

2.2 Aquaculture area

In the 21st century, with the development of sea cucumber culture technology and the exploration of aquaculture model, the range of sea cucumber culture has been gradually expanded from the Bohai Sea and the Yellow Sea, which are suitable for sea cucumber growth, to the East China Sea and South China Sea.

2.3 Present situation

According to official statistics in 2018, we can see:

A. The total production reached 1.7×10^5 t (Fig. 1-12-14).

B. The surface areas cultivated reached 2.3×10^5 hm^2 (Fig. 1-12-15a).

C. The seeds number has reached 56 billions in 2018 (Fig. 1-12-15b).

D. Valued at over 30 billion *yuan*.

E. The culture of *A. japonicus* has become one of the most important new aquaculture industries in China.

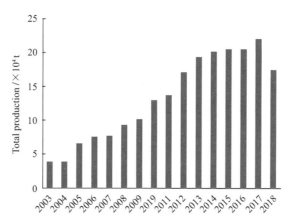

Fig. 1-12-14　Production of sea cucumber in China

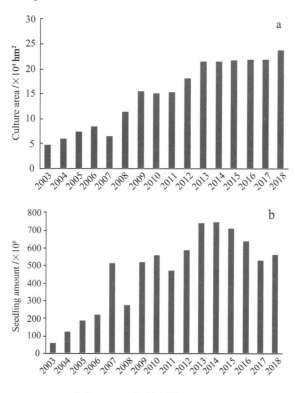

a. Culture area; b. Seedling amount.

Fig. 1-12-15　Culture area and seedling yield of sea cucumber in China

3　Artifical breeding techniques of *A. japonicus*

3.1　Site selection principles for nursery farms

A. Convenient seawater extraction.

B. Away from the estuary.

C. Good water quality, small wind and pollution-free coast or inner bay.

D. Places with convenient fresh water, electricity and transportation.

3.2 Facilities for nursery farms

A. Water intake and drainage systems.

B. Water treatment systems.

C. Heating systems.

D. Aeration systems.

E. Power supply systems.

3.3 Construction of nursery room

A. The nursery room is mostly with brick-concrete structure, roof curved and steel frame structure, generally 50–100 m long and 12–15 m wide.

B. The nursery room should be equipped with light control, temperature control and inflation system. The light is controlled at 500–1,500 lx to avoid direct light entering the room (Fig. 1-12-16).

Fig. 1-12-16   Nursery room for sea cucumber

3.4 The layout of the nursery workshop

The layout of the nursery workshop is mainly composed of the nursery ponds, leaving a certain area for disinfection and bait configuration. The middle walkway is 1–1.2 m wide and has a cover plate. The underside of the cover is a trench. The depth of the trench is 1.2–1.5 m. The nursery pond is preferably rectangular, with a depth of 1.2–1.5 m and volume of 10–50 m^3. The bottom of the pond should have a slope of 2°–3° to the outlet to facilitate thorough drainage. Each nursery pond should be equipped with an independent water supply, drainage system and gas supply channel.

3.5 Breeding stage

Artificial breeding of *A. japonicus* for the production of seedlings includes the following stages:

A. Collection and cultivation of broodstock (Fig. 1-12-17).

B. Spawning and hatching.

C. Larval rearing.

D. Larval settlement.

E. Juvenile rearing.

Fig. 1-12-17 Broodstock collection of sea cucumber

3.6 Collection and cultivation of broodstock

A. Collection: *A. japonicus* adults are gametogenically mature between the end of June and the end of July, with slight shifts based on local seawater temperature. The broodstock collected by divers is maintained in culture pond. The weight of adult broodstock should be above 200 g.

B. Cultivation: The cultured density should be no more than 30 sea cucumbers per cubic meter. The quality of the holding environment is maintained through regular water changes and close monitoring of temperature, according to the production plan (Fig. 1-12-18).

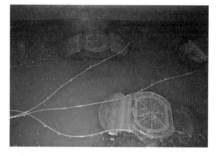

Fig. 1-12-18 Cultivation of the broodstock of sea cucumber

3.7 Spawning and hatching

Release of spermatozoa and oocytes by mature *A. japonicus* occurs when seawater temperature reaches 16-20 ℃ (Fig. 1-12-19). When this is not possible, spawning may be triggered using various techniques, such as desiccation, running water and temperature shock.

Spawning males should be removed from the pond to avoid excessive accumulation of sperm and polyspermy. For fertilization, spermatozoa density should be 20–100 individuals per milliliter.

For hatching, the temperature should be kept at 20–22 ℃ and light intensity should be 500–1,000 lx with moderate aeration. The hatching water should be agitated to avoid egg sinking.

a. Oviposition; b. Spawning.
Fig. 1-12-19　Oviposition and spawning of sea cucumber

3.8　Larval rearing

A. Time for larvae selection: 30 h after fertilization (early auricularia).

B. Larval density: 2×10^5–6×10^5 individuals per cubic meter.

C. The auricularia feed: Unicellular algae, such as *Phaeodactylum tricornutum*, *Nitzschia closterium*, *Chaetoceros muelleri*, *Dicrateria inornata* and *Dunaliella salina*, partially substituted with marine red yeast.

D. Feeding dose: The daily feeding dose increases from 1×10^4 cells per milliliter in early auricularia stage to 2.0×10^4 cells per milliliter in late auricularia stage.

E. Water change: It is generally advised to change 1/10–1/5 volume of fresh seawater daily to the culture tank daily.

F. Conditions of the surrounding: Continuous aeration and dark conditions are essential during the rearing period.

G. Regular observation: Larvae under a microscope is necessary to monitor development, growth rate, average body length, presence of algae in the stomach, organogenesis, and so on.

3.9　Larval settlement

A. Time: It takes about 10 d for auricularia larvae of *A. japonicus* to develop into doliolaria larvae. When doliolariae account for about 1/3 of the total population, settlement substrata can be added.

B. Settlement: The most commonly used substratum is a framework of about 30 cm × 50 cm × 50 cm made up of 10–15 layers of corrugated polyethylene sheets, which are spaced out every 6–7 cm (Fig. 1-12-20a).

C. Condition of the settlement: Settlement plates must be conditioned to ensure the growth of benthic diatoms on their surfaces (expected survival rate is 1%–30%) (Fig. 1-12-20b).

a. Settlement; b. Nursery tank with settlement.
Fig. 1-12-20　Laying of settlement in *A. japonicus* nursery tank

3.10　Juvenile rearing

It takes about 12 d from fertilization for *A. japonicus* to settle. Hard smooth substrata are required for the pentactula larvae of *A. japonicus* to settle.

A. Food: Juveniles of *A. japonicus* mainly feed on benthic diatoms present on the settlement plates, supplemented by unicellular algae and other food, such as the filtrate of ground fresh *Sargassum thunbergii*.

B. The feed ration: It is determined based on juvenile development, growth rate and body size.

C. Water change: It is recommended to change 1/2 volume water every day.

D. Rearing period: It would take about 100 d for juveniles of *A. japonicus* to reach up to 1 cm (Fig. 1-12-21).

Fig. 1-12-21　Juvenile rearing of *A. japonicus*

4　Culture techniques of *A. japonicus*

4.1　Main aquaculture patterns of sea cucumber

Main aquaculture patterns of sea cucumber are shown in Fig. 1-12-22.

a. Pond culture; b. Indoor culture; c. Suspended culture; d. Polyculture.

Fig. 1-12-22   Main aquiculture patterns of sea cucumber

4.2   Pond culture: one of the most important farming methods

4.2.1   Pond models

A. Standard pond: Made from modified shrimp ponds (Fig. 1-12-23a).

B. Cofferdam pond: Use natural tidal water on fenced shoals and shallow water lowland (Fig. 1-12-23b).

4.2.2   Pond construction

A. Area and layout: Square or rectangular, 1.3-2.0 hm^2 area and 1.5-3 m deep.

B. Dam: Dam foundation built with reef and rock or hard sand, the slope ratio is 1 : (1.5-2.5);

C. Facilities: The water supply and drainage system; aeration systems and substratum (stones or titles) (Fig. 1-12-23c).

a. Standard pond; b. Cofferdam pond; c. Substratum in pond.

Fig. 1-12-23   Pond culture of sea cucumber

4.2.3 Techniques and management of pond culture

A. japonicus pond cultivation involves disinfection, water preconditioning, introduction of seedlings and management.

A. Disinfection: 1–2 months before seedling broadcasting. The procession includes solarization, tillage, temper, disinfection using quicklime and immersion with sea water.

B. Water preconditioning: 15–20 d before seedling introduction. The pond water should be fertilized using fertilizer. The water transparency should be kept at 60–100 cm depth.

C. Introduction of seedlings:

① Seedlings quality: Healthy seedlings with good adherence.

② Seedling time: April to May in spring or September to October in autumn.

③ Seedling density: For 1–3 g per individual seedlings, stocking is around 10–20 individuals per square meter; For 3–5 g per individual seedlings, stocking is around 7–15 individuals per square meter.

4.2.4 Management

The management of *A. japonicus* facilities focuses on monitoring and controlling of water quality, feeding, and prevention and control of diseases. What is perhaps unique to this culture is that the pond bottom environment must also be monitored and adapted because of the benthic deposit-feeding lifestyle of *A. japonicus*.

Water quality management involves careful control of water temperature, salinity, dissolved oxygen level, water transparency, water color and pH. Water change is crucial in controlling water quality and it should be adjusted according to the season and particular situation.

The addition of food in exterior ponds is rarely needed to meet the demands of growing *A. japonicus*. Natural food supply is often enough to satisfy their needs. In farms that culture *A. japonicus* in high density and that recently started up, artificial feed containing fish powder mixed with marine mud has become widely used. The amount of artificial feed used is about 3% of total wet weight of *A. japonicus*, and the frequency is once every 2–3 d.

The annual feeding cycle is as follows: once a week before mid-June; no feeding from mid-June to early October because *A. japonicus* is in a dormant stage (aestivation); once every 3 d after early October.

The amount of feed must be closely monitored and adjusted to avoid decay and deterioration of water quality. Change in the presence of enemy planktonic species (*e.g.*, copepods) and their numbers must also be closely monitored. The technical staff should inspect the growth, feeding (food intake) and defecation (fecal pellet output) of *A. japonicus* regularly.

4.3 Indoor industrial culture: a kind of intensive culture pattern

4.3.1 Workshop

The culture facilities are modified using fish culture facilities and the substrata are made by tiles or net mesh (Fig. 1-12-24a). Tanks with a volume of 30-50 m^3 and a depth of 1 m are shown in (Fig. 1-12-24b).

a. Culture tank with substrata; b. Workshop.
Fig. 1-12-24 Indoor industrial culture of sea cucumber

4.3.2 Management

A. Water temperature: Maintain at 12-20 ℃. Usually utilize underground water to control the water temperature.

B. Stocking density: Depend on the seedlings' initial sizes. Seedlings weighted 50–70 g per individual are kept around 30–40 individuals per square meter.

C. Feeding: Artificial feed containing fish powder mixed with marine mud has become widely used. The amount of artificial feeds is 1%–6% of total wet weight of *A. japonicus*, feeding once every day.

4.4 Suspended culture: new kind of intensive culture pattern

4.4.1 Area

A. Well-protected inner bays where waves are small, and water is clear and nutrient-rich.

B. Water depth should be less than 10 m.

C. Zhejiang and Fujian provinces, southern China.

4.4.2 Layout and Facilities

A. Layout: Cages are tied to the suspended ropes with buoys floating on the surface of the sea (Fig. 1-12-25).

B. Cage: Traditional abalone culture cage consists of six combined separate compartments.

Fig. 1-12-25 Layout and cage for suspended culture

4.4.3 Management

A. Culture time: 4-month culture period from November to March.

B. Seedlings size and density: 15–25 g per individual with a density of 5–6 individuals per cage.

C. Feeding: Macroalgae mixed with fish power 1%–6% of total wet weight of *A. japonicus*, feeding once every 3 d.

4.5 Polyculture: candidate for marine IMTA systems

4.5.1 Advantage and mechanism for polyculture

A. Mix animals that occupy different trophic levels.

B. Mitigate the accumulation of particulate organic waste produced by other species, due to its deposit-feeding behavior.

4.5.2 Species used for co-culture with sea cucumber

A. Pond: Shrimp, fish, jellyfish (Fig. 1-12-26).

B. Cage: Scallop and abalone.

Fig. 1-12-26 Species used for co-culture with sea cucumber

5 Diseases of *A. japonicus*

5.1 Diseases in *A. japonicus* larvae

5.1.1 Rotting edges syndrome

A. Epidemiology: Occurs during the auricularia stages from June to July, causing a high mortality of up to 90%; is widely detected in commercial sea cucumber.

B. Clinical signs: Infected larvae are recognized by the darkening of the body edges. Diseased specimens undergo autolysis and the bodies completely disintegrate within 2 d (Fig. 1-12-27).

C. Pathogen: *Vibrio lentus*.

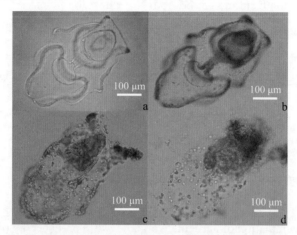

a. Normal aunicularia; b. Thicken and darken edge at the early stages; c. Edges of the larva decayed and the stomach shrank when infection developed; d. The whole body disintegrated and autolyzed and eventually the larva died.

Fig. 1-12-27 Autolyzing process of larvae with rotting edges syndrome

5.1.2 Stomach ulceration symptom

A. Epidemiology: Occurs in summer at high temperatures. It is associated with pathogenic bacteria and triggered by unsuitable feeds and high stocking density. It appears that the auricularia is susceptible to the infection. The mortality of affected larvae may rise up to 90% in certain case.

B. Clinical signs: The stomach walls of juveniles are thick, rough and visibly atrophic in the latter stages. The ulceration of the stomach usually results

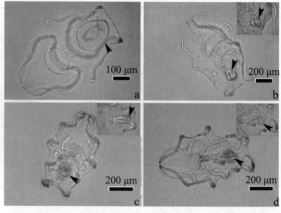

a. Normal auricularia; b. Stomach wall becoming thick and rough; c. Shrinking of stomach; d. Stomach lysis.

Fig. 1-12-28 Pathogenesis of auricularia with stomach ulceration symptom

in reduced growth and a low metamorphosis rate (Fig. 1-12-28).

C. Pathogen: *Vibrio splendidus*.

5.2 The disease in *A. japonicus* juveniles

The disease commonly found in *A. japonicus* juveniles is off-plate syndrome.

A. Epidemiology: Occurs in juveniles that have settled (normally on PVC plates) after the completion of metamorphosis from the doliolaria stage to the pentactula stage. Often the mortality can reach 100%.

B. Clinical signs: The infected juveniles shrink and gradually lose the ability to remain attaching onto the available substrate (Fig. 1-12-29). Meanwhile, the epidermis of infected individuals disappears; the whole body can even dissolve with the autolyzing process, then autolyzes.

C. Pathogen: *Vibrio parahaemolyticus*, *Vibrio lentus* and other *Vibrio* sp.

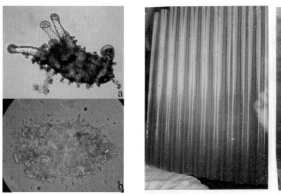

Healthy and diseased juveniles   Juveniles on the settlement

Fig. 1-12-29   Healthy juveniles (a) and diseased juveniles with off-plate syndrome (b)

5.3 Diseases in *A. japonicus* seedlings and adults

5.3.1 Bacterial ulcer syndrome

Epidemiology

Juveniles smaller than 5 mm are susceptible to this infection, which tend to occur as a result of high temperature and stocking density. The infection spreads rapidly from diseased individuals to healthy ones, making it difficult to control. Occasionally, an entire population can be wiped out in a short time (Fig. 1-12-30a).

Adults are susceptible to this disease during the warm season. It usually results in chronic mortalities, with cumulative rates of 30%. Generally, the infected sea cucumbers die after 15 d when the clinical signs appear (Fig. 1-12-30b).

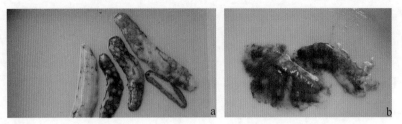

Fig. 1-12-30 Diseased juveniles (a) and adults (b) with bacterial ulcer syndrome

Clinical signs

A. Infected individuals are weak and not able to feed.

B. Their bodies shrink and eventually adopt a rounded shape and become white.

C. Skin ulceration begins with the appearance of small white patches that enlarge, and eventually exposes the underlying muscle and ossicles.

D. Finally, the whole body disintegrates and only white dots are clearly visible in the substrate (Fig. 1-12-31).

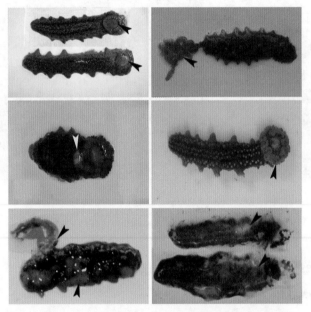

Fig. 1-12-31 Pathological process of bacterial ulcer syndrome

Pathogens

Many kinds of bacteria can cause the disease. The pathogens include *Pseudoalteromonas nigrifaciens*, *Vibrio splendidus*, *Aeromonas veronii*, *Bacillus cereus*, *Pseudoalteromonas* sp., *Pseudomonas putida*, *Vibrio parahaemolyticus*, *Vibrio harveyi*, *Vibrio alginolyticus*, ciliates, etc (Fig. 1-12-32).

 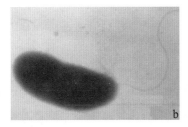 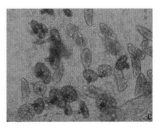

a, b. *Pseudoalteromonas nigrifaciens*; c. Ciliates.

Fig. 1-12-32 Pathogens for bacterial ulcer syndrome

5.3.2 Fungal disease

A. Epidemiology: Frequently it occurs in pond cultured sea cucumbers from April to August. Both juveniles and adults can be infected by the fungi, and result in an unhealthy appearance and poor quality of the final product.

B. Clinical signs: The papillae of the sea cucumbers become white during the early stage of the infection. With the development of the infection, large areas of body wall appear bluish white as the skin is eroded by the fungi. Unlike bacterial infections, there is no obvious mucus around the lesions. In some cases, the whole body surface becomes discoloured and transparent; the body wall becomes thinner and the infected individuals develop oedema (Fig. 1-12-33).

C. Pathogen: Two fungal species.

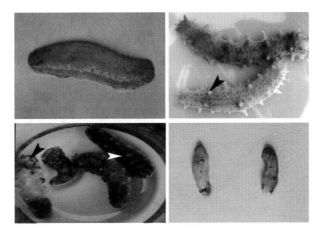

Fig. 1-12-33 The appearance of *A. japonicus* fungal disease

5.3.3 Enteritis disease

A. Epidemiology: It occurs in indoor culture system during all nursing period. The mortality is about 5% but the juveniles grow quite slowly with little feeding.

B. Clinical signs: Dark body, weak feeding ability, weak motility and adhesion ability. The food in the intestine of sick sea cucumber is discontinuity and there is plenty of yellow-white mucus in the intestine. The intestinal wall is brittle, poor toughness and easy to fracture (Fig.

1-12-34).

C. Pathogen: *Vibrio harveyi*.

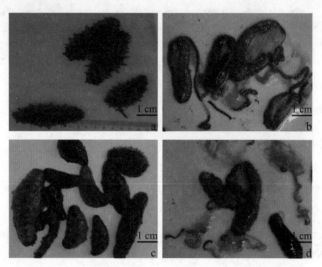

Fig. 1-12-34 The appearance and anatomic image of healthy (a-b) and sick sea cucumbers (c-d) with enteritis disease

6　New progress in *A. japonicus*

6.1　Selective breeding: a new variety of *A. japonicus* "Shen You No. 1"

New variety of *A. japonicus* "Shen You No. 1" was selected for 4 generations by means of mass selective breeding targeting at fast growth rate and high survival rate (Fig. 1-12-35).

Fig. 1-12-35 Adult and seedlings of a new variety of *A. japonicus* "Shen You No. 1"

The survival rate at 6-month old after *Vibrio splendidus* infected is 11.68% higher than that of unselected sea cucumber.

Using pond culture type, the body weight and the survival rate at harvesting time is 24.46% and 23.52% higher than that of unselected sea cucumber, respectively.

6.2 Settlement and device improvement

6.2.1 Settlement device

A. Different shapes (Fig. 1-12-36).

B. Different materials.

Fig. 1-12-36 New settlement device for sea cucumber culture

6.2.2 New culturing device

Fig. 1-12-37 shows new culturing devices for sea cucumber culture.

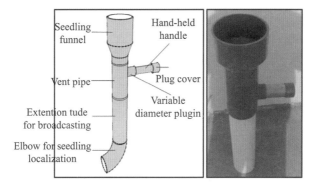

Seed broadcasting device

Auto-feeding device   Feeding device for cage

Fig. 1-12-37 New culturing devices for sea cucumber culture

References

Liao Y L. Fauna of China: Echinodermata Holothuroidea [M]. Beijing: Science Press, 1997.

Purcell S W , Samyn Y , Conand C. Commercially Important Sea Cucumbers of the World [M]. Rome: FAO, 2012.

Wang Y G , Rong X J , Liao M J. Culture and Disease Control of Sea Cucumber (*Apostichopus japonicus*) [M]. Beijing: Agriculture Press, 2014.

Theory and technology of kelp cultivation in China

Author: Liu Fuli
E-mail: Liufl@ysfri.ac.cn

1 Overview of China seaweed cultivation

1.1 Economic and ecological value

1.1.1 Food

Many seaweeds are cooked as traditional sea food, such as salad, noodle, sushi and instant vegetable (Fig. 1-13-1).

Fig. 1-13-1 Some common seaweed food

1.1.2 Raw industry materials

A. Alginate (Fig. 1-13-2): *Laminaria*, *Macrocystis*.

B. Agar, carrageenan (Fig. 1-13-3): *Gracilaria*, *Gelidium*, *Euchuma/Kappaphycus*.

C. Iodine: *Laminaria*, *Macrocystis*.

D. Mannitol: *Laminaria*, *Macrocystis*.

E. Other bioactive metabolites.

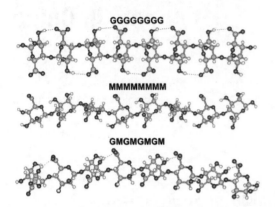 

Fig. 1-13-2 Schematic diagram of alginate molecular structure

Fig. 1-13-3 Agar used for jelly production

1.1.3 Aquatic animals feed

Seaweed can also be used as feed for aquatic animals, such as abalone, urchin and sea cucumber (Fig. 1-13-4).

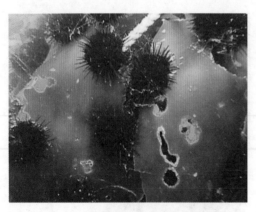 

Fig. 1-13-4 *Saccharina japonica* used as feed for abalone and sea urchin

1.1.4 Ecological value

Seaweeds play very important roles in shore ecology system.

A. Primary producer.

B. Providing food for other organism.

C. Bio-remediation: decrease eutrophication (nitrogen, phosphorus), sequester CO_2, increase O_2 and bio-adsorb the heavy metals.

D. Habitat forming to increase biodiversity (Fig. 1-13-5).

Fig. 1-13-5　Kelp forest

Regarding the production and value of the world's major seaweed producing countries in 2018, please refer to Fig. 1-13-6 and Fig. 1-13-7.

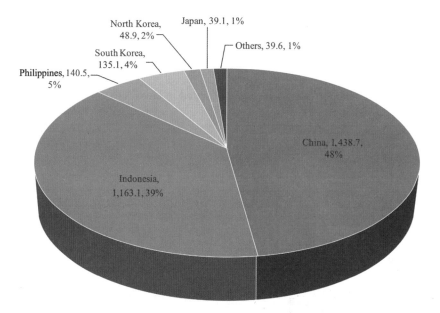

Fig. 1-13-6　Yields and proportions of the world's major seaweed producing countries (million ton)

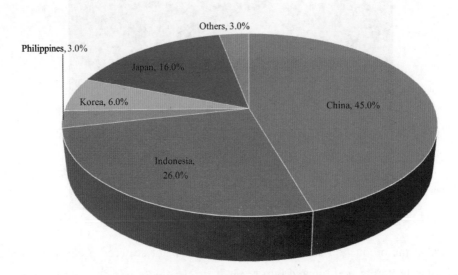

Fig. 1-13-7 Proportions of output value of the world's major seaweed producing countries

1.2 Seaweed species cultivated in China

Roughly, ten species of nine genera have been commercially cultivated in China.

There are five species of brown seaweed cultivated in large scale in China (Fig. 1-13-8).

a. *Undaria pinatifida*; b. *Macrocystis pyrifera*; c. *Saccharina japonica*; d. *Sargassum fusiformis*; e. *Sargassum thunbergii*.

Fig. 1-13-8 Representative species of brown seaweeds cultivated in China

There are five species of red seaweed cultivated in large scale in China (Fig. 1-13-9).

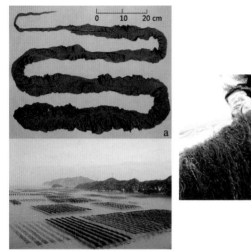

a. *Pyropia* sp.; b. *Gracilaria* sp.; c. *Eucheuma/Kappaphycus* sp.; d. *Gelidium amansii*.
Fig. 1-13-9 Representative species of red seaweeds cultivated in China

1.3 Cultivation scale and distribution

1.3.1 China seaweed production in 2018

Table 1-13-1 shows the production and area of China's major cultured seaweeds in 2018.

Table 1-13-1 Production and area of China's major cultured seaweed in 2018

Categories	*Saccharina*	*Undaria*	*Pyropia*	*Gracilaria*	*Eucheuma/ Kappaphycus*	*Sargassum*
Yield/t	1,522,537	175,503	201,779	330,334	1,820	23,246
Area/hm²	43,619	6,908	65,766	9,912	402	1,277

1.3.2 Distribution of seaweed farm in China

A. *Saccharina japonica*: mainly in Liaoning, Shandong and Fujian provinces.

B. *Undaria pinnatifida*: mainly in Liaoning and Shandong provinces.

C. *Pyropia*: mainly in Jiangsu and Fujian provinces.

D. *Gracilaria lemaneiformis*: mainly in Shandong, Fujian and Guangdong provinces.

E. *Sarggassum fusiformis*: mainly in Zhejiang Province.

F. *Sargassum thunbergii*: mainly in Zhejiang and Shandong provinces.

G. *Eucheuma/Kappaphycus*: mainly in Hainan Province.

1.4 Cultivation technology and model in China

1.4.1 Primary enhancement on the natural substrate

Porphyra enhancement on the rocks by cleaning away the other organisms (animals and seaweeds) has been developed as early as 200 years ago in Fujian Province.

1.4.2 Scale-up enhancement on the artificial substrate

Collect the seedlings on the artificial substrate (stone, bamboo, etc.), and place the substrate onto the sea bottom.

1.4.3 Commercial farming using the artificial seeding

Seedlings can be obtained under artificial control. Floating raft system was developed (Table 1-13-2).

Table 1-13-2 Cultivation technology and model in China

Culture method	Species	Technology
Floating long-line	*Saccharina/Laminaria*; *Undaria*; *Gracilaria*; *Sargassum*; *Eucheuma*; *Gelidium*, etc.	Artificial seeding: *Saccharina*; *Undaria*; *Sargassum*; New varieties; Natural seeding: *Gelidium*; Vegetative propagation: *Eucheuma*, *Gracilaria*
Floating net	*Porphyra/Pyropia*	Artificial seeding; New varieties
Indoor tanks	*Caulerpa lentillifera*; *Ulva* spp.	Introduced from Japan and Vietnam; Vegetative propagation

1.4.4 Commercial farming with genetic improved cultivars

New varieties with excellent performance (faster growth, stronger resistance, better quality, etc.) play significant roles in the culture, increasing the cultivation efficiency.

Seaweed enhancement

There are two methods used for seaweed enhancement in recent years. The first one is throwing stones without seaweed seedlings and these stones can provide as substrate for seaweed growing and propagating (Fig. 1-13-10). The second method is reading the seaweed seedlings on the stones (namely artificial reefs) firstly, and then throwing the stones into the sea (Fig. 1-13-11).

Fig. 1-13-10 Stones were threw into the sea as artificial reefs   Fig. 1-13-11 Seedling raised on the artificial reefs

2 Theory and technology for *S. japonica* cultivation

S. japonica is the first species cultivated commercially in China. *S. japonica* is also the first species whose seedling can be obtained artificially, and the summer seedling rearing system represents the highest level of seaweed seedling rearing technology.

The culture technologies have been widely used in the cultivation of other species; The annual yield of *S. japonica* ranks first.

2.1 Basic biology of *S. japonica*

Originally the taxonomic status of *Laminaria japonica* is: Phaeophyta: Phaeosporeae: Laminariales: Laminariaceae: *Laminaria*. Then the genus was changed from *Laminaria* into *Saccharina*.

The sporophyte is a thallus which is also called the frond, composed of three parts: the blade or lamina, the holdfast/rhizoid and the stipe (Fig. 1-13-12). It grows to a length of 2-6 m and a width of 35-50 cm at maturity. The blade or lamina is ribbon-like, with a thick central band area and thinner lateral edges appearing somewhat wavy.

The life cycle of *S. japonica* involves an alternation of two heteromorphic generations, the diploid, macroscopic sporophyte generation (Figs. 1-13-13) and the haploid, microscopic gametophyte generation (Fig. 1-13-14).

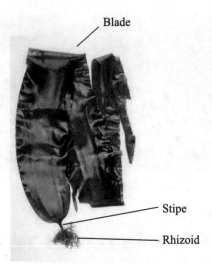

Fig. 1-13-12　Morphological characteristics of *S. japonica* sporophyte

Fig. 1-13-13　The life cycle of *Saccharina japonica* (a new variety named "Huangguan No. 1" of *S. japonica*)

Fig. 1-13-14　The sporophyte gametophyte generation of *S. japonica*

2.2 The culture of summer seedlings

The rearing of summer seedlings is the crucial base for the *Saccharina* industry.

A. Parent *Saccharina* stock selection and cultivation.

B. Sporeling curtain preparation.

C. Zoospores collection.

D. Gametophyte and sporelings cultivation.

Schematic diagram of the seedling-rearing factory is as Fig. 1-13-15, including water treatment system and the greenhouse with tanks.

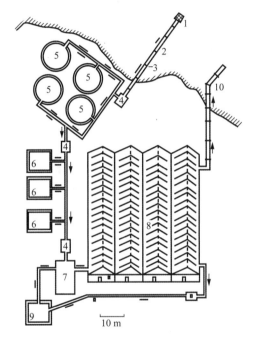

1. Wire cage; 2. Seawater intake pipe; 3. Cement support;
4. Pumping station; 5. Settling tanks; 6. Filter tanks;
7. Refrigeration station; 8. Glasshouse;
9. Recycled water circulation system; 10. Waste water outlet pipe.

Fig. 1-13-15 Schematic diagram of the seedling hatchery of *S. japonica*

Outside and inside pictures of greenhouse are shown in Figs. 1-13-16 to 1-13-18. Tank are 8-10 m in length and 1.2 m in width.

Fig. 1-13-16　The outside picture of the seedling hatchery of *S. japonica*

Fig. 1-13-17　The inside picture of the seedling hatchery of *S. japonica*

Fig. 1-13-18　Production management during seedling hatchery of *S. japonica*

Parent *Saccharina* stock screening and raising is as Fig. 1-13-19.

A. Time: in the end of June, when the sporangial sori begin to form.

B. Selection criteria: excellent agronomic traits (thick, wide and long frond; robust stipe, etc.); healthy; with large areas of sporangial sori no epiphytes.

C. The selected parent *Saccharina* stock will be cultivated in the special sea area with lower temperature. After cultivation of 1-2 months, they will release the zoospores.

D. Preparation of sporeling curtains.

Sporeling curtain or spores collector, is the substrate used to collect spores.

Fig. 1-13-19　The selected parent stock of *S. japonica*

Two types of sporeling curtains:

A. Brown curtain: It is made of palm fiber rope, brown in colour, used widely traditionally in North China (Fig. 1-13-20).

B. White curtain: It is made of vinylon fiber rope, white in colour, developed in South China, used as the substitute of brown curtain (Fig. 1-13-21).

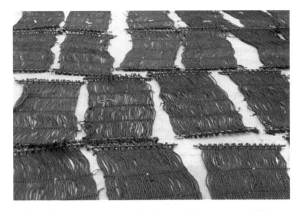

Fig. 1-13-20　The brown curtain for the seedling raising of *S. japonica*

Fig. 1-13-21　The white curtain for the seedling raising of *S. japonica*

Brown sporeling curtains are weaved with palm rope, and then are trimmed and autoclaved to clear away harmful materials (Fig. 1-13-22).

Fig. 1-13-22 The preparation of brown curtains

When the sporeling curtains were prepared well, they were arranged in the cistern in the greenhouse (hatchery). The parent *Saccharina* stock will be transferred into the cistern and release the zoospore, which will attach to the curtains and develop into young sporelings (Fig. 1-13-23).

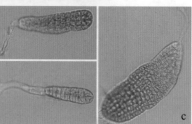

a. The inside picture of greenhouse; b. The brown curtains used for seedling attaching; c. Early development of young seedlings.

Fig. 1-13-23 Seedling raised in the hatchery

2.2.1 Hatching procedures in greenhouse

A. Parent kelp cleaning: Thoroughly clean and remove surface attachments of kelp stock with soft towel (Fig. 1-13-24).

B. Dry in dark to stimulate spore release: The parent stocks were placed on clean gauze in the hatchery for more than 1 hour, with no water droplets on the surface of the blade (Fig. 1-13-25).

C. Spore water preparation: Put the kelp into a sink filled with refrigerating water, let the spores release, occasionally stir, promote the release, take some water at intervals, examine the spore density under the microscope, control the spore density at 5-7 spores in every view of field 400 times, then stop the release (Fig. 1-13-26).

Fig. 1-13-24 Rinse the parent stock to clean off the epiphytes or other attachments

Fig. 1-13-25 Desiccate the blade to stimulate the release of zoospores

Fig. 1-13-26 The detection of zoospores released and density using microscope

D. Spore attachment: Put the curtains into the water tank, let the spores attach, and put the slides at the same time, observe the spore attachment density on the slide, 15 spores per field at 400 times, then stop the attachment.

E. Sporeling curtain rearing: Scatter curtains into the tanks in the hatchery for further rearing.

2.2.2 Sporelings management in greenhouse

A. Water temperature: It should be maintained 8-10 ℃, with a maximum of 12 ℃.

B. Light intensity and period: When zygote formation to the sporelings of a length 0.1 cm, optimum light intensity is 1,000-2,000 lx; Sporeling of 0.5-1.0 cm, 2,000-3,000 lx; Sporeling of 1-2 cm, 3,000-4,000 lx. About 10 hours of daylight is sufficient for sporeling growth in early developmental stages.

C. Nutrient requirements: Phosphorous (about 0.3 mg/L) and nitrogen (about 3 mg/L) based fertilizers (ammonium nitrate) should be added to the indoor culture system.

D. Monitoring stages of sporeling development: Careful daily observations of sporelings should be made during all stages of germination, growth and development (Fig. 1-13-27).

Fig. 1-13-27　The different development stages of the young seedling

When the sea temperature is lower than 20 ℃, the young sporeling will be transferred out of the greenhouse to the nearshore sea area for further culture. When seedlings reach about 15 cm, they will be clamped onto the rope and cultivated on the sea (Fig. 1-13-28).

Fig. 1-13-28　The seedlings clapped onto the culturing rope on the sea

2.3　Technology for grow-out

There are three types of floating rafts of cultivating technology (Fig. 1-13-29).

A. Vertical rope method;

B. "One dragon" rope method;

C. Horizontal rope method.

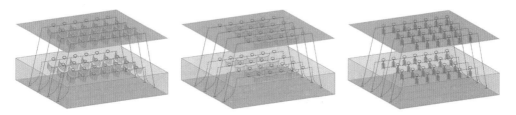

Fig. 1-13-29 Three types of floating rafts for *S. japonica* cultivation

The first method is the vertical method. Weight is tied to the sporeling rope (Fig. 1-13-30). Sporeling rope is vertical to the floating rope.

A. Advantage: Very simple; use the space vertically.

B. Disadvantage: The seaweed at bottom can not receive enough light.

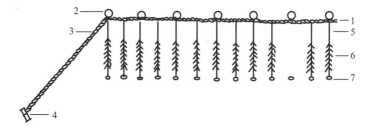

1. Floating raft rope; 2. Floating ball; 3. Anchor rope; 4. Anchor; 5. Connecting rope; 6. Sporeling rope; 7. Stone weight.

Fig. 1-13-30 The vertical rope method for *S. japonica* cultivation.

The seedling rope and the floating ropes are parallel to each other (Fig. 1-13-31).

The second method is the "one dragon" method.

A. Advantage: The seaweed can receive light equally and it is resistant to strong wave.

B. Disadvantage: The seaweed can not use the space efficiently.

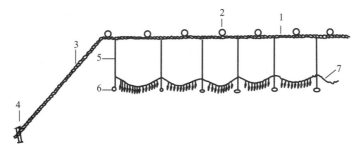

1. Floating raft rope; 2. Floating ball; 3. Anchor rope; 4. Anchor; 5. Hanging connecting rope; 6. Stone weight; 7. Sporeling rope.

Fig. 1-13-31 The "one dragon" rope method for *S. japonica* cultivation

The third method is horizontal rope method (Fig. 1-13-32). The kelp can receive the light evenly and utilize the space efficiently, resulting in the yield higher than the two previous methods. So it is the most popular method in China.

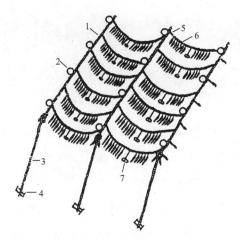

1. Floating raft rope; 2. Floating ball; 3. Anchor rope; 4. Anchor; 5. Connecting rope; 6. Kelp culture rope; 7. Stone weight.

Fig. 1-13-32　The horizontal rope method for *S. japonica* cultivation

Attentions should be paid in this method:

A. Management of cultivation density: Density of kelp plants on culture ropes.

B. Distance between culture ropes: Distance between adjacent rafts and blocks of rafts.

C. Adjusting water depth during grow-out: Adjustments are made in response to many interrelated factors at the raft site, such as the stage of kelp growth, current flow, turbidity, water temperature, etc (Fig. 1-13-33).

Fig. 1-13-33　The photos of different culturing stages on the sea

D. Fertilization application: Minimum nutrient demand level of about 20 mg/m^3 of nitrogen. Ammonium nitrate was applied in porous plastic bags sometimes.

References

Bureau of fisheries and fisheries administration. China fisheries statistical yearbook [M]. Beijing: China Agriculture Press, 2018.

Buschmann A H, Camus C. An introduction to farming and biomass utilization of marine macroalgae [J]. Phycologia, 2019, 58: 5, 443-445.

Fao. The state of world fisheries and aquaculture 2018: meeting the sustainable development goals [R]. Rome, 2018.

Tseng C K. Algal biotechnology industries and research activities in China [J]. Journal of Applied Phycology, 2001, 13: 375-380.

Theory and technology of *Pyropia* culture in China

Author: Wang Wenjun
E-mail: wjwang@ysfri.ac.cn

1. Background

1.1 Classification of Bangiaceae

Fig. 1-14-1 shows the classification of Bangiaceae.

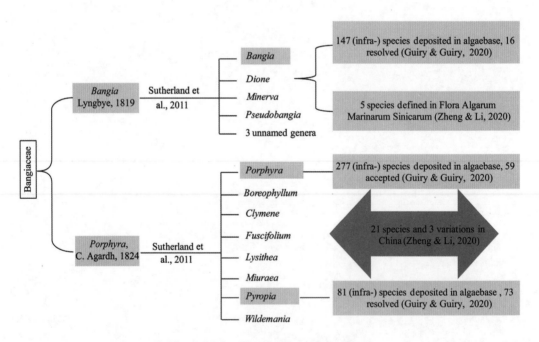

Fig. 1-14-1　Classification of Bangiaceae

1.2 Nutrient content of *Pyropia* and various kinds of *Pyropia* food

A. High protein, essential amino acids (EAAs), delicious amino acids (DAAs) content, EAA/TAA (Total amino acid) and DAA/TAA(Wang et al., 2019).

B. Low lipid content but high ratio of unsaturated fatty acids (TUFAs) to total fatty acids (TFAs), with eicosapentaenoic acid (EPA) accounting for up to 80% of TUFAs in some species (Wang et al., 2019).

The following data in Fig. 1-14-2 come from a publication of *Pyropia yezoensis* (Jung et al., 2016). Kinds of nutritious and delicious food have been made of *Pyropia*, such as dried *P. haitenensis* and *P. yeroensis* soup, fried with eggs, sauce roll, flavored laver broken, sandwich, and sushi.

a. Dried *P. haitanensis* and *P. yezoensis*; b. Soup; c. Fried eggs; d. Sauce; e. Roll; f. Flavored laver broken; g. Sandwich; h. Sushi made of/with laver.

Fig. 1-14-2　Food made of *Pyropia*

1.3　Statistics of seaweed farming in China

Pyropia has the largest culture area and *Saccharina* has the largest annual yield (Fig. 1-14-3). The culture area of *Pyropia* has been increased and doubled in the recent 10 years and has become the highest among all the cultured seaweeds in China (Wang et al., 2019).

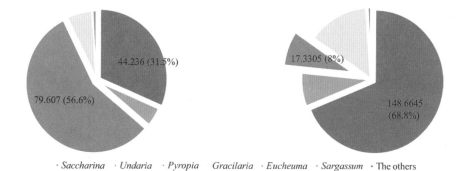

a. Scale of cultivated seaweeds in China in 2017 ($\times 10^3$ hm^2); b. Yield of cultivated seaweeds in China in 2017 ($\times 10^4$ t).

Fig. 1-14-3　Culture area and annual yield of the main cultured seaweeds in 2017 in China

1.4 Distribution of *Pyropia* and *Porphyra* resourse and industry in China

The wild *Pyropia* and *Porphyra* resources are widely distributed along the coast of China, from Hainan Province to Liaoning Province. The farmed *Pyropia* industries are mostly distributed from Guangdong Province to Shandong Province with the largest culture area in Jiangsu Province (Fig. 1-14-4a) and the highest yield in Fujian Province (Fig. 1-14-4b).

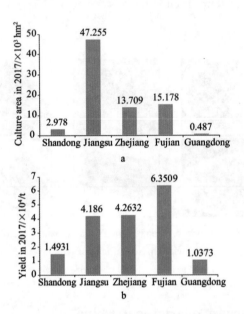

Fig. 1-14-4 Culture area (a) and annual yield (b) of *Pyropia* in 2017 in China

1.5 History of *Pyropia* utilization in China

The earliest record of *Pyropia* culture dates back to 960-1279 AD in Pingtan, Fujian Province, China. The ancient Chinese used lime to remove attached algae and invertebrates from the rocks in autumn before releasing of "seeds", which was named as "Caitan" farming method (Fig. 1-14-5) (Jung et al., 2016).

Fig. 1-14-5 Ancient mariculture and harvesting of *Pyropia* in China

1.6 "Caitan" farming in China

The ancient "Caitan" method is still used in some coastal areas in South China (Fig. 1-14-6).

Fig. 1-14-6 Heavy *P. haitanensis* beds are distributed on the rocks in South China (a) and the local farmer is harvesting them by hands (b)

1.7 The landmark for modern *Pyropia* industry

A. KM Drew's finding was in 1949: Conchocelis has been determined as a life stage of *Porphyra* other than a separate alga species (Drew, 1949).

B. Ecological and developmental investigation on the life history made it possible to manipulate the life history in culture.

  a. Long preservation of free-living conchocelis has been developed for germplasm.

  b. Large-scale culture of shell conchocelis has been achieved for industrial producing (Fig. 1-14-7).

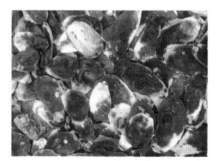

Fig. 1-14-7 Shell conchocelis of *Pyropia* (*Porphyra*)

C. Raft-aquaculture has been established. Two traditional culture methods for *Pyropia* are different from those for other seaweeds, *i.e.*, standing bamboo poles (Fig. 1-14-8) and semi-floating rafts.

Fig. 1-14-8 Standing bamboo poles culture system for *Pyropia* and *Porphyra*

2 Theory and technology of seedling culture

2.1 Life history of *Pyropia*

Most *Pyropia* species have sexual cycle with an alternation between the gametophyte and the sporophyte generation. A few species have only asexual cycle. Some have both cycles (Fig. 1-14-9).

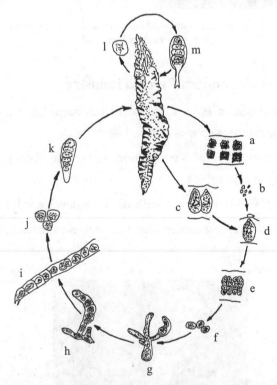

a. Spermatangium; b. Spermatium; c. Carpogonium; d. Fertilized carpogonium; e. Zygotosporangium; f. Zygotospore; g. Filamentous conchocelis; h. Conchosporangial branch; i. Conchospore formation and release; j. Conchospore; k. Seedling from conchospore; l. Archeospore; m. Seedling from archeospore.

Fig. 1-14-9 The life history of *Pyropia* and *Porphyra*

2.2 Culture of *Pyropia* shell conchocelis

Conchocelis penetrates into shell matrix, grows and develops into conchospore, which is the seed of laver for cultivation. All the other algae can't penetrate and live in shell matrix. Shell conchocelis is applied for industrial seedling culture (Fig. 1-14-10).

The contaminated algae are blocked outside the shell matrix and washed off periodically.

a, b. The shell conchocelis is growing; c. Plat culture of shell conchocelis; d. Vertical culture of shell conchocelis; e. Management of shell conchocelis culture: washing.

Fig. 1-14-10　Culture of *Pyropia* and *Porphyra* shell conchocelis

2.3　Disease occurs during *Pyropia* shell conchocelis culture

Disease occurs when the environment conditions are not suitable, *e.g.*, the temperature is very high, the light is too high or too low and the water quality is not so good. Yellow spot disease is the most commonly occured disease for *Pyropia* shell conchocelis at high temperature (Fig. 1-14-11).

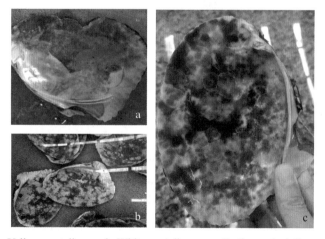

a. Yellow spot disease; b. White spot disease; c. Turtle cracked disease.

Fig. 1-14-11　Kinds of diseases during shell conchocelis culture

Treatment: immersed for several hours to several days with freshwater, $KMnO_4$ (Fig. 1-14-12), ClO_2, and so on.

Fig. 1-14-12 The shell conchocelis with yellow spot disease is immersed with $KMnO_4$

2.4 Conchocelis development and conchospore germinating into sporeling

Three morphologically distinct stages occur during conchocelis culture: vegetative conchocelis, chonchosporangial branches and conchospore formation and release (Fig. 1-14-13). Different culture conditions are required for different stages of conchocelis.

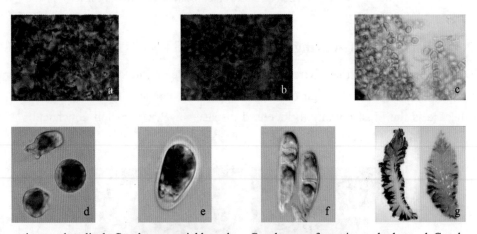

a. Vegetative conchocelis; b. Conchosporangial branch; c. Conchospore formation and release; d. Conchospore; e. Conchospore germinating; f. Two-cell sporeling; g. Gametophyte blade.

Fig. 1-14-13 Development of the shell conchocelis, conchospore releasing and developing

2.5 Seeding conchospores onto culture nets

In autumn, mass formation and release of conchospores occur. The released conchospores are collected and attached onto culture nets and germinated into gametophytes. The collecting was done completely by hand (Fig. 1-14-14a). Nowadays, in most cases, the collecting process is done by machine (Fig. 1-14-14b, c). In the afternoon, the shell conchocelis is washed and water is renewed (Fig. 1-14-14d).

a. Seeding conchospores onto culture nets by hand; b, c. Seeding conchospores onto culture nets by machine; d. Washing shell conchocelis and renewing water.

Fig. 1-14-14　Conchospores collecting process

3　Theory and technology of *Pyropia* aquaculture

3.1　Therory of *Pyropia* aquaculture: based on the endurance to dehydration

Most *Pyropia* species can endure desiccation. The blade becomes crisp sheet and recoveries after rehydration (Fig. 1-14-15).

During laver cultivation, the competitive algae green algae (*Ulva*, Enteromorpha) and plankton cannot endure extreme desiccation. The periodically drying culture protocols are applied for laver cultivation so that the contaminated algae are killed by drying.

Fig. 1-14-15　*Pyropia* blades are naturally dried to crisp sheets on the coastal rocks during ebb tide

3.2 "Green nets" and treatment

The nets full of green algae are called green nets (Fig. 1-14-16) which usually occurr when the seawater temperature is high and the culture nets fail to be dehydrated effectively and completely. The dominant competitive alga species is *Enteromorpha prolifra*. On the one hand, long period of drying under blazing sun can effectively prevent the burst of green algae. On the other hand, the green algae can be killed by low concentration acid and a period of freezing at around −20 ℃.

Fig. 1-14-16   *Pyropia* culture nets are full of green algae (a) and the green nets are dried on ground under blazing sun (b)

3.3 *Pyropia* is usually farmed in near-shore shallow water

Due to the highly dessication tolerance of *Pyropia* gametophyte blades and their vulnerability to the other competitive algae such as green algae and diatoms, the culture systems periodically emerging from and submerging into seawater are used for *Pyropia* cultivation. The semi-floating system (Fig. 1-14-17a) and the supporting bamboo pole system (Fig. 1-14-17b) have been the most widely applied for *Pyropia* culture. The former system is applicable in shallow water with large tide range and the later system is applicable in shallow water with depth of less than 20 m. Therefore, *Pyropia* is usually farmed in near-shore shallow water.

Fig. 1-14-17   The semi-floating system (a) and the supporting bamboo pole system (b) for *Pyropia* culture

3.4 Farming is extending to offshore deep seawater

The *Pyropia* culture areas have been expanding in recent years, however, the suitable coastal shallow areas are nearly completely exploited for *Pyropia* farming. It is necessary to extend *Pyropia* culture to offshore and deeper seawater. The semi-floating system and the supporting bamboo pole system are not applicable for deeper seawater cultivation. Culture systems have been improved to support *Pyropia* culture in deeper seawater (Fig. 1-14-18).

Fig. 1-14-18 An improved culture system for *Pyropia* culture

3.5 Farming systems for *Pyropia* culture

3.5.1 Semi-floating culture system

When tide rises, the culture nets are floating and immerged in seawater. When tide ebbs, the culture nets are submerged and dried in sun (Fig. 1-14-19). The system includes bamboo supports, anchors and culture nets (Fig. 1-14-20).

a. The system is immerged at tide rising; b. The system is submerged at ebb tide.
Fig. 1-14-19 The semi-floating culture system

Fig. 1-14-20 The basic structure of semi-floating culture system

The advantage of this culture system is that the culture nets are periodically dried along with the tide cycle and independent of labor or machine. So this system is very economical and easily managed.

It has some disadvantages, *e.g.*, it is limited to shoal with large tidal range where there should be enough time for drying the culture nets. Besides, it is limited to the areas where the low tide during spring tide should be daytime so that the nets can be effectively dried. What's more, the time and period for drying cannot be manipulated, so the nets can only be dried naturally.

3.5.2 Supporting bamboo pole culture system

Fig. 1-14-21 shows the supporting bamboo pole culture system.

a. The system is emerged at low tide; b. The system is hanged up and submerged at ebb tide.
Fig. 1-14-21 The supporting bamboo pole culture system

The structure is also simple with bamboo supports, anchors, culture nets and big and small floaters (Fig. 1-14-22). The advantage of this system is that the nets are dried when necessity. The nets are hanged up and down on the supporting poles. It had disadvantages: it is limited to shallow water because the length of the poles is limited; the nets are dried mostly by hand, so it is labor-consuming.

Fig. 1-14-22 The basic structure of the supporting bamboo pole culture system

3.5.3 Full floating culture system

Fig. 1-14-23 shows the full floating culture system.

Fig. 1-14-23 The full floating culture system

Advantages

It is applicable for deep water.

Disadvantages

A. The nets cannot be submerged and dried.

B. Limited to large current and clear water.

3.5.4 Improved full-floating culture method

This culture system can be turned over. The nets are growing under seawater (Fig. 1-14-24a,

Fig. 1-14-25a). When it is turned over, the nets are dried upon the big floaters (Fig. 1-14-24b, Fig. 1-14-25b). This technique has gained Chinese patent authorization.

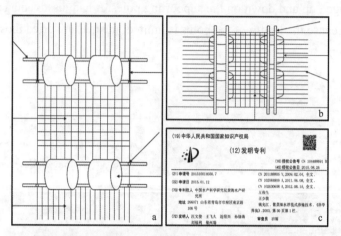

a. Emerged; b. Submerged; c. The Chinese patent authorization.
Fig. 1-14-24 Diagram of the improved full-floating culture system

a. Growing under seawater; b. Turning over to dry out.
Fig. 1-14-25 Application of the improved full-floating culture system

Advantages

A. It is applicable for deep water.

B. The nets are dried if necessity.

C. Easy management and labor saving.

D. Low cost for the facilities.

Disadvantages

The net is difficult to be turned over when laver grows up and the quality of product is reduced.

3.6 Harvesting *Pyropia* products

When the *Pyropia* blades grow up to 15–20 cm in length, they are harvested by ships (Fig. 1-14-26). They are harvested for several times in a farming season. Generally, they can be harvested every 10–20 d.

a. The harvesting ship; b. The big bamboo baskets used to hold the harvested *Pyropia*; c. The ship is harvesting *Pyropia*.

Fig. 1-14-26　Harvesting *Pyropia* products

References

Drew K. Conchocelis-phase in the life-history of *Porphyra umbilicalis* (L.) Kütz [J]. Nature, 1949, 164: 748–749.

Guiry M D, Guiry G M. AlgaeBase[EB/OL]. [2020-02-17]. http://www.algaebase.org.

Jung S M, Kang S G, Son J S, et al. Temporal and spatial variations in the proximate composition, amino acid, and mineral content of *Pyropia yezoensis* [J]. Journal of Applied Phycology, 2016, 28(6): 3459–3467.

Sutherland J E, Lindstrom S C, Nelson W A, et al. A new look at an ancient order: generic revision of the Bangiales (Rhodophyta) [J]. Journal of Phycology, 2011, 47: 1131–1151.

Wang W J, Li X L, Sun T Q, et al. Effects of periodical drying and non-drying on nutrient content and desiccation tolerance of an intertidal *Pyropia yezoensis* strain subject to farming conditions[J]. Journal of Applied Phycology, 2019, 31(3): 1897-1906.

Zheng B F, Li J. Flora Algarum Narinarum Sinicarum, Tomus II. Rhodophyta, no 1 Porphyricliales Erythropeltidales Goniotrichales Bangiales [M]. Beijing: Science Press, 2009.

Chapter 2
Mariculture modes

Mariculture mode
—Recirculating aquaculture system (RAS)

Author: Liu Baoliang
E-mail: liubl@ysfri.ac.cn

1 What is RAS?

Recirculating aquaculture systems (RAS) represents a new way to farm fish: high stocking densities; in indoor tanks with a "controlled" environment; treatment and reuse of the water (Fig. 2-1-1). New water is added to the tanks only to make up for splash out and evaporation and for that used to flush out waste materials. Fish farmed in RAS must be supplied with all the conditions necessary to ensure health and growth (Badiola et al., 2012).

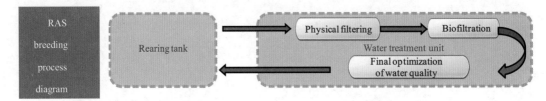

Fig. 2-1-1　A RAS breeding process diagram

2 The advantages of RAS

In brief, compared to traditional pond culture, the advantages of RAS are mainly reflected in the following points (Fig. 2-1-2).

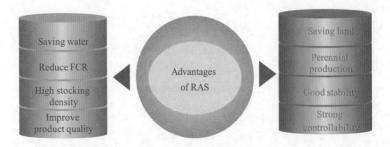

Fig. 2-1-2　List of advantages of RAS

2.1 Intensive production

RAS is applied to rearing large numbers of fish in a relatively small space. This is an excellent alternative to open pond culture where low densities of fish are reared free in large ponds and tend to be free from diseases, pollutants, stress, and seasonally suboptimal growing conditions (Fig. 2-1-3).

a. RAS; b. Open pond.
Fig. 2-1-3 The fish in RAS and open pond

2.2 Water and land conserving

RAS conserve both water and land (Fig. 2-1-4). They maximize production in a relatively small area of land using a relatively small volume of water. Similarly, since water is reused, the water volume requirements in RAS are only about 20% of what conventional open pond culture demands.

a. Eel; b. Schematic view of RAS.
Fig. 2-1-4 The schematic view of RAS and eel

2.3 Location flexibility

RAS are particularly useful in areas where land and water are expensive and not readily available. They require relatively small amounts of land and water. They are most suitable in northern areas where a cold climate can slow fish growth in outdoor systems and prevent year-round production (Fig. 2-1-5).

a. RAS base near the sea; b. RAS base in the farm.
Fig. 2-1-5 The location of RAS

2.4 Species and harvest flexibility

A. Rearing a wide diversity of fish species.

B. Rearing a number of different species simultaneously in the same or different tanks with in the same system.

C. Raising a different sizes of fish depending on market demand and price.

D. Harvest at the most profitable time.

2.5 Note

Constant supervision and skilled technical support are required to manage and maintain the relatively complex circulation, aeration and biofilter systems, and to conduct water quality analysis.

The danger of mechanical or electrical power failure would cause fish loss when rearing fish in high densities and in small water volumes. Continuous vigilance and quick reaction (15 minutes or less) are needed to avert total mortality.

However, the higher risk factor, capital investment, and operating costs can be offsetting by continuous production, reduced stress, improved growth, and the production of superior products in the RAS.

3 The application and prospects of RAS in China

In the first 15 years of the new millennium, the aquaculture contribution to the world production of aquatic animals has increased from 25.7% to 46.4% and the finfish farming counted for 66.6% of total aquaculture output of aquatic animals in 2017 (Fig. 2-1-6).

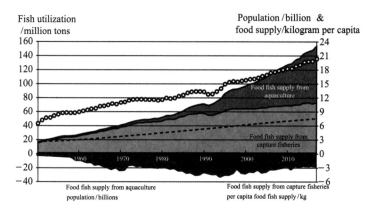

Fig. 2-1-6 World aquaculture production of food fish and aquatic plants (1950–2017)

3.1 Aquaculture in China

China is one of the most important contributors to world aquaculture production. Since 1993, aquaculture has accounted for more than half of the country's aquatic production, making China the largest producer of aquatic products in the world. In 2018, the farming fish production in China was 49.9 million tons, which occupied 62.3% of the world total aquaculture production.

3.2 Aquaculture in China's models

Currently, two main aquaculture models are developed in China, eco-culture (such as eco-pond and rice-fish ecosystem) and intensive farming (sea cage and factory farming). But the goal is to be unified, which is to achieve responsible and sustainable development.

Over the past 30 years, the Chinese aquaculture has gradually established the intensive breeding technological system and obtained successful experience. Up till now, the total area of are about 30 million square meters industrial aquaculture in China, including three main models of flow-through system, semi-closed recirculating system and RAS. The structure of the industrial aquaculture area is shown below:

A. Area: about 30,000,000 m².

B. Flow-through system: 85%–90%.

C. Semi-closed recirculating system: 10%.

D. Recirculating aquaculture system: <5%.

3.3 Marine RAS in China

The development of marine recirculating aquaculture industry has experienced nearly 40 years, and can be divided into 3 stages.

A. 1980s: 9 RASs were introduced from Germany and Denmark.

B. 1980–2000: Research and exploration.

C. 2001 to now: Rapid application in farm. Over 200 corporations used RAS to produce fish and shrimp, etc.

Since the new century, China has made considerable progress in research and application of marine industrial aquaculture, and established the suitable development mode of China's RAS. During this period, many species, such as grouper, sole fish, fugu and abalone were firstly cultured in RAS. Turbot, Atlantic salmon and scallop were also bred in RAS successfully. The successful application of RAS has improved the efficiency of the aquaculture, protected the ecological environment, and promoted the development of marine economy (Liu et al., 2015).

3.4 Present status

RAS has developed rapidly in recent years. Techniques are being transformed from "imitate & follow stage" into "independent innovation stage" incentive policy (Fig. 2-1-7). Although the RAS is still at primary stage, it has good prospects for development.

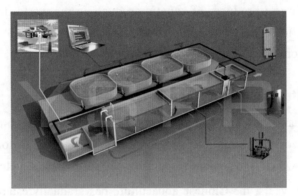

Fig. 2-1-7 The schematic view of RAS

3.5 The problems of Marine RAS in China

Despite the rapid developent of the basic technology RAS in recent years, there are many technical innovations needed to ensure the good performance of the systems performing well for a broader range of aquatic species and culture conditions The problems are as follows.

A. Late beginning, small scale.

B. Lag in technology, waste resources.

C. High running cost (energy cost).

D. Poor practicability and reliability of equipment.

E. Low intelligent and automatic level.

3.6 Typical RAS of marine aquaculture

3.6.1 Typical RAS for turbot

Turbot is one of the aquaculture species for industrialized farming in the northern coastal areas of China with superior high stocking density (Fig. 2-1-8). The RAS for turbot is the basic type, consisting of greenhouse, tank, foam fraction, biofilter, filtration, sterilization and monitoring system. Generally, the stocking density of turbot could reach about 60 kg/m³.

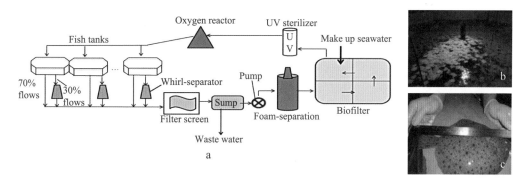

a. The process diagram of RAS; b: Turbot fish in RAS; c. Turbot fish.
Fig. 2-1-8 The process diagram of RAS for turbot

3.6.2 Typical RAS for Atlantic salmon

Atlantic salmon was first introduced to China for RAS culture in 2010. Until now, 78 RASs are running, with the scale of 100,000 m² and 36,800 m³ water spaces. This farm could produce 1,000 t of salmon every year, and is the largest RAS base of Atlantic salmon in the world. All the systems are designed and built independently, with high degree of automation and stability. In addition, the technology of automatic feeding and internet of things were also successfully introduced. To some extent, this system represents the top level land-based marine aquaculture in China. Generally, the stocking density of Atlantic salmon could reach about 45 kg/m³ (Fig. 2-1-9).

a. Salmon in the RAS; b. The inside view of workshop.
Fig. 2-1-9 Atlantic salmon farming workshop

4　The RAS design

Schematic diagram of land-based RAS is shown as Fig. 2-1-8a: Water flows from rearing tanks ⟶ Whirl-separator (30% flows) ⟶ Filter screen ⟶ Sump ⟶ Pump ⟶ Foam-separation ⟶ Biofilter ⟶ UV sterilizer ⟶ Oxygen reactor ⟶ Rearing tanks (Odd-Ivar et al., 2007).

The design includes:

A. Rearing tank.

B. Solid waste removal device.

C. Water circulation pump.

D. Foam-separation.

E. Biofilter.

F. Disinfection.

G. Dissolved oxygen regulator.

4.1　Fish culture tanks

What is important when choosing a design is that the new water should be uniformly distributed throughout the entire tank volume. Round or polygonal (six to eight edges) tanks with a circulating flow pattern are suitable because no dead zones are provided that the inlet and outlet are correctly designed (Fig. 2-1-10).

Square tanks have dead zones in each corner and the effective farming volume is therefore not so large. For this reason square tanks are not recommended.

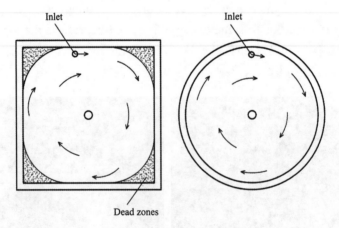

Fig. 2-1-10　Flow pattern simulation in fish culture tanks

The height of the tank compared with the diameter will also affect the water exchange. For tanks with a circular flow pattern, a ratio of diameter to height of between 2 and 5 has been successfully used.

The size of the tank depends on a variety of factors including stocking rates, species selected, water supply, water quality and economic considerations.

Tanks can be constructed of plastic, concrete, FRP and many other materials.

In China, rearing ponds constructed with concrete and FRP materials are more common (Fig. 2-1-11).

Fig. 2-1-11　Concrete and FRP tanks

4.2　Physical filtering

The removal of solid waste is one of the key aspects of RAS water treatment. Common physical filtration devices typically include a microfiltration machine and a foam separator device. The microfiltration machine is used to remove solid particles larger than 80 microns in diameter, generally including both drum-type and crawler-type. Foam separation is based on the principle of adsorption to remove tiny suspended particles and organic waste from water.

4.3　Biofiltration

The biological filter (biofilter) is the heart of RAS.

As the name implies, it is a living filter composed of a media (corrugated plastic sheets or beads or sand grains) upon which a film of bacteria grow. The bacteria provide the waste treatment by removing pollutants.

The two primary water pollutants that need to be removed are as follows.

A. Fish waste (toxic ammonia compounds) excreted into the water.

B. Uneaten fish feed particles.

The biofilter is the site where beneficial bacteria remove (detoxify) fish excretory products, primarily ammonia.

4.3.1　Ammonia and nitrite toxicity

Ammonia and nitrite are toxic to fish. Ammonia in water occurs in two forms: ionized ammonium (NH_4^+) and unionized (free) ammonia (NH_3) (Fig. 2-1-12). The latter, NH_3, is highly toxic to fish in small concentrations and should be kept at levels below 0.05 mg/L. The total amount of NH_3 and NH_4^+ remain in proportion to one another for a given temperature and pH, and a decrease in one form will be compensated by conversion of the other.

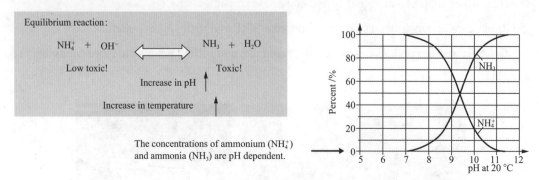

Fig. 2-1-12 The balance of ammonium and ammonia is affected by temperature and pH

4.3.2 Calculating ammonia loading

Ammonia loading can be roughly estimated from the biomass (weight) of fish in the tank or it can be based on the weight of feed fed each day. On the average, about 250 mg of ammonia per day is produced for every 1,000 g of fish in the tank. For example, in a tank containing 1,000 striped bass fingerlings each weighing 75 g (75,000 g total fish weight), the daily ammonia load produced by all the fish would be 18,750 mg.

Ammonia loading can also be estimated based on the total amount of feed fed. For manufactured fish feed with standard protein levels of 30% to 40%, simply multiply the total weight of the feed (in grams) times (25). For example, if the fish are fed 1 kg of pelleted feed per day, the amount of ammonia produced per tank would be about 25 g per day.

Nitrification refers to the bacterial conversion of ammonia nitrogen (NH_3) to less toxic NO_2 and finally to non-toxic NO_3.

Two groups of aerobic (oxygen requiring), nitrifying bacteria are needed for this job. *Nitrosomonas* bacteria convert NH_3 to NO_2, the *Nitrobacter* bacteria convert NO_2 to NO_3 (they oxidize toxic nitrite to largely nontoxic nitrate).

$$NH_4^+ + 1.5O_2 \xrightarrow{\textit{Nitrosomonas} \text{ bacteria}} 2H^+ + H_2O + NO_2^- + \text{energy}$$

$$NO_2^- + 0.5O_2 \xrightarrow{\textit{Nitrobacter} \text{ bacteria}} NO_3^- + \text{energy}$$

Nitrification is an acidifying process, but will become most efficient when the pH is maintained between 7 and 8 and the water temperature is about 27-28 ℃. Acid water (less than pH 6.5) inhibits nitrification and should be avoided.

4.3.3 Biofilter design and materials

A biofilter, in its simplest form, is a wheel, barrel, or box that is filled with a media that provides a large surface area on which nitrifying bacteria can grow.

The biofilter container can be constructed of a variety of materials, including plastic, wood, glass, metal, concrete, or any other nontoxic substance (Fig. 2-1-13).

The size of the biofilter directly determines the carrying capacity of fish in the system. Larger biofilters have a great ammonia assimilation capacity and can support greater fish production.

Fig. 2-1-13　Some kinds of ideal biofilter media

The ideal biofilter media has high surface area for dense bacterial growth, sufficient pore spaces for water movement, clog resistance, easy cleaning and maintenance characteristics. We use and recommend plastic because it is lightweight, flexible, and easy to clean, but it may be expensive.

4.3.4　Biofilter sizing

Important factors that must be considered in designing a biofilter are media surface area (square feet of surface for bacteria attachment), ammonia leading and hydraulic loading.

In general, the volume of the biofilter is not less than 30% of the total aquaculture water, and the amount of biofilter media is about 25%-50% of the volume of the biofilter (Fig. 2-1-14).

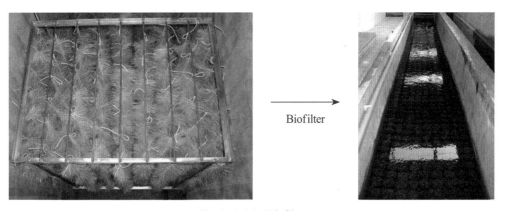

Fig. 2-1-14　Biofilter

4.3.5　Biofilter start-up and recirculation rates

Biofilter start-up

Efficient biofiltration depends on colonization of the biofilter media by nitrifying bacteria. Complete colonization of a biofilter is dependent on a number of environmental conditions and may take one to three months. Inoculation of a new tank with seed bacteria from a existing system can reduce start-up times and facilitate full efficiency. Cooler water temperatures, below 21 ℃, can reduce bacterial activity, slow the bacterial colonization, and diminish the effectiveness of the biofilter.

Recirculation rates

The recirculation rate (turnover time) is the amount of water exchanged per unit of time. Recirculation systems are designed to provide at least one complete turnover per hour (24 cycles per day). Increasing the number of turnovers per day would provide increased biofiltration, greater nitrification (bacterial contact) and reduced ammonia levels.

4.4 Oxygen management

The addition of oxygen is used for the survival of high-density fish, the survival rate of aerobic, nitrifying bacteria on biological filters and for the decomposition of organic waste.

Atmospheric oxygen can be added to the tanks by surface agitation with aerators or by large blowers. Pure oxygen can be delivered and stored in a tank as liquid oxygen or it can be produced on-site by an oxygen generator. Oxygen cones are often used to increase the dissolved oxygen level in RAS (Fig. 2-1-15).

a, b. Pure oxygentanks; c. Oxygen cones.
Fig. 2-1-15   Pure oxygen storage tank and oxygen cones

4.5 Sterilization

Ozone (O_3) is a powerful oxidizing agent that can be used to break down compounds.

UV sterilization mainly uses immersion UVB lamps. UV sterilization efficiency is higher, but water needs better light transmittance and a certain retention time. The disadvantage of UV is that it has high energy consumption, is easy to attenuate aging, and cannot effectively kill parasites.

References

Badiola M, Mendiola D, Bostock J. Recirculating Aquaculture Systems (RAS) analysis: Main issues on management and future challenges [J]. Aquacultural Engineering, 2012, 51(none): 26-35.

Liu Y, Li B L, et al. Recirculating Aquaculture Systems in China-current application and prospects [J]. Fisheries and Aquaculture Journal, 2015, 2: 12-20.

Odd-Ivar L. Aquaculture engineering [M]. Oxford: Blackwell Publishing Ltd. UK, 2007: 99-200.

Advances in the research of recirculating aquaculture systems (RAS) in China

Authors: Cui Zhengguo, Qu Keming
E-mail: cuizg@ysfri.ac.cn

1 Introduction

1.1 Definition

Recirculating aquaculture systems (RAS) are a set of systems with aquaculture water treatment and recycling system in which the wastewater is purified and treated by physical, chemical and biological methods, and then all or most of the water is reused. RAS represent an exciting, eco-friendly and unique way to farm fish or shrimp. Instead of the traditional method of mariculture outdoors in open ponds, net cages or tanks, this system rears fish at intensive densities within controllable conditions (Fig. 2-2-1).

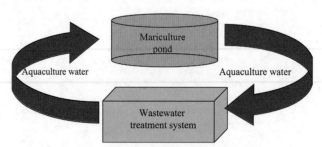

Fig. 2-2-1 Recirculating aquaculture systems

1.2 Mechanism

Generally, RAS include mechanical filter system, nitrogen removal system, bio-filter system, sterilizing system (Ozone concentrator & UV lights), thermostatic system (keeping temperature stable) and oxygen aerating system (Fig. 2-2-2).

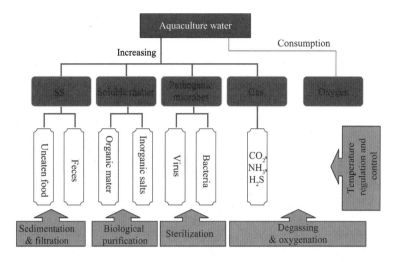

Fig. 2-2-2　Removal mechanism of recirculating aquaculture systems

1.3　RAS's advantages

RAS have many advantages such as intensive culture density, low feed coefficient, ecology friendly, water-saving, land, and energy efficient (Fig. 2-2-3).

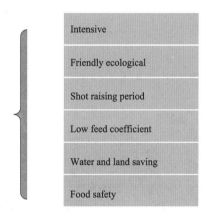

Fig. 2-2-3　Advantages of RAS

2　Research progress of RAS in China

2.1　Overview

Since the middle of the 1990s, the wastewater recycling technology of marine RAS in China has experienced four important development stages. At each stage, some problems that limited the development of RAS were solved (Fig. 2-2-4).

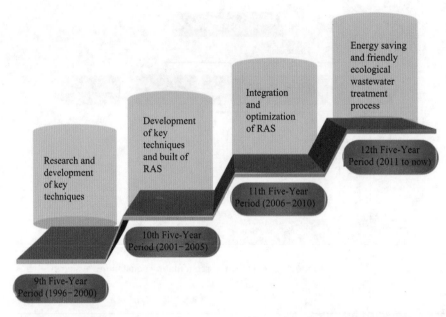

Fig. 2-2-4 Four important development stages of research progress of RAS in China

2.2 Development stages

Stage I: 9th Five-Year Period (1996–2000)

Technical characteristics: At this stage, the key techniques of rapid filtration, sterilization and oxygenation were solved, and some wastewater equipments such as microstrainer, rapid sand filter tank and high efficiency oxygen tank were studied and developed (Fig. 2-2-5).

a. Microstrainer; b. Rapid sand filter tank; c. High efficiency oxygen tank.

Fig. 2-2-5 Wastewater equipments

Stage II: 10th Five-Year Period (2001–2005)

Technical characteristics: At this stage, some other key equipments including protein skimmer, biofilter and UV disinfector were created. Basically, RAS were built (Fig. 2-2-6).

a. Protein skimmer; b. Biofilter; c. UV disinfector.

Fig. 2-2-6 Key equipments

Stage III: 11th Five-Year Period (2006-2010)

Technical characteristics: At the third stage, the focus of research were on the integration and optimization of RAS, and industrial, efficient RAS were built (Fig. 2-2-7).

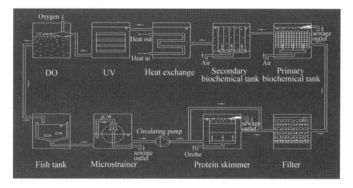

Fig. 2-2-7 An integrated and optimized RAS

Stage IV: 12th Five-Year Period (2011 to now)

Technical characteristics: At the fourth stage, modern engineering techniques and biological technologies were applied in RAS. Its research objects were to be standard, energy saving and ecological (Fig. 2-2-8).

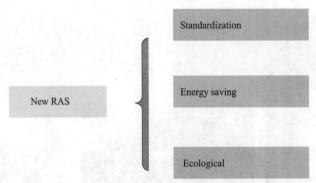

Fig. 2-2-8　The research objects of new RAS

The figures below show the standardized farming and the new energy saving and ecological RAS process (Figs. 2-2-9 to 2-2-10).

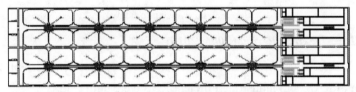

Fig. 2-2-9　Farm standardization

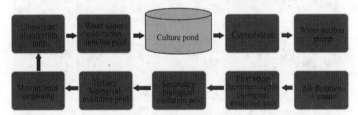

Fig. 2-2-10　New RAS process

In Stage IV, most of "facilities" of RAS were substituted by the "equipments" of RAS to reduce the cost and save energy. Table 2-2-1 shows the comparison between the two types of RAS.

Table 2-2-1　The comparison between the two types of RAS

	Facilities	Equipments
Design		
Advantages	Scale of design: 150 m^3; Running stable; High removal rate	Scale of design: 350 m^3; Low cost; Saving energy
Disadvantages	High cost and energy	Low removal rate, unstable
Applications	Most in developed countries	Most in China

Compared with RAS in 10th Five-Year Period, the costs and energy consumption of RAS in 12th Five-Year Period were obviously reduced, by 95% and 70%, respectively (Table 2-2-2).

Table 2-2-2 The comparison of cost, energy consumption and characteristics at different periods

	10th Five-Year Period	11th Five-Year Period	12th Five-Year Period
Cost	3,000 *yuan*/m^2	800 *yuan*/m^2	150 *yuan*/m^2
Energy consumption	43 W/m^2	19.5 W/m^2	12.9 W/m^2
Characteristics	Frequent damage, bad coupling, sometimes hypoxia, only for one culture species	Frequent damage, bad coupling, only for one culture species	Most are "facilities", running stable, good coupling, for one or several species

3　YSFRI's RAS research and application

3.1　Filtration equipments

The filtration equipments are used for removing suspended solids and some organic matter. The research team studied and developed the filtration equipments including microstrainer, filter tank, protein skimmer, rotary curved screen, and so on (Fig. 2-2-11).

a. Microstrainer; b. Rapid sand filter tank; c. Rotary curved screen.

Fig. 2-2-11　The filtration equipments

3.2 Sterilization and disinfection equipments

YSFRI has developed various sterilization and disinfection equipments with UV, Ozone or Protein skimmer (Fig. 2-2-12).

Fig. 2-2-12 Sterilization and disinfection equipments

3.3 Oxygenation equipments and facilities

Oxygenation equipments include molecular sieve oxygen generating device, high DO tank, pipeline/cone DO tank, DO pond and convection type DO pond (Fig. 2-2-13).

Fig. 2-2-13 Oxygenation equipments and facilities

3.4 Degassing facilities and equipments

The degassing facilities and equipments include degassing pond, degassing tower and micropore aeration tank (Fig. 2-2-14).

Fig. 2-2-14　Degassing facilities and equipments

3.5　Biofilm & biofilter

The choice of bilfilter materials, the cultivation of biofilm and analysis of microbial community were studied (Fig. 2-2-15).

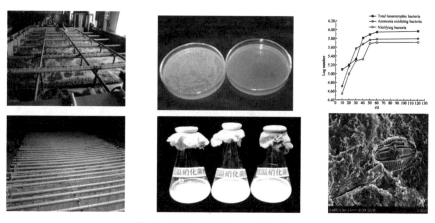

Fig. 2-2-15　Biofilm and biofilter

3.6　Online water quality auto-monitoring system

The online water quality auto-monitoring system can monitor various parameters such as temperature, salinity, conductivity, pH, DO and oxidation reduction potential (Fig. 2-2-16).

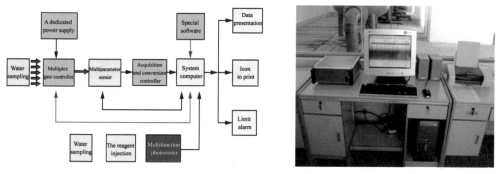

Fig. 2-2-16　Online water quality auto-monitoring system

3.7 Integration and optimization of RAS

The RAS used for fish culture consist of aquaculture system and waste water treatment system. The RAS is monitored by an online monitoring system (Fig. 2-2-17).

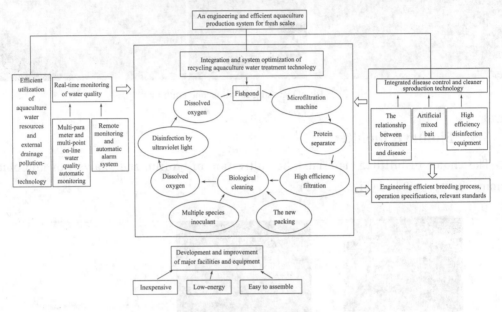

Fig. 2-2-17 RAS of fish

3.8 Applications

Our RAS process has been successfully used in 23 aquaculture factories. The total demonstration area is 2×10^5 m^2 and occupies an area of 39% in the whole country (Fig. 2-2-18).

Fig. 2-2-18 Applications and demonstrations

3.9 Some demonstration bases

Figs. 2-2-19 to 2-2-20 are some pictures of our RAS in the demonstration bases.

Fig. 2-2-19　Demonstration base of Yantai Tianyuan Aquatic Co., Ltd.

Fig. 2-2-20　Demonstration base of Laizhou Mingbo Aquatic Co., Ltd.

4　Challenges and prospects

Although RAS which are highly efficient, ecological, energy-saving and cost saving for aquaculture, have been successfully used, there are still challenges and problems worthy of further studies (Fig. 2-2-21).

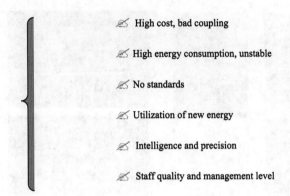

Fig. 2-2-21 Challenges and prospects

References

Chen J, Xu H, Ni Q, et al. Research report on the development of China's factory recycling aquaculture [J]. Fishery Modernization, 2009, 36(4): 1-7 (in Chinese).

Hu J C, Yu X Q, Xin N H, et al. Research Status and Application Prospect of industrial recycling aquaculture[J]. China Fisheries, 2017(6): 94-97 (in Chinese).

Lei Q L, Huang B, Liu B, et al. Strategic research on the construction of high-end farming industry in China based on the concept of aquatic animal welfare[J]. China engineering science, 2014(3): 14-20 (in Chinese).

Qu K M, Du S E, Cui Z G, et al. Engineering technology for the construction of efficient seawater factory farming system (revised edition) [M]. Beijing: Ocean Press, 2018 (in Chinese).

Zhang Y Q, Li X Y, Qin F, et al. Advances in the application of macroalgae to recycling aquaculture [J]. Food Science, Technology and Economics, 2018, 43(10): 115-117 (in Chinese).

Development of coastal integrated multi-trophic aquaculture (IMTA) in China

Author: Zhang Jihong
E-mail: zhangjh@ysfri.ac.cn

1 What's IMTA? Why is IMTA?

1.1 Key roles of aquaculture

China is the largest aquaculture producer in the world. In 2017, its aquaculture production was 49 million tons, among which, mariculture occupied 51.5%.

Seafood products contribute to approximately one-third of the total animal-based protein uptake in china.

The proportion of various activities in China's fishery economy in 2017 was showed in Fig. 2-3-1, among which, marine fisheries economy occupied 18%. Mariculture economic value accounted for 13% of the total fishery economy in China. Almost 13.6 million people were engaged in fishery.

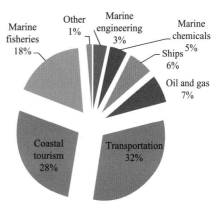

Fig. 2-3-1 The proportion of marine fisheries in the marine economy

1.2 Challenges of mariculture

At present, mariculture is mainly concentrated in shallow seas and inner bays where the water exchange is weak.

Waste nutrients discharged from mariculture activities become serious: eutrophication, organic matters accumulate in sediment, and then dissolved oxygen (DO) decreases and the content of harmful substances, such as ammonia and hydrogen sulfide, increases.

With multi-pressure of global climate change and human activity, the level of coastal environmental deterioration is alarming. Fig. 2-3-2 shows some natural ecological disasters, such as green tides and starfish disasters. The deteriorating environment causes frequent diseases outbreak, which leads to declined quality and quantity of seafood products and severe economic losses.

Fig. 2-3-2　Green tides and starfish disasters

1.3　Sustainable development — common problem in the world

The mariculture industry worldwide is searching for methods to support further growth of aquaculture production in a sustainable manner.

Integrated multi-trophic aquaculture (IMTA) has been proposed as a potential effective way of sustainable farming.

1.4　What is IMTA?

Early talks about what would become known as IMTA in 1995 were by Thierry Chopin.

IMTA refers to the combination of different trophic levels in the same culture system with the intention of increasing sustainability and profitability through mitigating environmental impacts. By-products (wastes) from higher trophic levels are reused by lower trophic (Fig. 2-3-3).

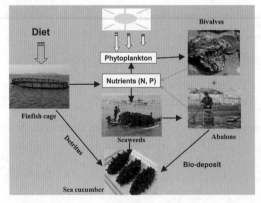

Fig. 2-3-3　Conceptual diagram of an IMTA operation including fish, shellfish and seaweeds

1.5 Why is IMTA?

IMTA has several benefits, such as:

A. Nutrient recovery;

B. Bioremediation;

C. Enhanced growth;

D. Enlarge maricultural carrying capacity;

E. Products diversification;

F. Disease prevention;

G. Carbon sequestration;

In short, IMTA combines the cultivation of fed aquaculture species (mainly finfish, shrimp) and/or inorganic extractive aquaculture species (seaweed) and/or organic extractive species (bivalves) in equilibrium with the site conditions, economic balance, social interest, and environmental concerns (Troell et al., 2009).

2 How to setup IMTA in coastal area? —Case study of IMTA in Sanggou Bay

2.1 IMTA of seaweed, abalone and sea cucumber

The simple schematic diagram of mutually beneficial relationship of seaweed, abalone and sea cucumber was showed in Fig. 2-3-4.

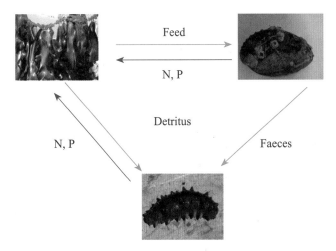

Fig. 2-3-4 Seaweed, abalone and sea cucumber

2.1.1 Principle of seaweed + abalone + sea cucumber

Nutrients fluxes within the IMIA-system were showed in Fig. 2-3-5 (Zhang et al., 2018).

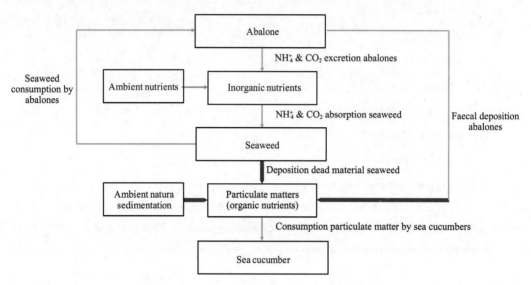

Fig. 2-3-5 Schematic representation of nutrient fluxes within an IMTA-system including abalone, kelp and sea cucumber cultures

2.1.2 Problems in kelp and abalone IMTA mode

Kelp and abalone IMTA mode has large-scale applications in China and the mode design in northern China is as follows (Fig. 2-3-6).

Per cultural unit (4 longline frames, about 1,600 m^2): 400 ropes for kelp, 100 net cage for abalone.

Total: kelp 12,000 individuals, abalone 20,000 individuals.

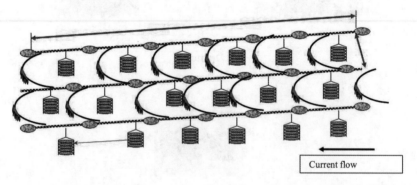

Fig. 2-3-6 Cultivation units of the IMTA of kelp and abalone

Daily management of abalone mainly includes the following aspects:

A. Feeding;

B. Removing residual seaweed.

In the integrated cultivation system of seaweed and abalone, feeding abalone and removing residual seaweed is a very heavy work (Fig. 2-3-7). With the increase of labor costs, the breeding profit space is squeezed. The feeding effect can effectively use residual bait and feces and reduce labor intensity.

Fig. 2-3-7 Heavy labor

To use seaweed as diet of abalone, the farmer needs to complete the following works:

A. Feeding abalone;

B. Removing residual seaweed.

2.1.3 Can sea cucumber survive and grow? What about the economic benefit?

Sea cucumber as a detrivorous species, can be an interesting candidate in suspended longline cultural of seaweed and abalone. The schematic diagram of longline seaweed and abalone cultivation was showed in Fig. 2-3-8.

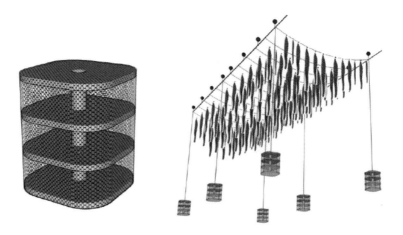

Fig. 2-3-8 Longline co-culture system (abalone + seaweed)

Production technique includes the following three aspects:

A. Kelp longline culture;

B. Abalone *Haliotis discus hannai* net cages hanging vertically from long lines;

C. Sea cucumber *Apostichopus japonicus* added directly to the abalone cages.

The cages were suspended from kelp longline and stocked with abalone (250 individuals per cage) and four densities of sea cucumbers (1, 2, 4 and 6 individuals per layer), 1 cage without abalone, and 1 cage without sea cucumber as control treatment (Fig. 2-3-9).

Fig. 2-3-9 Long cage culture feeding with kelp

Experiment design for the seaweed, abalone and sea cucumber:

Sea cucumber *A. japonica* was added directly to abalone cages without any modification of the culture equipment, in order to allow simple and low cost production.

Fig. 2-3-10 showed that for all treatments, the survival rates of sea cucumber was 100%, the body weight of sea cucumber increased from 65 g in Oct. to 129 g in May of next year, and the average specific growth rate (SGR) was 0.33% each day.

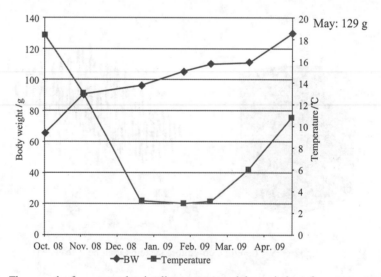

Fig. 2-3-10 The growth of sea cucumber in all treatments and the variation of sea water temperature
(Data from Jon Funderud, 2010)

Fig. 2-3-11 showed a student was weighting a sea cucumber.

The total value of kelp, abalone and sea cucumber was significantly higher than monoculture of kelp or co-culture of kelp and abalone. Food value occupied over 90% of the total value (Fig. 2-3-12).

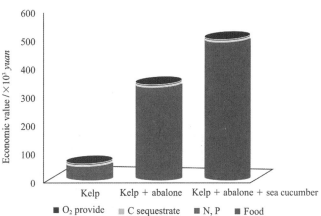

Fig. 2-3-11 Weighting the body weight of sea cucumber

Fig. 2-3-12 Comparing the economic value of different cultivated mode
(Data from Zhang, 2016)

2.1.4 Could we increase the density of sea cucumber?

Sea cucumbers in all treatment could grow and there was no significant difference in SGR (Fig. 2-3-13).

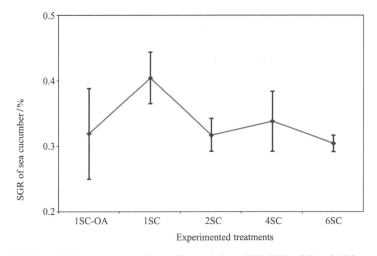

ISC-OA: ind./layer sea cucumber, without abalone ISC, 2SC, 4SC and 6SC sea cucumber densities of 1, 2, 4, 6, ind./layer respectively.

Fig. 2-3-13 The final SGR of sea cucumber in all treatments
(Data from Jon Funderud, 2010)

2.1.5 Principle of kelp + abalone + sea cucumber

Food supply and demand is the key to determine the density of sea cucumbers in the IMTA system. Fig. 2-3-14 showed the possible food sources of sea cucumbers. Firstly, it needs to know the food requirements of sea cucumbers and the effects of food quality and water temperature on their ingestion rate.

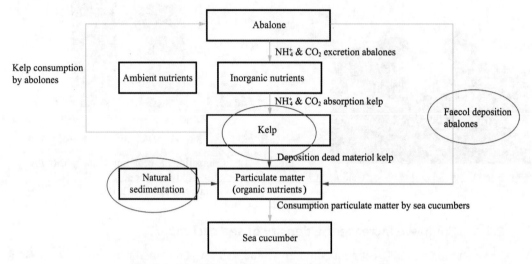

Fig. 2-3-14   The possible food sources of sea cucumber

Effect of food quality and temperature on ingestion rate of sea cucumber—indoor experiment

The feeds with different organic contents were got by mixing different ratio of kelp powder and sea mud (Table 2-3-1).

Table 2-3-1   Ingredient composition and proximate nutrient of the experimental diets

Content	Group			
	I	II	III	IV
Sea mud	100	88	76	64
Kelp powder	—	12	24	36
Protein/(kJ · g^{-1})	0.75	1.84	2.93	4.01
Fat/(kJ · g^{-1})	2.80	2.99	3.17	3.36
Ash/(kJ · g^{-1})	87.6	82.8	78.0	73.1
Water content/(kJ · g^{-1})	5.80	5.81	5.82	5.84
Total energy/(kJ · g^{-1})	0.25	1.07	1.88	2.69
Organic content/(kJ · g^{-1})	4.71	8.43	12.20	15.90

Organic content of diet had significant influence on the ingestion rate of *A. japonicas* (Fig. 2-3-15). The sea cucumbers were divided into 4 experimental groups by the wet weight.

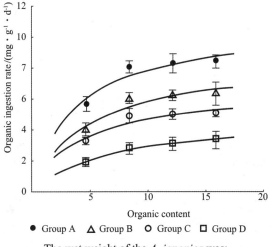

The wet weight of the *A. japonica* was:
A. 4.77 ± 0.95 g; B. 15.12 ± 1.14 g; C. 34.77 ± 7.95 g; D. 78.13 ± 4.99 g.

Fig. 2-3-15 Effects of feed organic content on ingestion rate of different size *A. japonicas*

Water temperature (T) and wet weight (W) had a significant influence on feeding behavior of *A. japonicas* (Fig. 2-3-16).

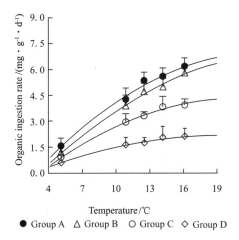

Fig. 2-3-16 Effects of water temperature on the ingestion rate of *A. japonicas* with different sizes

Stepwise regression got the relationship of OIR, T, W.

$$\text{OIR} = 2.2 \times W^{-0.384} + 0.033 \times W^{-0.384} \times T^2 + 0.077 \times T$$

$$(R^2 = 0.939, P < 0.01) \qquad \text{(Equation 1)}$$

Effects of water temperatures and diets on the ingestion of different size sea cucumber *A. japonicas* were measured. In order to understand the feeding eco-physiological characteristic of sea cucumber *A. japonicas*, the response of feeding behavior when changing organic content (OC) of feed or temperature was studied in lab oratory. There are 3 possible food sources for sea cucumber: deposition of dead kelp, faecal deposition of abalone and natural sedimentation. We measured the amount of the 3 possible food and then found that the potential density of sea cucumber in one net cage to be 55 individuals in theory.

Field experiment

During the experiment, the abalone could grow normally. Adding sea cucumber to the cage had no significant effect on the abalone growth (Fig. 2-3-17).

Abalone shell length: 4-5 cm, 200 individuals per cage;

Sea cucumber: A. 2 individuals per layer (6 individuals per cage); B. 6 individuals per layer (18 individuals per cage); C. 10 individuals per layer (30 individuals per cage); D. Control: abalone only. I: Initial body weight of abalone.

Fig. 2-3-17 Initial and finial body weight of abalone in all treatments

The survival rate of sea cucumber decreased with the increase of density, and the survival rate of group A was the highest, reaching 67%. During the experiment, the body weight of sea cucumber continued to increase, and the specific growth rate of group A was the highest, reaching 0.43 (Fig. 2-3-18).

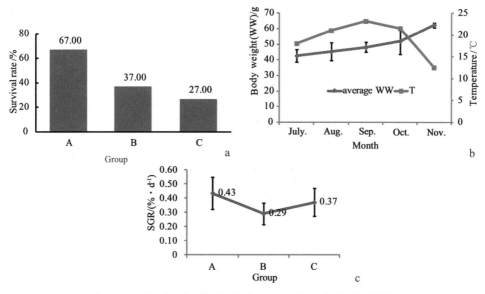

Sea cucumber density (individuals per cage): A. 6; B. 18; C. 30.

a. Survival rate of sea cucumber in different density; b. Seasonal change of sea cucumber body weight and sea water temperature; c. Special growth rate of sea cucumber in different density.

Fig. 2-3-18　The survival rate and grouth rate of field experiment

Food sources of A. japonicus

Stable isotope technique was used to quantify relative contributions of natural POM, abalone faeces and kelp to sea cucumber (Fig. 2-3-19).

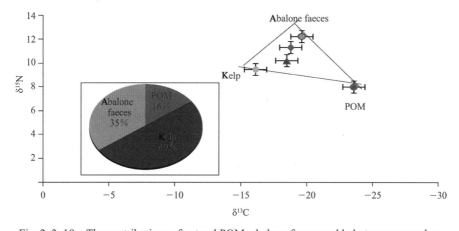

Fig. 2-3-19　The contributions of natural POM, abalone faeces and kelp to sea cucumber

Environmental impact of the IMTA—Sediment condition

MOM (monitoring, ongrowing fish farms, modelling) is Norwegian environmental regulation system (Hansen et al., 2001), MOM-B is performed in the local impact zone and combines three groups of parameters (Group Ⅰ, Ⅱ and Ⅲ). The environmental condition of the sediment is determined by combining the conditions parameters of the three groups and dividing then into four conditions (1, 2, 3 or 4), in which condition 4 is equivalent to unacceptable sediment conditions and the lower the score the better the environmental condition (Fig. 2-3-20).

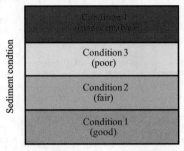

Fig. 2-3-20 The sediment condition varies from 1 to 4. the lower the number, the better the condition

A. Group Ⅰ—biological parameters: Presence or absence of macrofauna;

B. Group Ⅱ—chemical parameters: pH, redox potential;

C. Group Ⅲ—sensory parameters: outgassing, colour, smell, consistency (grab volume), thickness of deposits.

In Sanggou Bay, the large scale mariculture industry of kelp and abalone has lasted over 10 years. But the quality of benthos ecosystem is still kept in good or fair status (measured by MOM-B).

2.1.6 What we should think about the IMTA?

If you have already cultured abalone, it is also very easy for you to culture sea cucumbers!

Co-use of infrastructure has lower impact than monoculture.

The bioremediation potential depends on the relative culture densities.

2.2 How to setup "fish+shellfish+seaweed" IMTA in coastal area? —Case study of IMTA in Sanggou Bay

"Fish + shellfish + seeweed" IMTA mode was showed in Fig. 2-3-21.

Fig. 2-3-21 Experimental case and longline for "fish + shellfish + seaweed" IMTA

2.2.1 The current mariculture status in coastal of China

Mariculture production in China is still primarily based on molluscs and seaweeds, especially long-line culture, which is the main culture mode in northern China (Figs. 2-3-22 to 2-3-23).

Fig. 2-3-22　Suspended long-line culture in Sanggou Bay

Fig. 2-3-23　Proportion of various cultured organisms in marine aquaculture production

Searching for methods

A. Low carbon emission;

B. Low environmental impact;

C. High production;

D. High economic value.

Chinese finfish cage maricultural production

There were two kinds of typical cage in China (Fig. 2-3-24). The annual production increased from 2008 to 2013 (Fig. 2-3-25). During these years, the average increased rate for traditional cage and anti-wave cage were 8% and 18%, respectively.

Fig. 2-3-24　Two kinds of typical cage in China

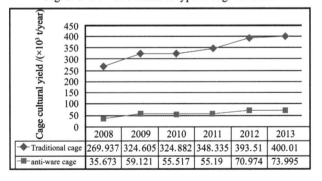

	2008	2009	2010	2011	2012	2013
Traditional cage	269.937	324.605	324.882	348.335	393.51	400.01
anti-ware cage	35.673	59.121	55.517	55.19	70.974	73.995

Fig. 2-3-25　Cage culture production during 2008-2013

The cage farming industry faces serious environmental challenges

With the blind growth of culture scale (Fig. 2-3-26), cage culture is over carrying capacity, and a large number of diet residues and fish feces are released into the sea. The environmental pollution causes frequent diseases outbreak.

Fig. 2-3-26　Picture of fish farms in China

2.2.2　The Schematic diagram of IMTA finfish, bivalve and seaweed

Fig. 2-3-27 showed the schematic diagram of IMTA finfish, bivalve and seaweed.

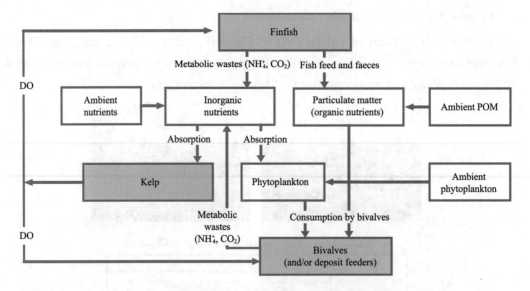

Fig. 2-3-27　The schematic diagram of IMTA finfish, bivalve and seaweed

How much inorganic and organic bound nutrients are released?

Fig. 2-3-28 showed the nitrogen budget in marine fish cage culture system in Sanggou Bay.

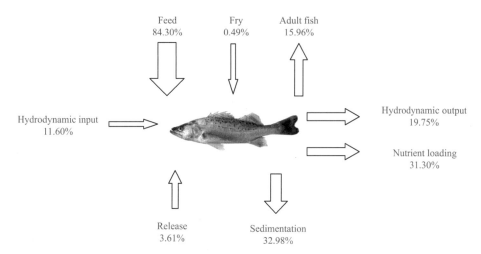

Fig. 2-3-28　Nitrogen budget in marine fish cage culture system in Sanggou Bay

In May, the initial growth rate was negative. The maximum growth rates were in September and August for *L. japonicus* and *S. fuscescens*, respectively (Table 2-3-2).

Table 2-3-2　Special growth rate of main cultured fish (Ge et al., 2007)

Month	Temperature/℃	*Lateolabrax japonicus*		*Sebastodes fuscescens*	
		Weight/g	SGR	Weight/g	SGR
May	10.4	205	−0.640	438	−0.070
June	15.2	235	0.441	280	0.239
July	19.0	295	0.734	285	0.057
Aug.	21.0	335	0.410	335	0.521
Sep.	21.6	460	1.023	365	0.277
Oct.	18.0	525	0.426	375	0.087
Nov.	12.0	630	0.588	390	0.127
Dec.	7.5	664	0.171	425	0.278

July-September was the peak season of ammonia excretion of black rock fish, and August-October was the peak season of ammonia excretion of sea bass (Table 2-3-3).

Table 2-3-3　Total ammonia released per month from anti-wave cage in Sanggou Bay (Ge et al., 2007)

	Black rock *Sebastodes fuscescens*		Sea bass *Lateolabrax japonicas*	
Month	Biomass/t	Ammonia excretion	Biomass/t	Ammonia excretion
Apr.	44.3	0.03	48.4	4.00×10^{-6}
May	49.0	0.11	22.0	2.10×10^{-4}
June	39.8	0.09	27.4	8.64×10^{-3}

(*to be continued*)

	Black rock *Sebastodes fuscescens*		Sea bass *Lateolabrax japonicas*	
Month	Biomass/t	Ammonia excretion	Biomass/t	Ammonia excretion
July	42.2	0.26	43.2	0.135
Aug.	42.9	0.20	46.9	0.27
Sep.	44.9	0.76	65.2	0.29
Oct.	25.3	0.08	81.2	0.09
Nov.	28.9	0.08	96.5	1.5×10^{-3}
Dec.	39.6	0.06	57.6	7.65×10^{-9}

From April to December, the sea bass exports 12.1 t of particulate matter to the water body through faeces released, with an average faces released of 1.34 t per month. The largest excretion amount occurred from August to September (Table 2-3-4).

Table 2-3-4 Fish faeces released per month from anti-wave cage in Sanggou Bay (Unit: t/m) (Ge et al., 2007)

Month	Apr.	May	June	July	Aug.	Sep.	Oct.	Nov.	Dec.
Sea bass *L. japonicas*	0.61	0.40	0.51	1.63	2.43	3.15	1.72	0.95	0.68
Black rock *S. fuscescens*	0.01	0.05	0.05	0.06	0.04	0.03	0.03	0.03	0.02

Distribution of nutrients in fish culture area

Under the net cage, the organic matter (the sum of residual bait and feces) reached 87.88%. There was significantly negative relationship between organic matter and distance from the cage ($P < 0.01$) (Fig. 2-3-29).

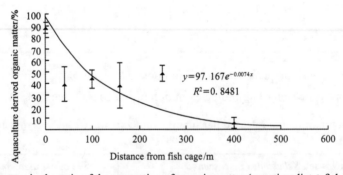

Fig. 2-3-29 Changes in the ratio of the proportion of organic matter (uneating diet + fish faeces) from cage culture sources to cages
(Data from Jiang et al., 2012)

For the bivalves, in the present study (Jiang et al., 2012), assimilation efficiency of the oysters for fish aquaculture derived organic matter was estimated at 54% (10% waste feed and 44% fish faeces) (Fig. 2-3-30).

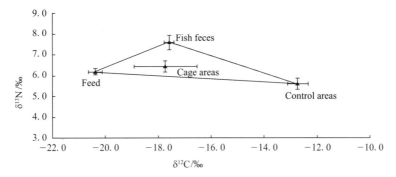

Fig. 2-3-30 Stable carbon and nitrogen isotope maps of sediments and their potential sources in the cage area
(Data from Jiang et al., 2012)

Can bivalves utilize nutrients from fish feed and faeces?

Feeding behaviors of Japanese scallop *Patinopecten yessoensis* feeding on *Paralichthys olivaceus* feed, fecal, sediment particulates from farming cage and microalgae were showed in Fig. 2-3-31. Results showed *P. yessoensis* could feed on fish faece, residual fed, deposition and microalgae. Absorption efficiency (AE) among microalgae, fish feces and uneaten feed groups were not significantly different, but they were significantly higher than those on sediment particulates. There was positive correlation between AE and OC; linear relationship between absorption rate (AR) and particle organic material (POM).

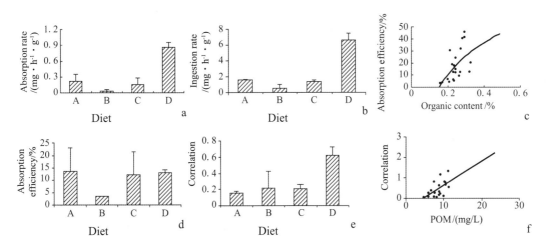

a. Absorption rate; b. Ingestion rate; c. Absorption efficiency; d. Clearance rate; e. Correlation of AE and organic content; f. Correlation of AR and POM. The 4 type of diets were: A. Residual diet; B. Sediment particulates; C. Fish feces; D. Microalgae.

Fig. 2-3-31 Clearance rate, ingestion rate, absorption rate and absorption efficiency of scallop *Patinopecten yessoensis* to the 4 type of diets and the correlation of feeding behavior with food quality

Appropriate integrated species election

5 species bivalves' feeding behavior on different food, including flounder (*P. olivaceus*) faeces, residual fed and sediment particulates were studied so as to discuss the potential IMTA based on filtering feed bivalves.

Ingestion rates of oyster, clam and scallop were significantly higher than the other 2 species (ANOVA, $P < 0.01$), so they are relatively good candidate species (Fig. 2-3-32).

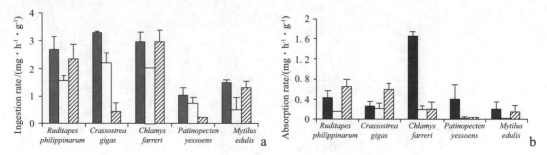

a: Ingestion rate; b: Absorption rate.

Fig. 2-3-32 The ingestion rate and absorption rate of 5 species of bivalve on the uneaten diet and fish faeces

Nutrient utilization of seaweed

The total inorganic nitrogen (TIN) uptake rate was measured for discs taken from different parts of the kelp under two temperature treatments of 4 ℃ and 10 ℃ in laboratory.

The uptake rates of the upper part of the middle band (60-110 cm) and the base of the plant (20-50 cm) were faster than the lower part of the middle band (150-200 cm) and the marginal part of the plant (Fig. 2-3-33).

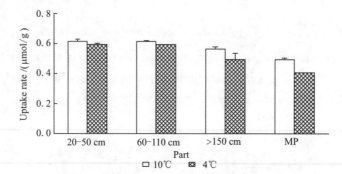

Fig. 2-3-33 The TIN uptake rate in different temperature of kelp discs at different parts
(Data from Mao et al., 2018)

Integrated culture of fish, bivalve, seaweed in Sanggou Bay

Fig. 2-3-34 shows the integrated culture of fish, seaweed and bivalve in Sanggou Bay.

a. Kelp harvest; b. Monitoring the growth of bivalves; c. Integrated culture of fish and longline.

Fig. 2-3-34　Picture of integrated culture of fish, seaweed and bivalve in Sanggou Bay

The schematic of finfish, bivalve and seaweed integrated mariculture is showed in Fig. 2-3-35.

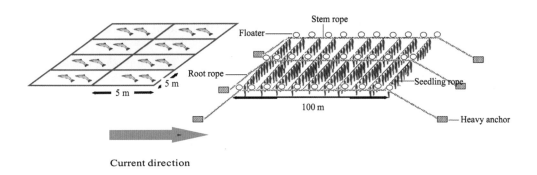

Fig. 2-3-35　Distribution of cage and longline *in situ* area

The soft tissue weight increased 35% at fish cage (Fig. 2-3-36). The contribution rate of fish feces and uneaten feed to oyster was 29.27% and 5.59% respectively (Fig. 2-3-37).

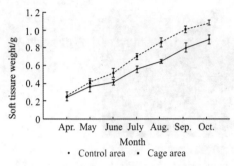 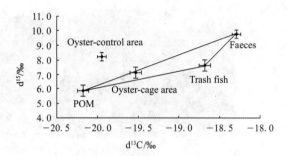

Fig. 2-3-36 Growth in soft tissue dry weight of oyster *Crassostrea gigas* at fish cage and in control area

Fig. 2-3-37 Dual isotope plot of oyster tissue and potential food sources
(Data from Jiang et al., 2013)

For the bivalves, assimilation efficiency of the oysters for fish aquaculture derived organic matter was estimated at 54% (10% waste feed and 44% fish faeces). If, for example, 42% of the total solid nutrient loads from fish cages are within the suitable size range (Elberizon et al., 1998), the oysters will theoretically be able to recover approximately 23% of the total particle organic matter released from fish cages.

The soft tissue weight of scallop in cage area increases 16% compared with the control area.

The assimilation efficiency of scallops for fish aquaculture-derived organic matter is estimated at 26%.

The scallops will theoretically recover nearly 10% of the total particle organic matter released from fish cages (Fig. 2-3-38).

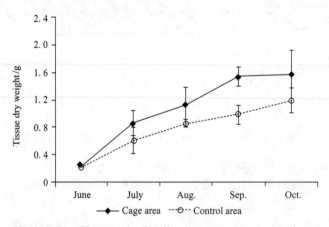

Fig. 2-3-38 The growth of scallop in cage area and control area

Based on the total amount of dissolved nutrients released from cage culture and the absorption and utilization of different species of macroalgae in different seasons, we calculated the appropriate ratio of macroalgae to fish (Table 2-3-5).

Table 2-3-5 Bioremediation strategy of fish cage culture

Season	Macroalgae (DW)/Fish (WW)
Winter	0.94 (*Laminaria*)
Spring	
Summer	1.53 (*Gracilaria lemaneiformis*)
Autumn	

3 Discussion and summary

3.1 Challenges

Adding sea cucumber into the net may make it more difficult to clean the detritus.

Is the IMTA possible without manual handling?

There are still challenges in the "seaweed + abalone + sea cucumber" IMTA mode. For example, adding sea cucumbers in the cage will increase the difficulty for farmers to clean up the residual diet (Fig. 2-3-39). The question we need to solve is: Is it possible to get the most suitable sea cucumber density and feeding strategy so that there is as little residual diet as possible, with no need to clean the residual diet manually?

Fig. 2-3-39 Picture of a farmer cleaning up the residual diet

3.2 Candidate species in IMTA

For candidate species of IMTA, it is necessary to consider many factors, such as the coupling of growth season, nutrients removal efficiency, economic benefit, local market etc. For sea cucumber, we could consider the seeding size, for that different size of sea cucumber has different characteristics to high temperature:

A. Larva which weight was less than 10 g was never going into aestivating.

B. When temperature was above 22 ℃ sea cucumber (weight was over 30 g) aestivation occurred.

C. When temperature was less than 1 ℃ in winter sea cucumber (weight was 80-200 g) stopped ingesting and felled into hibernation.

3.3　In coastal ecosystem, multi-culture system is complicated and strongly dependent on the local environment

The distribution of longline around finfish cage will depend on the waste mass and distance of waste dispersion.

For the differences in sea conditions, culture species and ratios, we cannot copy a model completely. We need to accept the idea of IMTA and understand the framework of establishing IMTA (Fig. 2-3-40).

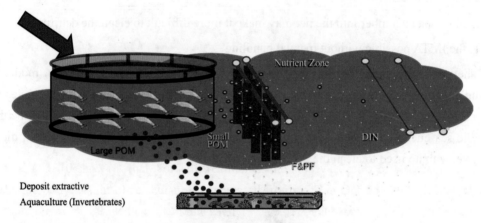

Fig. 2-3-40　Multi-culture system
(Data from Chopin et al., 2009)

3.4　Gaps to fill

Particle size is an important parameter in bivalve feeding (The diet includes ice-fresh small fish, artificial diet, re-suspension of decaying residual diet, phytoplankton).

Nutrient recovery efficiency is a function of technology, management, spatial configuration, species selection, trophic level, biomass ratios, natural food availability, particle size, digestibility, season, light, temperature and water flow.

Multi-species numerical model (Fig. 2-3-41) is being studied now.

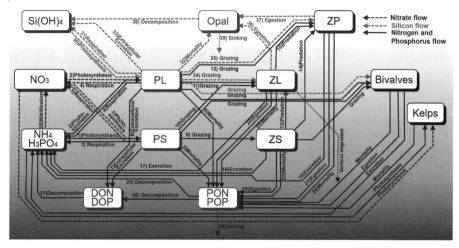

Fig. 2-3-41　The conceptual model diagram of IMTA
(Data from Sun et al., 2020)

References

Ge C Z, Fang J G, Guan C T, Wang W, Jiang Z J. Study on key biological process of cage mariculture: Metabolism of *Lateolabrax japonicas* [J]. Marine Fisheries Research, 2007, 28(2): 45-50.

Hansen P K, Ervik A, Schaanning M, et al. Regulating the local environmental impact of intensive, marine fish farming Ⅱ. The monitoring programme of the MOM system (modelling-ongrowing fish farms-monitoring) [J]. Aquaculture, 2001, 194: 75-92.

Sun K, Zhang J H, Lin F, et al. Evaluating the influences of mariculture on pelagic ecosystem by a numerical approach: A case study of Sungo Bay, China [J]. Ecological Modelling, 2020: 415.

Zhang J H, Ge C Z, Fang J G, Tang Q S. Multi-Trophic mariculture practices in coastal waters, [M]// Gui J F, et al, Aquaculture in China: Success Stories and Modern Trends. Bei. 2018: 541-554.

Zhang J H. Integrated multi-trophic aquaculture of fish, bivalves and seaweed in Sanggou Bay [M]//Miao W and Lal K K, Sustainable intensification of aquaculture in the Asia-Pacific region. Documentation of successful practices. Bangkok, Thailand, FAO, 2016: 110-121.

MOM-B system and its application in IMTA

Author: Wu Wenguang
E-mail: wuwg@ysfri.ac.cn

1 MOM system introduction

1.1 The origin of MOM system

The full name of MOM is modelling-ongrowing fish farms-monitoring.

MOM system was established by Ervik et al. in 1997, and a series of research papers have been published (Ervik et al., 1997; Hansen et al., 2001; Stigebrandta et al., 2004).

1.2 The function of MOM system

It is a management systems for monitoring and assessing the environmental impact of aquaculture activities (first used in fish farming in 1997).

With the improvement of the MOM system, it has been widely applied in the monitoring of other mariculture activities as shown in Fig. 2-4-1 (Zhang et al., 2001, 2009).

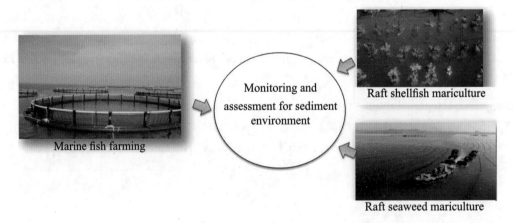

Fig. 2-4-1　The application of MOM system

1.3 The composition of MOM system

The MOM system is composed of A, B and C parts. MOM-A refers to measurement of sedimentation under and around the cage; MOM-B refers to assessment of sedimentary environmental status; MOM-C refers to survey of benthic community structure (Fig. 2-4-2).

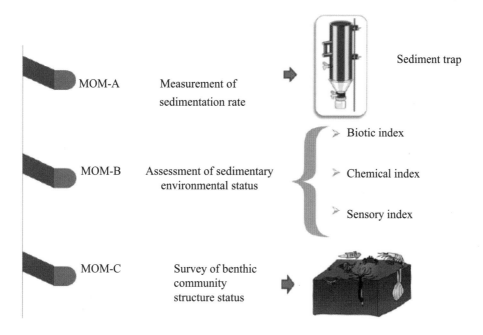

Fig. 2-4-2 The composition of MOM system

1.4 The advantages and application prospect of MOM-B system

1.4.1 Advantages of MOM-B system

The advantages of MOM-B include: simple to operate and low monitoring cost; result; and easy to learn through simple training.

1.4.2 Application prospect

MOM-B is operated primarily at site level but may be linked to more comprehensive management systems which are operated at higher geographic levels concerning the coastal zone and involving different activities. It is a very effective tool to assess the sediment environment pressure. Firstly, MOM-B can monitor the sediment environment of mariculture regularly and systematically. Secondly, it can help people understand the response of the sediment environment to mariculture timely. Finally, it can spot and correct the harmful mariculture activities timely.

2 The monitoring content and standards of MOM-B

The three groups of parameters used in the B-investigation are: biological parameters, chemical parameters and sensory parameters. A scoring system (Fig. 2-4-3) has been developed for the three groups of parameters and the lower the score, the better the environmental condition is.

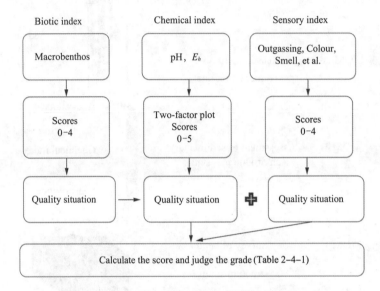

Fig. 2-4-3　The composition of MOM-B system

2.1　MOM-B composition and awarded points rule

The biotic index is a determination of whether the sediment contains a macro-infauna and is directly linked to the environmental quality objective, which states that a viable macro-infauna must be present under the fish farm. The chemical index parameters are based on direct measurements of pH and redox potential (Eh) by electrodes inserted in the sediment immediately after sampling in transparent corers. The sensory index parameters are a number of sensory sediment variables, which change with organic enrichment (Table 2-4-1).

Table 2-4-1　MOM-B composition and awarded points rule

MOM-B	Awarded points rule	Scores		Judge the grade (Mean ± S.D.)
Biotic index	Macrobenthos	Exist	Exist	1, 2, 3
		None	None	4
Chemical index	pH	pH & Eh two-factor plot		1, 2, 3, 4
	Eh			

(*to be continued*)

MOM-B	Awarded points rule	Scores		Judge the grade (Mean ± S.D.)
Sensory index	Outgassing	None	0	1, 2, 3, 4
		Exist	1	
	Colour	Grey	0	
		Brown or black	2	
	Smell	None	0	
		Medium	2	
		Pungent	4	
	Thickness of deposits	< 2 cm	0	
		2 – 8 cm	1	
		> 8 cm	2	
	Consistency	Solid	0	
		Soft	2	
		Loose	4	
	Grab volume	< 1/4	0	
		1/4 – 3/4	1	
		> 3/4	2	

2.2 pH & *Eh* two-factor plot

As shown in Figs. 2-4-4 to 2-4-5, the *y*-coordinate is *Eh*, the *x*-coordinate is pH, and the measured value of Example-1 and Example-2 is also given.

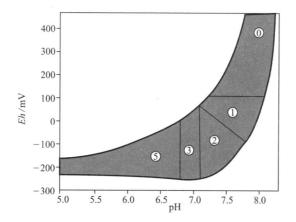

Fig. 2-4-4 pH & *Eh* Two-factor plot

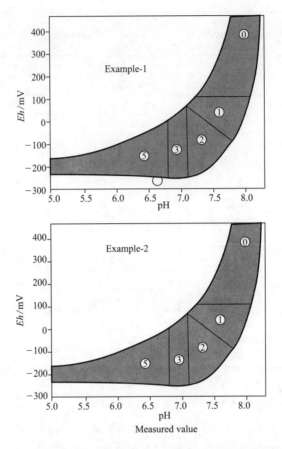

Fig. 2-4-5 pH & Eh two-factor plot and measured value of the example

2.3 The principle of grading of MOM-B

The lower the score, the better the sedimentary environmental status is (Table 2-4-2).

Table 2-4-2 The condition standards of MOM-B

Index	The condition standards of MOM-B			
Average value (Chemical & sensory) (AV)	< 1.1	1.1 ≤ AV < 2.1	2.1 ≤ AV < 3.1	> 3.1
Condition	1	2	3	4
Evaluation result	Very good	Good	Bad	Very bad

3 The application of MOM-B in Sanggou Bay

Fig. 2-4-6 shows the application of MOM-B in Sanggou Bay.

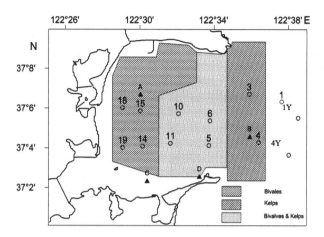

Fig. 2-4-6 The survey stations in Sanggou Bay

3.1 The calibration of monitoring instrument

The monitoring instruments needs to be calibrated before investigation as shown in Fig. 2-4-7.

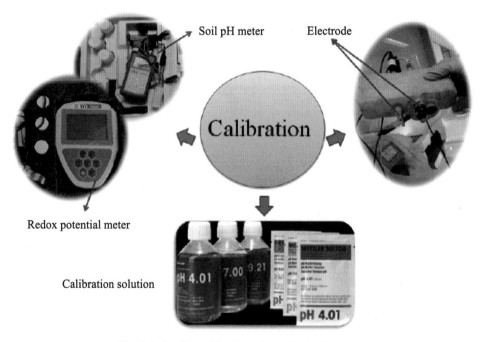

Fig. 2-4-7 The calibration of monitoring instrument

3.2 The sampling and monitoring of sediment

The sampling and monitoring process of sediment is shown in Fig. 2-4-8.

Fig. 2-4-8　The sampling and monitoring of sediment

3.3 The survey results in 2006

3.3.1 The chemical index in 2006

The two-factor plot for pH-Eh in summer and autumn of 2006 is shown in Fig. 2-4-9 and Table 2-4-3.

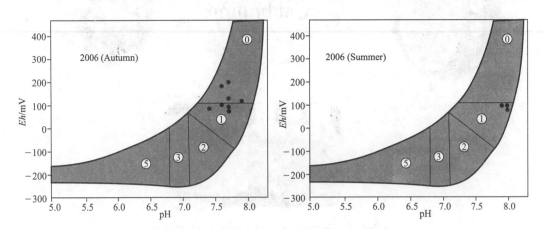

Fig. 2-4-9　Two-factor plot for pH-Eh in 2006
(Data from Zhang et al., 2009)

Table 2-4-3 The evaluation results of chemical index in 2006

Season	Parameter	Sample number										Index
		1	5	8	10	12	13	14	16	18	19	
Summer	Score	0	1	1	1	0	0	0	0	1	1	0.5
	Condition	1	1	1	1	1	1	1	1	1	1	1
Autumn	Score	1	1	1	1	0	1	1	1	1	1	0.87
	Condition	1	1	1	1	1	1	1	1	1	1	1

3.3.2 The survey results of sensory index in 2006

The evaluation results of sensory index in summer and autumn of 2006 are as shown in Table 2-4-4.

Table 2-4-4 The evaluation results of sensory index in 2006

Group Parameter	Summer										Autumn							
	1	5	8	10	12	13	14	16	18	19	1	5	8	10	12	14	16	18
Outgassing	0	0	0	0	0	0	0	0	0	0	0	0	0	0	0	0	0	0
Colour	0	0	0	0	0	0	0	0	0	0	0	0	0	0	0	0	0	0
Smell	0	0	0	0	0	0	0	0	0	0	0	0	0	0	0	0	0	0
Consistency	2	2	2	2	2	2	2	2	2	2	2	2	2	2	2	2	2	2
Grab volume	1	1	1	1	1	1	1	1	1	1	1	2	2	1	2	1	2	1
Thickness of deposits	0	0	0	0	0	0	0	0	0	0	0	0	0	0	0	0	0	0
Sum side (scores)	3	3	3	3	3	3	3	3	3	3	3	4	4	3	4	3	4	5
Sediment condition	1	1	1	1	1	1	1	1	1	1	1	1	1	1	1	1	1	2

3.3.3 Evaluation results of the sedimentary environment of MOM-B

The evaluation results of the sedimentary environment of MOM-B in Sanggou Bay in summer and autumn of 2006 are shown in Table 2-4-5 and Table 2-4-6. The results show that all station evaluation levels are very good for stations 18 and 19.

Table 2-4-5 The evaluation results of MOM-B in July, 2006

MOM-B	Stations									
	1	5	8	10	12	13	14	16	18	19
Chemical index	0	1	1	1	0	0	0	0	1	1
Sensory index	0.66	0.66	0.66	0.66	0.66	0.66	0.66	0.66	0.66	0.66
Mean value	0.33	0.83	0.83	0.83	0.33	0.33	0.33	0.33	0.83	0.83
Judge the grade	1	1	1	1	1	1	1	1	1	1
Evaluation result	Very good	Very good	Very good	Very good	Very good	Very good	Very good	Very good	Very good	Very good

Table 2-4-6 The evaluation results of MOM-B in November, 2006

MOM-B	Stations									
	1	3	5	8	10	12	13	14	16	18	19
Chemical index	1	1	0	1	0.5	1	0.5	1	1	1	1
Sensory index	0.44	1.54	1.32	1.32	1.32	1.21	1.54	1.54	1.32	1.76	1.54
Mean value	0.72	1.27	0.66	1.16	0.91	1.11	1.02	1.27	1.16	1.38	1.27
Judge the grade	1	2	1	2	1	2	1	2	2	2	2
Evaluation result	Very good	Very good	Very good	Very good	Very good	Very good	Very good	Very good	Very good	Good	Good

3.4 The survey results in 2016

3.4.1 The chemical index in 2016

The evaluation results of chemical index in autumn 2006 are shown in Fig. 2-4-10 and Table 2-4-7.

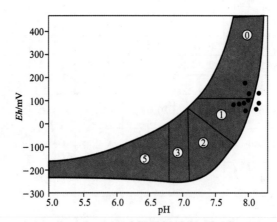

Fig. 2-4-10 The Two-factor plot for pH-Eh in 2016
(Data from Yang et al., 2018)

Table 2-4-7 The evaluation results of MOM-B in 2016

Parameter	Sample number											Index
	1	3	5	8	10	12	13	14	16	18	19	
Score	1	1	0	1	0.5	1	0.5	1	1	1	1	0.82
Condition	1	1	1	1	1	1	1	1	1	1	1	1

3.4.2 The survey results of sensory index in 2016

The survey results of sensory index in 2016 are shown in Table 2-4-8.

Table 2-4-8 The survey results of sensory index in 2016

Sensory index	Station										Mean value
	3	5	8	10	12	13	14	16	18	19	
Outgassing	0	0	0	0	0	0	0	0	0	0	0
Colour	1	1	1	1	1	1	1	2	1	1	1.10
Smell	1	0	0	0	0	0	0	0	0	0	0.10
Thickness of deposits	1	1	1	1	1	0	1	1	1	1	0.90
Consistency	2	2	2	2	2	4	3	2	4	3	2.60
Grab volume	1	1	1	1	0.5	1	1	1	1	1	0.95
Sum	6	5	5	5	4.5	6	6	6	7	6	5.65
Calibration	1.32	1.10	1.10	1.10	0.99	1.32	1.32	1.32	1.54	1.32	1.24
Contidion	2	2	2	2	1	2	2	2	2	2	2

3.4.3 Evaluation results of the sedimentary environment

The evaluation results of the sedimentary environment MOM-B in Sanggou Bay in summer and autumn of 2006 and autumn of 2016 are shown in Table 2-4-9 and Table 2-4-10.

Table 2-4-9 The evaluation results of MOM-B in 2016

MOM-B	Station										Mean value
	3	5	8	10	12	13	14	16	18	19	
Chemical index	0	0	0	0	1	0	0	0	0	1	0.2
Sensory index	1.32	1.10	1.10	1.10	0.99	1.32	1.32	1.32	1.54	1.32	1.24
Mean value	0.66	0.55	0.55	0.55	0.995	0.66	0.66	0.66	0.77	1.16	0.72
Judge the grade	1	1	1	1	1	1	1	1	1	2	1
Evaluation results	Very good	Very good	Very good	Very good	Very good	Very good	Very good	Very good	Very good	Good	Very good

Table 2-4-10 The evaluation results of MOM-B in 2006 and 2016

Time	MOM-B	The survey stations									
		1	3	5	8	10	12	13	14	16	18	19
2006.07	Chemical index	1	1	1	1	1	1	1	1	1	1	1
	Evaluation results	Very good	Very good	Very good	Very good	Very good	Very good	Very good	Very good	Very good	Very good	Very good
2006.11	Chemical index	1	1	1	1	1	1	1	1	1	2	2
	Evaluation results	Very good	Very good	Very good	Very good	Very good	Very good	Very good	Very good	Very good	Good	Good

(to be continued)

Time	MOM-B	The survey stations									
		1	3	5	8	10	12	13	14	16	18	19
2016.09	Chemical index	1	2	1	2	1	2	1	2	2	2	2
	Evaluation results	Very good	Good	Very good	Good	Very good	Good	Very good	Good	Good	Good	Good

The influence of large-scale mariculture on benthic environment is good or very good in the ten years, and it is benefited from the development of coastal integrated multi-trophic aquaculture (IMTA) in Sanggou Bay.

3.5 The survey results in abalone + kelp mariculture area in Sanggou Bay

The MOM-B evaluation was conducted in May, July, August and October of 2010. The samping stations were set in Sanggou Bay: abalone + kelp mariculture area is shown in Fig. 2-4-11.

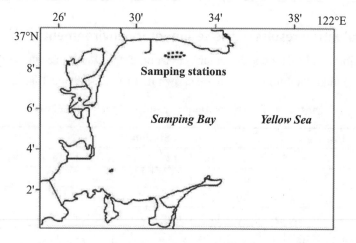

Fig. 2-4-11 Samping stations in Sanggou Bay: abalone + kelp mariculture area
(Data from Zhang et al., 2001)

3.5.1 The survey results of chemical index in abalone + kelp mariculture area

The survey results of chemical index in abalone + kelp mariculture area are shown in Table 2-4-11.

Table 2-4-11 Seasonal variation of sediment pH and *Eh* values at sample stations in Sanggou Bay

Station	May		July		August		October	
	pH	*Eh*/mV	pH	*Eh*/mV	pH	*Eh*/mV	pH	*Eh*/mV
1	7.4 ± 0.2	64 ± 9	7.3 ± 0.2	−43 ± 12	7.3 ± 0.1	−142 ± 34		
2	7.4 ± 0.1	66 ± 12	7.3 ± 0.1	34 ± 6	7.3 ± 0.1	−152 ± 21	7.1 ± 0.1	−135 ± 26
3	7.5 ± 0.1	8 ± 3	7.2 ± 0.1	91 ± 21	7.4 ± 0.3	−155 ± 43	7.3 ± 0.2	112 ± 8

(*to be continued*)

Station	May		July		August		October	
	pH	Eh/mV	pH	Eh/mV	pH	Eh/mV	pH	Eh/mV
4	7.5 ± 0.3	61 ± 24	7.1 ± 0.1	124 ± 18	7.4 ± 0.3	−167 ± 44	7.1 ± 0.1	109 ± 17
5	7.2 ± 0.1	73 ± 16	7.3 ± 0.2	119 ± 34				
6	7.5 ± 0.2	74 ± 12	7.2 ± 0.1	118 ± 11	7.3 ± 0.2	−165 ± 34	7.1 ± 0.1	−109 ± 15
7	7.4 ± 0.2	−34 ± 13	7.2 ± 0.0	121 ± 9	7.2 ± 0.1	−127 ± 12	7.1 ± 0.1	−166 ± 29
8	7.5 ± 0.1	96 ± 22	7.2 ± 0.1	106 ± 22	7.3 ± 0.2	−174 ± 46	7.0 ± 0.0	−108 ± 18
9	7.6 ± 0.3	99 ± 16	7.2 ± 0.2	116 ± 42	7.4 ± 0.1	−134 ± 9	7.2 ± 0.1	71 ± 26
10	7.4 ± 0.1	68 ± 25	7.1 ± 0.1	111 ± 23	7.4 ± 0.2	−134 ± 24		

3.5.2 The survey results of sensory index in abalone + kelp mariculture area

The survey results of sensory index in abalone + kelp mariculture area are shown in Table 2-4-12 and Table 2-4-13.

Table 2-4-12 Calculated results for group 2 parameters of MOM-B system of abalone longline culture area in Sanggou Bay

Month	Parameter	Station										Mean ± S.D.	Condition
		1	2	3	4	5	6	7	8	9	10		
May	Score	3	3	7	4	5	4	3	4	3	1	3.7 ± 1.6	1
	Condition	1	1	2	2	2	2	1	2	1	1		
July	Score	1	5	3	2	3	1	3	1	1	0	2.0 ± 1.5	1
	Condition	1	2	1	1	1	1	1	1	1	1		
August	Score	5	8	4	2		4	5	6	2	2	4.5 ± 2.0	2
	Condition	2	2	2	1		2	2	2	1	1		
October	Score		6	4	4		2	5	6	5		4.6 ± 1.4	2
	Condition		2	2	2		1	2	2	2			

Table 2-4-13 Calculated results for the group 3 parameters of MOM-B system of abalone longline culture area in Sanggou Bay

Month	Parameter	Station										Mean ± S.D.	Condition
		1	2	3	4	5	6	7	8	9	10		
May	Score	3	3	7	4	5	4	3	4	3	1	3.7 ± 1.6	1
	Condition	1	1	2	2	2	2	1	2	1	1		
July	Score	1	5	3	2	3	1	3	1	1	0	2.0 ± 1.5	1
	Condition	1	2	1	1	1	1	1	1	1	1		
August	Score	5	8	4	2		4	5	6	2	2	4.5 ± 2.0	2
	Condition	2	2	2	1		2	2	2	1	1		
October	Score		6	4	4		2	5	6	5		4.6 ± 1.4	2
	Condition		2	2	2		1	2	2	2			

3.5.3 Evaluation results of MOM-B in abalone + kelp mariculture area

Evaluation results of MOM-B in abalone + kelp mariculture area are shown in Table 2-4-14.

Table 2-4-14 Evaluation results of MOM-B in abalone + kelp mariculture area in different months

MOM-B	May	July	August	October	Mean value
Chemical index	1	1	2	2	1.5
Sensory index	0.22	0.22	0.44	0.44	0.33
Mean value	0.61	0.61	1.22	1.22	0.92
Judge the grade	1	1	1	1	1
Evaluation result	Very good	Very good	Very good	Very good	Very good

Totally, the sediment environment was well in abalone + kelp mariculture area in Sanggou Bay, but it should be paid more attention that the redox potential values were negative in August and October which might be related to abalone longline culture. The sedimentary environment was in reductive state, whose stress may related of abalone long-line culture.

References

Ervik A, Hansen P K, Aure J, et al. Regulating the local environmental impact of intensive marine fish farming Ⅰ. The concept of the MOM system (modelling-ongrowing fish farms-monitoring) [J]. Aquaculture, 1997, 158(1): 85-94.

Hansen P K, Ervik A, Schaanning M, et al. Regulating the local environmental impact of intensive, marine fish farming Ⅱ. The monitoring programme of the MOM system (modelling-ongrowing fish farms-monitoring) [J]. Aquaculture, 2001, 194: 75-92.

Stigebrandta A, Aureb J, Ervik A, et al. Regulating the local environmental impact of intensive marine fish farming Ⅲ. A model for estimation of the holding capacity in the modelling-ongrowing fish farm-monitoring system [J]. Aquaculture, 2004, 234(1-4): 239-261.

Yang Y Y, Zhang J H, Wu, W G, et al. Macrobenthic community characteristics of different culture areas in Sanggou Bay [J]. Journal of Fisheries of China, 2018, 42(6): 922-931. (in Chinese)

Zhang J H, Hansen P K, Fang J G, et al. Assessment of the local environmental impact of intensive marine shellfish and seaweed farming-application of the MOM System in the Sungo Bay, China [J]. Aquaculture. 2009, 287(3): 304-310.

Zhang J H, Ren L H, Wu T, et al. Assessment of the local environmental impact of abalone suspended longline culture application of the MOM system in Sungo Bay [J]. Fishery Modernization, 2011, 38(1): 1-6. (in Chinese)

Cage mariculture in China

Author: Cui Yong
E-mail: cuiyong@ysfri.ac.cn

1 General description of cage mariculture in China

1.1 Introduction

Sea-based marine fish aquaculture in China started in the early 1970s and has developed rapidly since the late 1980s. At the beginning, small wooden net cages with size of 9 m² to 25 m² were commonly used for farming. Since the year of 2000, offshore or deep sea cage culture has gradually developed.

1.2 Classification

At present, there are more than 2 million net cages in all, among which about 10,000 cages are deep sea cages with total water volume up to 10.6 million cubic meters. Fish production from cage mariculture in 2016 was 623,900 t (Figs. 2-5-1 to 2-5-2) (Yuan & Zhao, 2016).

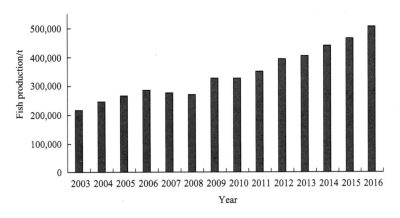

Fig. 2-5-1 Fish production from traditional cages in China (2003–2016)

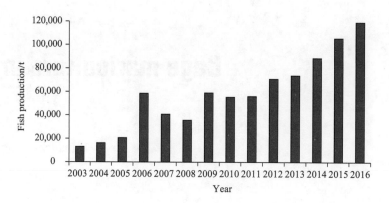

Fig. 2-5-2　Fish production from offshore cages in China (2003–2016)

1.3　Distribution

Cages are mostly congregated in the East China Sea. About 20% is in the South China Sea and 10% is in the Yellow Sea (Fig. 2-5-3) (Guo & Zhao, 2017).

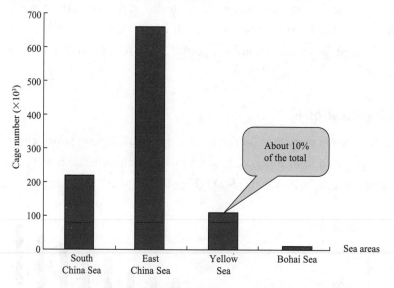

Fig. 2-5-3　Distribution of sea cages in China

2　Traditional cages and offshore cages

2.1　Structures, sizes and materials of small cages

The structures and materials of small cages are simple. The three types of cages shown in Fig. 2-5-4 are typical, and the sizes are 3 m × 3 m in the near shore.

a. Bamboos; b. Hardwood beams; c. Steel pipes.

Fig. 2-5-4　Structures, sizes and materials of small cages

2.2　Problems of traditional cages

These cages cannot withstand the strong wind and big waves in heavy sea. More and more cages have to be crowded in the near shore, shallow and sheltered areas, which brings in many problems. For instance, the stocking density is beyond the carrying capacity of the environment, and the lack of effective method to prevent net bag from fouling makes the exchange of water difficult.

The accumulation of waste metabolite and the waste feeds pollute the farming environment, and problems of self-pollution are serious. Moreover, most of the near shore waters are polluted by waste from on-land industries. All resulted in fish disease to break out and fish quality and economic efficiency to decrease, which have restricted the development of cage mariculture in China.

In August 2006, typhoon "Sangmei" caused major damage to Fuding's cage. More than 70,000 fish cages were destroyed, costing more than 600 million RMB (Fig. 2-5-5).

Fig. 2-5-5　Losses from pollution and fish diseases

In June 2011, typhoon "Milei" caused major damage to Rongcheng's cages and raft farming (Fig. 2-5-6) (Huang, et al., 2011).

Fig. 2-5-6   Major damage caused by typhoon "Milei"

To realize sustainable development, a new model of cage mariculture — offshore cage farming (Fig. 2-5-7) has to be developed.

Fig. 2-5-7   Offshore cage farming

2.3   Offshore cages

2.3.1   Advantages of offshore cage farming

A. High-tech and automation.

B. Long life span.

C. More resistant to wind and wave thus can be sited in more exposed waters.

D. Large carrying capacity and better economic benefits.

E. Friendly culture model to the environment.

F. Little risk of loss and lower cost.

G. Better fish quality and higher prices (Although the farming species are almost the same as those of traditional cages, the environment improved).

2.3.2 Major types and structures of deep sea cages

A. Typical offshore cage in China (Fig. 2-5-8). The cage circumference is from 40 m to 120 m. The cage volume is from 800 m^3 to 10,000 m^3. The cage is resistant to typhoon of 10–12 grade.

Fig. 2-5-8　Typical offshore cage farm in China

B. Submersible cage (Fig. 2-5-9). When the typhoon is coming, process of submergence can be remote controled by mobile phone message. The cage can rise and sink by releasing air or water.

a. High-density polyethylene (HDPE) submersible cage; b. Remote control.
Fig. 2-5-9　Submersible cage

C. Two-cone-shaped net cage (Fig. 2-5-10). The advantage of the cage is better performance of anti-current than HDPE cage. The disadvantages of the cage are higher cost (considering cost performance), installation difficulty and operation inconvenience. This cage is not used in China because of these disadvantages.

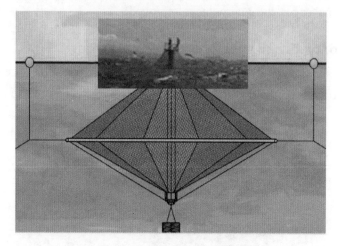

Fig. 2-5-10　Two-cone-shaped net cage

D. Cage made from ropes and floats. The advantage of the cage is lower cost. The disadvantages of the cage are lack of performance platform and large deformation. This cage is mostly used in Zhejiang Province (Fig. 2-5-11) (Cui et al., 2018).

Fig. 2-5-11　Cage made from ropes and floats

E. Steel structure cage (Fig. 2-5-12). The cage is resistant to current, not easy to be broken or fouled. But it is not easy to change and lift the net, since the net is made of metal (Cui et al., 2013).

Fig. 2-5-12 Steel structure cage

F. The flounder fish cage (Fig. 2-5-13). The characteristics of the cage are benthonic habit and bottom platform.

Fig. 2-5-13 Cage for flounder fish

G. Other types of net cage are newly developed. Several kinds of cages are developed for offshore farm in recent years as shown in Fig. 2-5-14.

a. Three floating pipes circular cage; b. Single pipe flexible joint square cage; c. Two floating pipes square cage; d. Steel structure submersible cage.

Fig. 2-5-14　Other types of net cage newly developed

2.4　Cage farm base in China

Several large cage farm companies are established in China such as cage farming base at Tianzheng company (Fig. 2-5-15) in Dalian, Liaoning Province.

Fig. 2-5-15　Cage farming base at Tianzheng company

New environment-friendly plastic materials are used for cage farming base in Ningde,

Fujian Province (Fig. 2-5-16).

Fig. 2-5-16 New environment-friendly plastic materials used in Ningde

3 Supporting facilities of cage aquaculture

3.1 Size grading of fish

The fish in cage can be divided into different group by the grating according to the fish size (Fig. 2-5-17).

a. Grating; b. Grading operation.
Fig. 2-5-17 Size grading of fish

3.2 Netting cleaning

After the cage is put into the sea, shellfish and algae will attach to the netting. We need to clean the net by the equipment shown in Fig. 2-5-18.

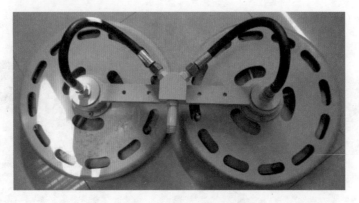

Fig. 2-5-18 Netting cleaning

3.3 Vacuum fish suction pump

Fish catching machine (Fig. 2-5-19) assembles two pots. The parameter of the machine are shown below.

Fig. 2-5-19 Vacuum fish suction pump

A. Pipe diameter: 200 mm.

B. Max flow: 55.48 m^3/h.

C. Volume: 561 L.

D. Pump power: 11 kW.

E. Vacuum degree: −0.08 MPa.

F. Period: 42 s.

G. Fish max weight: 2.0 kg.

H. Max capacity: 10–20 t/h.

3.4 Tuck net

The fish in cage can be caught by the tuck net (Fig. 2-5-20).

a. Norwegian salmon; b. *Fugu* catch ship in Dalian.

Fig. 2-5-20　Tuck net

3.5　Underwater monitoring system

Fig. 21b shows a remote operated vehicle called remoted operated vehicle (ROV). We can see the fish in cage underwater by the ROV (Fig. 2-5-21).

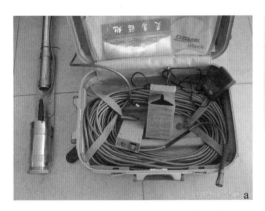 

a. Handheld single camera; b. Remote operated vehicle.

Fig. 2-5-21　Underwater monitoring system

3.6　Environment monitoring system

The system included 5 environmental factors, namely current, temperature, DO, salinity and pH (Fig. 2-5-22).

Fig. 2-5-22 Environment monitoring system

4 Main species farmed in China

Currently, there are about 30 kinds of marine fish cultured in China. Because the water temperature in different sea areas are different obviously, the mariculture fish in different areas are different. The main species farmed in China are divided into two categories: north farming and south farming (Table 2-5-1; Fig. 2-5-23).

Table 2-5-1 Main species farmed in China

Species	North farming	South farming
Pseudoscisena crocea		√
Sciaenops ocellatus		√
Rachycentron canadum		√
Lateolabrax japonicus	√	√
Epnephelus		√
Sebastodes fuscescens	√	
Hexagrammos otakii	√	
Pagrosomus major	√	√
Sparus macprocephahts	√	√
Trachinotus ovatus		√
Fugu rubripes	√	
Paralichthys olivaceus	√	
Scophthalmus maximus	√	√

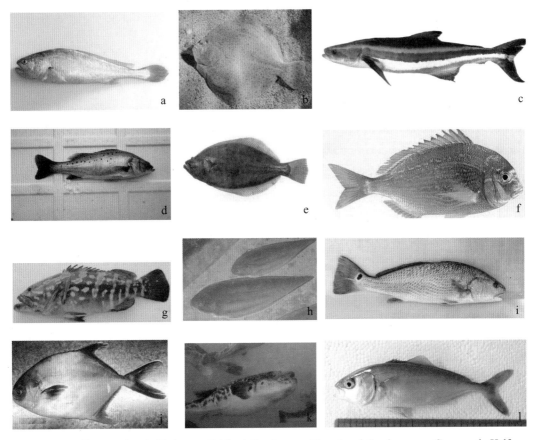

a. Large yellow croaker; b. Turbot; c. Cobia; d. Sea bass; e. Flounder; f. Sea bream; g. Grouper; h. Half smooth tongue sole; i. Red drum; j. Swallowtail; k. *Fugu*; l. Yellowtail.

Fig. 2-5-23   Main species farmed in China

5 Development strategy of cage aquaculture

5.1 Farming scale

Deep-sea cage culture is a scale-efficient industry. Suitable farming scale will not only reduce the management fee and labor cost, but also reduce the installation cost of mooring fittings and supporting facilities.

5.2 Farming fish species

The selection of farming species is closely related to the benefit, which mainly depends on market factors and sea area conditions. Salmon is the dominant species in Norway's cage culture. A fish forms a large industry and the farming species is specialized. This is a way for large enterprises to learn in the future. That is, the main farming species form a large industry, with fresh and processed products in the way of both domestic and foreign markets. In addition to develop the main species, small enterprises should develop and try out new species to meet the market demand.

5.3 Sea conditions

The selection of deep-sea cage aquaculture area should take into account not only its environmental conditions to meet the maximum survival and growth needs of farmed fish, but also the special requirements of breeding methods.

The tidal flow velocity of the sea area is generally 0.5-1 m/s. If the flow rate is greater than 1 m/s, it is easy to generate the deformation of the cage, and it is difficult to keep the shape of the cage and the volume of the cage. The waves have done great damage to the deep sea aquaculture cage. The long wave with regular period exerts less force on the cage. However, the force of flowering and breaking waves is very strong in the shallow water area near the shore. It acts on the cage frame and the netting directly, which often causes the frame to break unevenly and thus causes the damage of the netting.

5.4 Selection of cage specifications

HDPE cage specification and farming capacity are shown in Table 2-5-2.

Table 2-5-2 HDPE cage specification and farming capacity

Cage perimeter/m	Area/m²	Net height/m	Aquatic water /m³	Young fish number ($\times 10^4$)	Production/t
40	127	8	1,016	1.4–2.6	9–16
		7	889	1.2–2.3	8–15
		6	762	1.0–2.0	6–12
50	199	10	1,990	2.6–5.0	15–25
		8	1,592	2.0–4.0	12–20
		6	1,194	1.6–3.0	9–18
80	509	12	6,108	8.2–15.8	48–96
		10	5,090	6.9–13.1	40–80
		8	4,072	5.5–10.5	32–64
120	1,146	15	17,190	23.2–44.4	136–270
		12	13,752	18.6–35.5	108–210
		10	11,460	15.4–30	90–180

6 Equipment and facilities for deep sea or open ocean aquaculture

In recent years, with the strong support of relevant policies for developing deep sea or open ocean culture in China, new marine fish aquaculture models, equipment and facilities, such as the large steel structure pens, aquaculture vessels, large steel structure cages, deep sea farms or platforms, etc., have been developed, which greatly promotes the technology progress of sea-based marine fish aquaculture in China.

Norwegian semi-submerged deep sea aquaculture platform (deep sea farm; Fig. 2-5-24) was made in China. The cage was assembled completely in 2017, and then it was conveyed to Norway by ship. It left Qingdao on June 14, 2017 and arrived in Norwegian sea in September. It was officially delivered for use after completing offshore installation and commissioning on October 6, 2017.

Fig. 2-5-24　Deep sea farm

In May, 2018, a newly developed submersible aquaculture farm named "Shenlan No. 1" (Fig. 2-5-25) was used for salmon and trout farming in the cold water mass area in the Yellow Sea.

Fig. 2-5-25　"Shenlan No. 1"

Large steel structure pen (Fig. 2-5-26) was built in Laizhou, Shandong Province. It consists of 2 rings with the diameter of the outer ring up to 127 m. The total volume of the pen is about 160,000 m^3. Eight platforms built on top of the pen can be used for farming operation, management and recreation fishing.

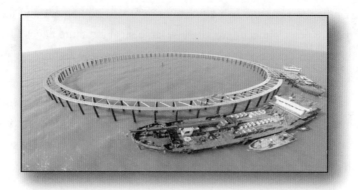

Fig. 2-5-26 Large steel structure pen

Other equipment and facilities are being developed for deep sea or open ocean culture including the aquaculture vessels, combined floating tanks, net cages for cold mass culture, various large steel structure cages, etc (Fig. 2-5-27).

a. Aquaculture vessel; b. Combined floating tanks; c. Special net cage for cold mass culture; d. Large steel structure cage.

Fig. 2-5-27 Other equipment and facilities

References

Cui Y, Guan C T, Huang B, et al. Study on hydrodynamic characteristics of the float-type net enclosure [J]. Fishery Modernization, 2018, 45(5): 14-18.

Cui Y, Guan C T, Wan R, et al. Numerical simulation of a flatfish cage system in waves and currents [J]. Aquacultural Engineering, 2013, 56: 26-33.

Guo Y F, Zhao W W. China fishery statistical yearbook [M]. Beijing: Agricultural Press of China, 2017.

Huang B, Guan C T, Cui Y, et al. The survey and technology resolving of influence on cage farming in Shan Dong caused by Meari typhoon [J]. Fishery Modernization, 2011, 38(4):17-21.

Yuan X C, Zhao W W. China fishery statistical yearbook [M]. Beijing: Agricultural Press of China, 2016.

Research and application of marine ranching technology

Author: Guan Changtao
E-mail: guanct@ysfri.ac.cn

1 General description of marine ranching in China

1.1 What is marine ranching?

Marine ranching is to build or remediate grounds based on the principle of marine ecosystem in certain sea areas for marine organisms to breed, grow, seek foods and dodge predators by means of deploying artificial reefs, stock enhancement and releasing, etc (Ministry of Agriculture and Rural Affairs of the People's Republic of China, 2017). It is proved that marine ranching is an effective way to enhance fishery resources, improve eco-environment and realize the sustainable utilization of fishery resources.

In my mind, however, the concept of marine ranching is to create a kind of fishery just like herding animals on grassland, as showed in Fig. 2-6-1.

a. Grassland for land animals; b. Ground for marine organisms.
Fig. 2-6-1 Marine ranching is to create a kind of fishery like herding animals on grassland

1.2 Why construct marine ranches?

As is known to all, China is the biggest fishery country in the world. So the sustainable development of fishery is very important for the economy and society along the coast of China. However, long-term overfishing, environment pollution and reclamation for land use have caused the bareness in seabed, desertification of water area and destruction of fish habitat, which accelerated the deterioration of marine ecosystem and the reduction of fishery resources (Fig. 2-6-2). Facing the problems, China has taken a series of measures to protect the marine environment and enhance the fishery resources. Construction of marine ranches is one of the most important measures in recent years.

a. Petroleum pollution; b. Overfishing; c. Reclamation for land use.
Fig. 2-6-2 Examples of environment pollution, reclamation and overfishing

1.3 General development of marine ranching in China

Marine ranching in China can be traced back to 1979, when the first set of artificial reefs were put into the sea in the Bay of Beibuwan in Guangxi. Generally, the construction of artificial reefs and the development of marine ranching in China can be divided into three periods.

The first period (1981-1987): This was the experimental stage and the construction of artificial reefs was in a small scale, while rich materials and experiences both in theory and in practice were acquired.

The second period (1988-1999): The construction of artificial reefs was stopped for insufficient understanding on the resources and environmental protection and lack of investment.

The third period (2000-present): Studies, experiments and large scale marine ranching based on the construction of artificial reefs have been carrying out.

By the year of 2018, the total investment from the government and companies on marine ranching in China had been over 50 billion RMB. More than 250 marine ranches had been constructed and the total reef volume had been over 60 million hollow cubic meters covered an area of 860 km^2 (Yang et al., 2019).

In the Bohai Sea and the Yellow Sea, 136 ranches covering a total area of 350 km^2 have been constructed with the total reef volume up to 1.8×10^7 m^3.

In the East China Sea, 23 ranches covering a total area of 240 km^2 have been constructed with the total reef volume up to 7×10^5 m^3.

In the South China Sea, 74 ranches covering a total area of 270 km^2 have been constructed with the total reef volume up to 4.2×10^7 m^3.

1.4 Types of artificial reefs used in China

Many types of artificial reefs have been used for the construction of marine ranching in China. Figs. 2-6-3 to 2-6-7 show some of the typical reef structures.

Fig. 2-6-3　Ship-shaped reef

Fig. 2-6-4　Roof-shaped reef

Fig. 2-6-5　Cubic structure reef

Fig. 2-6-6　Assembly cylinder reef

Fig. 2-6-7 Pyramid structure reef

2 Key technologies for marine ranching

Key technologies for marine ranching are shown in Fig. 2-6-8.

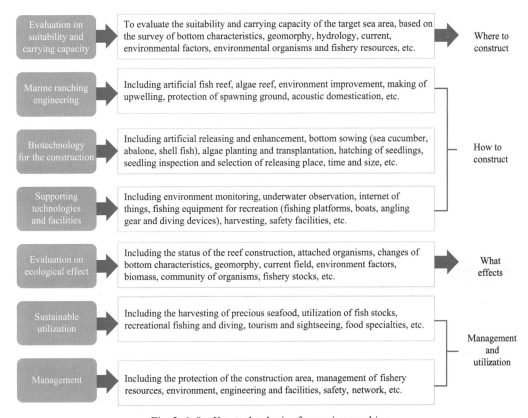

Fig. 2-6-8 Key technologies for marine ranching

3 Research progress in marine ranching

3.1 Evaluation on the suitability and carrying capacity of the sea area

Several research teams from the research institutions and universities have been carried out the investigations and studies on the suitability and carrying capacity of the environment for marine ranching. The evaluation methods have been established. Fig. 2-6-9 shows the site investigation of the marine ranching in Laizhou Bay.

Fig. 2-6-9　Site investigation of the marine ranching in Laizhou Bay

3.2　Investigations on the effects of artificial reefs set up in the early 1980s

Investigations on the effects of artificial reefs set up in the early 1980s were carried out (Fig. 2-6-10 to 2-6-12). The current status, effects and functions of these reefs stayed in the sea bed for more than 20 years were acquired, which provided us rich references and experiences for the further research and construction of artificial reefs.

Fig. 2-6-10　Investigations on the effects of artificial reefs set up in the early 1980s

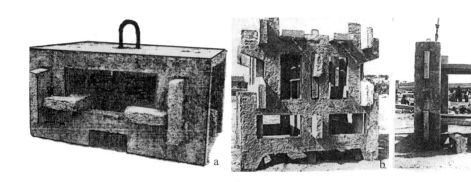

a. Sea cucumber reef in Penglai, Yantai; b. Fish reef of type 81-1; c. Fish reef of type 81-2; d. Cubic reef in Lingshan Island, Qingdao; e. Artificial reef assembly in Penglai, Yantai.

Fig. 2-6-11　Artificial reefs set up in the early 1980s

a. Star-shaped reef; b. Triangle reef; c. Double M-shaped reef.

Fig. 2-6-12　Lifting the reef up from the sea bed for investigation

3.3　Physical tests and numerical modeling on hydrodynamics of artificial reefs

Physical tests and numerical modeling on hydrodynamics of different artificial reef structures, such as star-shaped, concrete pipe, triangle, cubic and pyramid reefs have been carried out (Figs. 2-6-13 to 2-6-18). The results have been applied for the selection of reef structures and design of reef distribution (Liu et al., 2010; Zheng et al., 2015; Li et al., 2017).

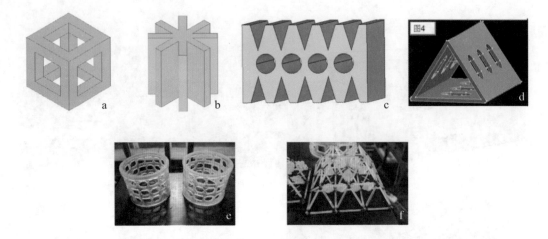

a. Cubic structure reef; b. Star-shaped reef; c. Double M-shaped reef; d. Triangle reef; e. Assembly cylinder reef; f. Pyramid structure reef.

Fig. 2-6-13　Physical tests and numerical modeling on hydrodynamics of different artificial reef structures

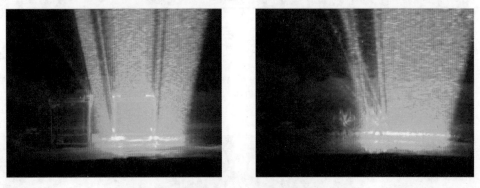

Fig. 2-6-14　PIV test on flow fields of artificial reefs

Fig. 2-6-15　Test on physical stability in waves and currents

Fig. 2-6-16 Test on hydrodynamics of combined reefs

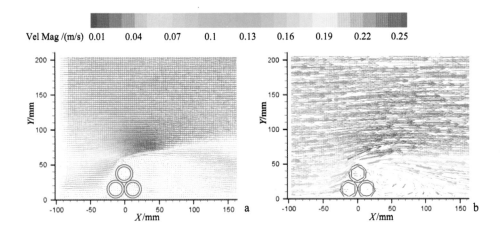

a. Physical tests; b. Numerical modeling.

Fig. 2-6-17 Numerical modeling on hydrodynamics of artificial reefs

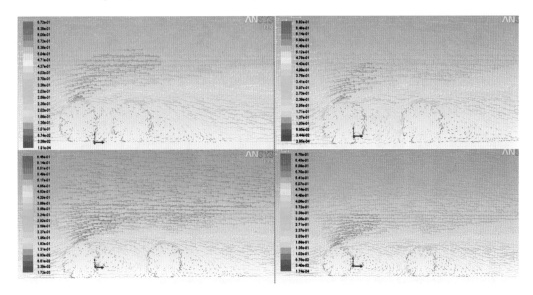

Fig. 2-6-18 Numerical simulation on flow field of combined star reefs in different distances

3.4 Experimental studies on the attractive effect and organisms attaching on artificial reefs

Experimental studies on the attractive effect of sea cucumber and attaching effect of organisms on artificial reefs with different shapes and materials were carried out (Fig. 2-6-19). The results provide basis for the selection of reef structures in accordance with different types of marine ranches constructed for different functions and in different sea conditions (Cui et al., 2010; Wang et al., 2019).

a. Attractive effect of sea cucumber on artificial reefs with different shapes and materials; b. Reefs for organisms attaching test; c. Reefs being lifted for observation.

Fig. 2-6-19 Experimental studies on the attractive effect and organisms attaching on artificial reefs

3.5 Development of new artificial reefs and supporting facilities for marine ranching

New types of artificial reefs and supporting facilities, such as the enhancement facilities, submersible net cage, multifunctional platform, remote control underwater camera and environment monitoring system have been developed.

3.5.1 New types of artificial reefs

With the rapid development of marine ranching in China, many new types of artificial reefs have been designed, manufactured and put into use. Fig. 2-6-20 shows some of the newly developed artificial reefs used in China.

a. PE pipe filled with concrete; b. Star-shaped reef; c. Cubic reef with heightened structure; d. Steel structure reef; e. Assembly cylinder reef.

Fig. 2-6-20 Some of the new types of artificial reefs developed in China

3.5.2 Facilities for enhancement of rare sea animals

Enhancement facilities, such as the facility for the rare sea animals (Fig. 2-6-21) and the submersible cage for sea cucumber farming (Fig. 2-6-22), have been developed and used for marine ranching.

Fig. 2-6-21　Enhancement facility for rare sea animals

Fig. 2-6-22　Submersible sea cucumber farming cage

3.5.3　Remote environment monitoring system

Environment monitoring system for marine ranching area has been developed. The system can continuously monitor the temperature, salinity, dissolved oxygen, pH value and current velocity, and transmit the data remotely (Fig. 2-6-23).

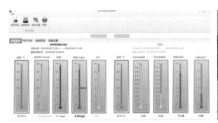  

Fig. 2-6-23　Products of remote environment monitoring system

3.5.4　Multifunctional platform for management and leisure fishing

To meet the need of management, monitoring and leisure fishing in marine ranching areas, multifunctional platforms (Fig. 2-6-24) have been developed.

Fig. 2-6-24　Multifunctional platforms used for marine ranching in China

3.6　Studies on planting and transplantation of sea weed and algae

In order to rebuild and remediate the coastal sea grass fields, we have developed technologies for the planting and transplantation of sea weed and algae including the species as eelgrass (*Zostera marina* L.), kelp (*Laminaria japonica*), *Sargassum thunbergii* and *Sargassum fusiforme* etc, and achieved obvious effects (Fig. 2-6-25).

Fig. 2-6-25　Effects of algae transplantation

3.7　Studies on stock enhancement and fish releasing

We carried out studies on stock enhancement and fish releasing with many species, such as shrimp (*Fenneropenaeus chinensis*), jelly fish (*Rhopilema esculenta*), swimming crab (*Portunus trituberculatus*), cuttlefish (*Sepia esculenta*) and other fishes (Luo et al., 2016). Species and sizes for fish releasing in the Bohai Sea and the Yellow Sea are given in Table 2-6-1.

Table 2-6-1 Species and sizes for fish releasing in the Bohai Sea and the Yellow Sea

Species	Size /mm		Releasing time/month	
	Shandong	Hebei and Tianjin	Shandong	Hebei and Tianjin
Paralichthys olivaceus	50	30, 50	7–9	6–7
Pseudopleuronectes yokohamae	50		9–10	
Cynoglossus semilaevis	50		70–10	
Liza haematocheila	50	30	6–7	6–7
Pagrosomus major	50	20	6–7	6
Sparus macrocephalus	50		7–9	
Trachidermus fasciatus	50		4–5	
Sebastes schlegeli			6–7	
Takifugu rubripes		40		6–7

3.8 Investigation and evaluation on ecological effects of marine ranching

More than 50 surveys on biological resources and ecological environment of marine ranches have been made, and the evaluation method on ecological effects of marine ranching has been established (Li et al., 2018; Zhao et al., 2019). Figs. 2-6-26 to 2-6-27 show some of the process of the surveys.

a. Sampling of organisms; b. Current survey; c. Water environment survey; d. Sampling of fishery resources.

Fig. 2-6-26 Surveys on the effects of marine ranching

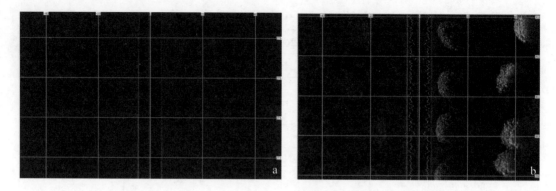

a. Sea bed without reef construction; b. Sea bed with artificial reefs construction.
Fig. 2-6-27 Survey on the deployment of artificial reefs by C3D-LPM sonar imaging system

4 Application of marine ranching technology

The above achievements and technologies have provided strong support for the construction and planning of marine ranches in the area of the Bohai Sea and the Yellow Sea, even all over the coast of China.

4.1 Marine ranch in Laizhou Bay

The demonstration area is located in Laizhou Bay, constructed by Laizhou Mingbo Aquatic Products Co., Ltd., with a total area of 3,300 hm^2 (Fig. 2-6-28).

Fig. 2-6-28 Marine ranch in Laizhou Bay

4.2 Marine ranch in Beidaihe, Hebei Province

The demonstration area is located in Beidaihe, Hebei Province, with a total area of 96.3 hm^2 (Fig. 2-6-29).

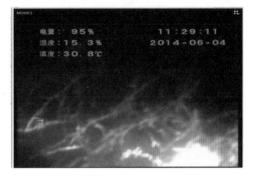

Fig. 2-6-29 Marine ranch in Beidaihe, Hebei Province

4.3 Marine ranch in Changdao Island, Shandong Province

The demonstration area is located in Changdao Island, Shandong Province, with a total area of 100 hm^2 and water depth of 18-24 m. Large reefs were constructed in the ranch to form a conservation area (Fig. 2-6-30).

Fig. 2-6-30 Marine ranch in Changdao Island, Shandong Province

4.4 Qingdao Luhaifeng Marine Ranch

The demonstration area is located in Qingdao, with a total area of 3,500 hm^2. The ranch has been approved as the National Marine Ranching Demonstration Areas and the Recreational Fishing Base of Shandong Province (Figs. 2-6-31 to 2-6-33).

Fig. 2-6-31　Qingdao Luhaifeng Marine Ranch

a. Yu Kangzhen, vice minister of the Ministry of Agriculture and Rural Affairs of China came to the ranch for investigation; b. Li Shumin, deputy director general, fisheries administration of the Ministry of Agriculture and Rural Affairs of China came to the ranch for investigation; c. Visitors in the angling platform; d. Sonar scan over the reefs in sea bed.

Fig. 2-6-32　The success of Luhaifeng Marine Ranch attracted wide attention of leaders, researchers and vistors

Fig. 2-6-33 On May 25, 2016, a fishery delegation from Western Australia visited the marine ranching base after participating the International Forum on Marine Fishery Resources Conservation and Artificial Reef held in Qingdao

4.5 Planning of marine ranching in Changhai County and construction of marine ranch in Zhangzidao Island, Dalian, Liaoning Province

The total planning area of marine ranching in Changhai County is up to 10,000 km² and the planning report was approved in April 2016. Moreover, the marine ranch around the Zhangzidao Island has been in scale (Figs. 2-6-34 to 2-6-35).

Fig. 2-6-34 Planning of marine ranching and the marine ranch in Zhangzidao Island, Dalian, Liaoning Province

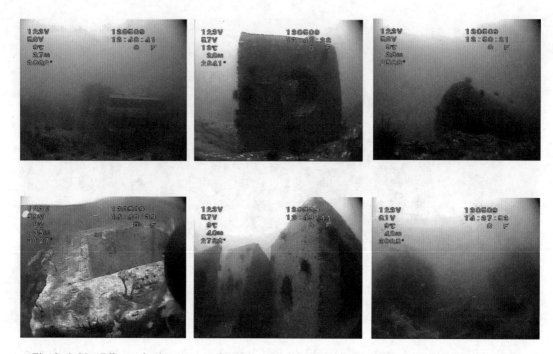

Fig. 2-6-35   Effects of enhancement of marine animals and algae around the reefs in Zhangzidao Marine Ranch

References

Cui Y, Guan C T, Wan R, et al. The study of attractive effects of artificial reef models on *Apostichopus japonicas* [J]. Progress in Fishery Sciences, 2010, 31(2): 109-113 (In Chinese).

Liu Y, Guan C T, Zhao Y P, et al. Experimental study on two-dimensional flow field of the star artificial reef in the water stream with PIV [J]. Chinese Journal of Hydrodynamics, 2010, 25(6): 777-783 (In Chinese).

Luo G, Zhuang P, Zhao F, et al. Development status, existing issues and countermeasure in the selection of suitable species for stock enhancement [J]. Marine Fisheries, 2016, 38(5): 551-560.

Li J, Zhang Y, Yuan W, et al. Research on artificial reef ecosystem health assessment based on fuzzy synthetic evaluation [J]. Progress in Fishery Sciences, 2018, 39(5): 10-19.

Li J, Zheng Y X, Gong P H, et al. Numerical simulation and PIV experimental study of the effect of flow fields around tube artificial reefs [J]. Ocean Engineering, 2017, 134: 96-104.

Ministry of Agriculture and Rural Affairs of the People's Republic of China. SC/T 9111-2017 Fishery industry standard of the People's Republic of China. Classification of marine ranching [S]. Beijing, 2017.

Wang Z, Gong P H, Guan C T, et al. Effect of different artificial reefs on the community structure of organisms in Shique Beach of Qingdao [J]. Progress in Fishery Sciences, 2019,

40(4): 163-171.

Yang H S, Zhang S Y, Zhang X M, et al. Strategic thinking on the construction of modern marine ranching in China [J]. Journal of fisheries of China, 2019, 43(4): 1255-1262 (In Chinese).

Zhao R R, Gong P H, Zhang Y, et al. Ecosystem health assessment of artificial reef area in Long Island [J]. Progress in Fishery Sciences, 2019, 40(6): 9-17.

Zheng Y X, Liang Z L, Guan C T, et al. Numerical simulation and experimental study of the effects of disposal space on the flow field around the combined three-tube reefs [J], China Ocean Engineering, 2015, 29(3): 445-458.

Chapter 3
Maricultural organism disease control and molecular pathology

Activities for WSD and IHHN OIE reference laboratory

Author: Yang Bing
E-mail: yangbing@ysfri.ac.cn

1 The world organization for animal health (OIE)

1.1 What is OIE ?

OIE is the intergovernmental organization responsible for improving animal health worldwide. It is recognized as a reference organisation by the World Trade Organization (WTO) and in 2019 it had a total of 182 members. The OIE maintains permanent relations with 71 other international and regional organizations and has regional and sub-regional offices on every continent.

As a global leader for animal health and welfare standards, the OIE plays an influential role on animal disease prevention, control and information sharing. It comprises the delegates of all member countries and meets at least once a year. The General Session of the Assembly lasts five days and is held every year in May in Paris.

Asia and the Pacific Regional Offices is in Tokyo, Japan. OIE Regional Representative for Asia and the Pacific is Dr Hirofumi Kugita.

1.2 The objectives of the OIE

The objectives of the OIE are to ensure transparency in the global animal disease situation; collect, analyse and disseminate veterinary scientific information; encourage international solidarity in the control of animal diseases; safeguard world trade by publishing health standards for international trade in animals and animal products; improve the legal framework and resources of national veterinary services; provide a better guarantee of food of animal origin; and promote animal welfare through a science-based approach.

1.3 International standards

The *Terrestrial Animal Health Code* and *Aquatic Animal Health Code* respectively aim to assure the sanitary safety of international trade in terrestrial animals, aquatic animals and their products.

The *Manual of Diagnostic Tests and Vaccines for Terrestrial Animals* and the *Manual of Diagnostic Tests for Aquatic Animals* provide a harmonized approach to disease diagnosis by describing internationally agreed laboratory diagnostic techniques.

The OIE regularly updates its international standards as new scientific information comes to light, following its established transparent and democratic procedures. The only pathway for adoption of a standard is via the approval of the World Assembly of Delegates Meeting in May each year at the OIE General Assembly.

1.4 Overview of OIE reference centers

1.4.1 OIE reference laboratories

The principal mandate of OIE reference laboratories is to function as world reference centres of expertise on designated pathogens or diseases.

1.4.2 OIE collaborating centers

The principal mandate of OIE collaborating centres is to function as world centres of research, expertise, standardization of techniques and dissemination of knowledge on a specialty. Two or several reference centres may be designated as an OIE Reference Centres Network, following the Guidance for the Management of OIE Reference Centre Networks.

The network of collaborating centres and reference laboratories constitutes the core of OIE scientific expertise and excellence.

2 OIE reference laboratories for aquatic animal diseases in China

The OIE reference laboratories for aquatic animal disease in China are the Infectious Haematopoietic Necrosis and Spring Viraemia of Carp, and the Infectious Hypodermal and Haematopoietic Necrosis and White Spot Disease.

3 OIE reference laboratories for WSD and IHHN

3.1 OIE expert for WSD and IHHN

The OIE expert for WSD and IHHV is Dr. Jie Huang, who was a former member of Aquatic Animal Health Standards Commission (2009–2012) and the former vice president of Aquatic Animal Health Standards Commission (2012–2015).

3.2 OIE reference laboratories for WSD and IHHN

OIE reference laboratories for WSD and IHHN were designated in May 2011 (Fig. 3-1-1).

Fig. 3-1-1 The opening ceremony of OIE reference laboratory

3.3 Quality management system of OIE reference laboratories for WSD and IHHN

The Quality management system (QMS) of OIE reference laboratories for WSD and IHHN has been accredited by China National Accreditation Service for Conformity Assessment (CNAS) in accordance with ISO/IEC 17025: 2005 in March 2018 (Fig. 3-1-2)

Fig. 3-1-2 Laboratory accreditation certificate

3.4 Works of OIE reference laboratories for WSD and IHHN

Annual reports of activities of OIE reference laboratories for WSD and IHHN have been submitted to OIE since 2011.

3.4.1 Application of WSD diagnostic methods

WSD diagnostic methods were used in OIE reference laboratories including PCR, qPCR, LAMP, histopathology, IAH, bioassay, TEM and T-E staining in 2011-2018 (Table 3-1-1) (OIE, 2019).

Table 3-1-1 WSD diagnostic methods (2011-2018)

Method	2011	2012	2013	2014	2015	2016	2017	2018
PCR	51	339	285	558	591	528	—	—
qPCR	540	30	6	—	—	—	—	—
LAMP	230	18	15	6	35	—	—	20
Histopathology	6	4	43	476	—	60	—	—
ISH	—	—	—	—	—	—	—	—
Bioassay	16	34	6	—	—	—	—	—
TEM	—	—	—	—	—	31	—	—
T-E staining	—	—	4	—	—	7	—	—

3.4.2 Application of IHHN diagnostic methods

IHHN diagnostic methods were used in OIE reference laboratories including PCR, qPCR, LAMP, histopathology and bioassay in 2011-2018 (Table 3-1-2) (OIE, 2019).

Table 3-1-2 IHHN diagnostic methods (2011-2018)

Method	2011	2012	2013	2014	2015	2016	2017	2018
PCR	51	194	278	—	—	—	—	—
qPCR	—	—	20	—	—	—	—	—
LAMP	225	—	16	476	—	—	—	—
Histopathology	6	4	40	—	—	—	—	—
Bioassay	—	—	20	—	—	—	—	—

3.4.3 Diagnostic reagents preparation and distribution

A series of field-used high sensitive detection kits (Fig. 3-1-3) were made for detection of WSSV and IHHNV.

A total of 3,318 and 769 field-used high sensitive detection kits for WSSV and IHHNV were provided for shrimp farms, hatcheries, aquaculture disease control station and some research laboratories.

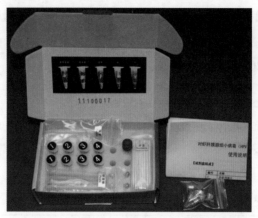

Fig. 3-1-3　High sensitive detection kit for HPV

3.4.4　Contribution for OIE standards

During 2011-2018, OIE reference laboratories contributed a series of comments and suggestions for OIE standards (Table 3-1-3).

Table 3-1-3　Comments and suggestions for OIE standards

Year	Number of comments and suggestions	Field of comments and suggestions
2011	10	WSD
	4	IHHN
	44	Other aquatic manual chapters
2012	17	OIE aquatic code
2013	2	OIE aquatic code
2014	6	OIE aquatic code
2015	6	WSD
	2	IHHN
	3	OIE aquatic code
2016	3	IHHN
	7	OIE aquatic code
	12	Other aquatic manual chapters
2017	4	IHHN
	8	WSD
	7	OIE aquatic code
2018	1	IHHN
	7	WSD
	2	OIE aquatic code

3.4.5 The reference materials of WSSV provided by OIE reference laboratory

Provision of the reference materials of WSSV

The reference materials of WSSV is shown in Table 3-1-4.

Table 3-1-4　Reference meterials of WSSV (2011-2018)

Year	Type of material	Quantity	Purpose
2011	WSSV-infected tissue	100 g	Positive infected shrimp tissue
	Purified WSSV preparation	1 tube	Positive control
	for VP24 and polyclonal antibodies VP28 of WSSV	2 tubes	Positive control
2012	WSSV-infected tissue	170 g	Infection experiment and positive control
2013	WSSV-infected tissue	30 g	Infection experiment
2014	WSSV-infected tissue	160 g	Infection experiment
2015	WSSV-infected tissue	18 g	Infection experiment and positive control
2016	WSSV DNA	1×10^6 reactions	positive control for PT
	WSSV-infected tissue	100 g	Infection experiment
2017	WSSV DNA	1×10^5 reactions	positive control for PT
	WSSV-infected tissue	15 g	Infection experiment
2018	WSSV DNA	1×10^4 reactions	Positive control
	WSSV-infected tissue	22 g	Infection experiment

Provision of the reference materials of IHHNV

The reference materials of IHHNV is shown in Table 3-1-5.

Table 3-1-5　Reference materials of IHHNV (2011-2018)

Year	Type of material	Quantity	Purpose
2011	IHHNV-infected tissue	30 g	Positive infected shrimp tissue
	IHHNV DNA	1 tube	Positive control
2012	IHHNV-infected tissue	20 g	Positive infected shrimp tissue
2013	IHHNV-infected tissue	1 g	Positive control
2014	IHHNV-infected tissue	2×50 g	Infection experiment
	IHHNV DNA	3×185 μL	Positive control
2015	IHHNV-infected tissue	6 g	Positive infected shrimp tissue
2016	IHHNV DNA	1×10^6 reactions	Positive control
2017	IHHNV DNA	1×10^4 reactions	Positive control
2018	IHHNV DNA	1×10^5 reactions	Positive control
	IHHNV-infected tissue	100 mg	Positive infected shrimp tissue

3.4.6 Training courses of disease diagnosis technology and detection method of pathogens

Training courses for Southeast Asian countries were held in 2016 and 2018 (Fig. 3-1-4).

a. Trainee of Southeast Asian countries in 2016; b. 2016 Southeast Asia shrimp disease diagnosis technology training course in 2018; c, d. China ASEAN Maritime Cooperation training of rapid detection for aquaculture pathogens in 2018.

Fig. 3-1-4　Training courses for Southeast Asian countries and African countries

3.4.7 Southeast Asia Marine Cultivation Technology Training Workshop

Operational training of rapid detection for aquaculture pathogens in Southeast Asia Marine Cultivation Technology Training Workshop was held in 2018 (Fig. 3-1-5).

Fig. 3-1-5　In Southeast Asia Marine Cultivation Technology Training Workshop

3.4.8 Regional proficiency testing programme for aquatic animal disease laboratories in the Asia-Pacific

A. Objectives: The objectives are to establish a regional laboratory proficiency testing programme and to strengthen the capability for molecular diagnosis of important aquatic animal diseases in Asia.

B. Expected outcomes: The programme is expected to improve diagnostic capability for significant aquatic animal diseases, increase confidence between trading partners, certify the disease status of commodity exports, meet quarantine requirements, ensure the sanitary safety of trade appropriate pre-border measures and reduce spread of disease.

C. Timetable: Timetable of regional proficiency testing program for aquatic animal disease laboratories in the Asia-Pacific was held in 2013 to 2015 (Table 3-1-6).

Table 3-1-6 Timetable of regional proficiency testing programme

Round	Milestone	Date
1	Test samples prepared and QC complete	Apr. 2013
1	Test panels distributed	May 2013
1	Report of test results sent to participants	Aug. 2013
2	Test samples prepared and QC complete	Oct. 2013
2	Test panels distributed	Nov. 2013
2	Report of test results sent to participants	Jan. 2014
3	Test samples prepared and QC complete	Apr. 2014
3	Test panels distributed	May 2014
3	Report of test results sent to participants	Jul. 2014
4	Test samples prepared and QC complete	Oct. 2014
4	Test panels distributed	Nov. 2014
4	Report of test results sent to participants	Jan. 2015
	Final project report	Apr. 2015

D. Sample selection: Sample selection of regional proficiency testing program for aquatic animal disease laboratories in the Asia-Pacific participated in 2013-2015 (Table 3-1-7).

Table 3-1-7 Sample selection of regional proficiency testing programme

Sample panel	Required
White spot syndrome virus (WSSV) PCR	√
Yellow head virus (YHV) PCR	√
Taura syndrome virus (TSV) PCR	√
Infectious myonecrosis virus (IMNV) PCR	√
Infectious hypodermal and haematopoietic necrosis virus (IHHNV) PCR	√
Megalocytiviruses (RSIV, ISKNV, GIV, etc.) PCR	√
Nervous necrosis viruses (NNV) PCR	√
Koi herpesvirus (CyHV-3) PCR	√

(*to be continued*)

Sample panel	Required
Macrobrachium rosenbergii nodavirus (MrNV) and extra small virus (XSV) PCR	√
Spring viraemia of carp virus (SVCV) PCR	√

E. Results: The sample results of regional proficiency testing program for aquatic animal disease laboratories in the Asia-Pacific are satisfactory.

3.4.9 National testing programme for aquatic animal disease laboratories in aquatic animal epidemic prevention system of China

WSSV and IHHNV testing for aquatic animal disease laboratories in China worked since 2014-2019 (Tables 3-1-8 to 3-1-9).

Table 3-1-8 WSSV testing for aquatic animal disease laboratories in aquatic animal epidemic prevention system of China (2014-2019)

Year	Local fisheries, technology promotion departments and aquatic animal epidemic prevention and detection laboratories	Laboratories in colleges and universities	Laboratories of scientific research institutes	Customs laboratory	Others	Total
2014	17	3	5	2	—	27
2015	34	4	10	5	—	53
2016	55	4	11	8	—	78
2017	106	4	12	7	2	131
2018	97	5	11	5	0	118
2019	82	7	13	6	17	125

Table 3-1-9 IHHNV testing for aquatic animal disease laboratories in aquatic animal epidemic prevention system of China (2014-2019)

Year	Local fisheries, technology promotion departments and aquatic animal epidemic prevention and detection laboratories	Laboratories in colleges and universities	Laboratories of scientific research institutes	Customs laboratory	Others	Total
2014	12	3	3	2	—	20
2015	22	4	8	4	—	38
2016	42	5	8	5	—	60
2017	60	3	7	8	5	83
2018	58	4	12	5	1	80
2019	—	—	—	—	—	—

3.4.10 Proficiency testing on IHHNV with University of Arizona in 2018

Proficiency testing on IHHNV with University of Arizona was held in 2018 and the evaluation results were satisfactory.

3.4.11 Diagnostic technology training with the National Fisheries Health Agency of Peru (SANIPES)

The diagnostic technology training with the National Fisheries Heath Agency of Peru is shown in Fig. 3-1-6.

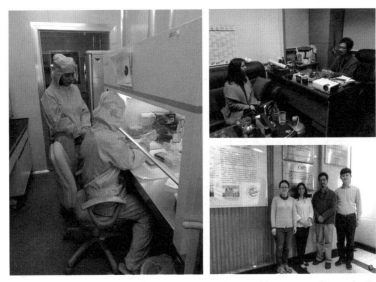

a. Diagnostic technology training with the National Fisheries Health Agency of Peru; b. Cooperation discussion; c. Experts of Peru visiting OIE reference laboratory.

Fig. 3-1-6　Diagnostic technology training with the National Fisheries Heath Agency of Peru

3.4.12 Twinning project on IHHN and WSD with Indonesia

The project was initially approved by OIE and the contract of project has been signed by its representatives of Yellow Sea Fisheries Research Institute, Chinese Academy of Fishery Sciences (YSFRI), China, the Fish Quarantine and Inspection-Standard Examination Laboratory (FQI-SEL), Indonesia, and Dr. Eloit the OIE Director General on August 20, 2019.

The project began with the project leader and one staff member from the Parent Laboratory travelling to the Candidate Laboratory for a period of five days (August 26 to August 30, 2019). During this visit, the Candidate Laboratory held a formal Project Launch Meeting on 27 August 2019 with Director and staff of the Centre (Fig. 3-1-7). A comprehensive audit of the Candidate Laboratory was carried out to examine the listing staff, their skills and also the current physical resources. It is recognized that the Candidate Laboratory has highly competent staff that can easily support all of the objectives of the project (Fig. 3-1-8).

Fig. 3-1-7　Project launch meeting in Jakarta, Indonesia in 2019

Fig. 3-1-8　Project meeting in Jakarta, Indonesia in 2019

References

OIE.Manual of diagnostic tests for aquatic animals. (2019) [M/OL]. https://www.oie.int/standard-setting/aquatic-manual/access-online/.

Competence criteria of aquatic animal disease testing laboratory

Author: Wan Xiaoyuan
E-mail: wanxy@ysfri.ac.cn

1 Laboratory accreditation

1.1 What is CNAS?

China National Accreditation Service for Conformity Assessment (CNAS) is the national accreditation body of China unitarily responsible for the accreditation of certification bodies, laboratories and inspection bodies, which is established under the approval of the Certification and Accreditation Administration of the People's Republic of China (CNCA) and authorized by CNCA in accordance with the Regulations of the People's Republic of China on Certification and Accreditation.

The purpose of CNAS is to promote conformity assessment bodies and strengthen their development in accordance with the requirements of applicable standards and specifications, and to facilitate the conformity assessment bodies to effectively provide service to the society by means of impartial conduct, scientific means and accurate results.

1.2 CNAS organizational structure

CNAS organizational structure is shown as Fig. 3-2-1.

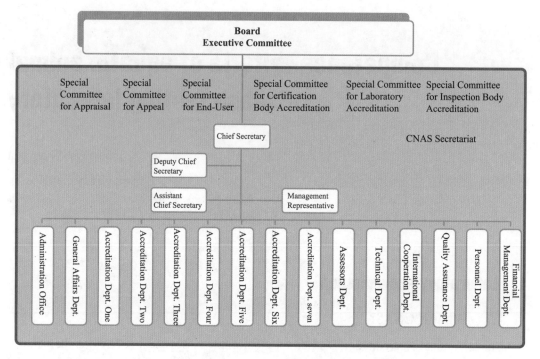

Fig. 3-2-1　CNAS organizational structure
(Date from https://www.cnas.org.cn/english/introduction/index.shtml)

1.3　Mutual recognition arrangement (MRA)

China national accreditation system for conformity assessment has been a part of international accreditation multilateral recognition system, and is playing an important role in it.

CNAS was the accreditation body member of the International Accreditation Forum (IAF) and the International Laboratory Accreditation Cooperation (ILAC), as well as the member of the Asia Pacific Laboratory Accreditation Cooperation (APLAC) and the Pacific Accreditation Cooperation (PAC). The Asia Pacific Accreditation Cooperation (APAC) was established on 1 January, 2019 by the amalgamation of two former regional accreditation cooperations—the APLAC and the PAC (Date from https://www.cnas.org.cn/english/intemation alcooperations/index.shtml).

1.4　Labortory and relevant bodies accreditation

Labortory and relevant bodies accreditation includes accreditation of testing laboratories (including accreditation of forensic unit), accreditation of calibration laboratories, accreditation of medical laboratories, accreditation of proficiency testing providers, accreditation of reference material providers, accreditation of laboratory safety, and accreditation of good laboratory practice.

1.5　What is laboratory accreditation?

Laboratory accreditation is a third-part attestation to a laboratory conveying formal demonstration of its competence to carry out specified testing, measurement and calibration. CNAS provides a means of determining, recognizing and promoting the competence of laboratories, thus providing a ready means for customers to access reliable testing and calibration services.

1.6　The criterion of laboratory accreditation

ISO/IEC 17025:2017 specifies the general requirements for the competence, impartiality and consistent operation of laboratories. ISO/IEC 17025:2017 is applicable to all organizations performing laboratory activities, regardless of the number of personnel. Laboratory customers, regulatory authorities, organizations and schemes use peer-assessment, accreditation bodies, and others use ISO/IEC 17025:2017 in confirming or recognizing the competence of laboratories.

1.7　The requirement of quality management for OIE reference laboratory

Five critical points have been identified for the evaluation of a labortory's performance:

A. Submission of an annual report;

B. Accreditation to the ISO 17025: 2017 or equivalent quality management system (from January 2018);

C. Response to requests from the OIE headquarters for scientific expertise (*e.g.*, revision of the *Aquatic Manual* chapters);

D. Diagnostic activity or production and supply of reference materials related to the disease or pathogen.

E. No response to requests from the OIE for administrative issues relating to transparency and confidentiality (*e.g.*, not renewing the potential conflicts of interest declaration or providing a confidentiality undertaking).

2　Quality management in aquatic animal disease testing laboratories

2.1　The guarantee of the quality of the test results

The guarantee of the quality of test results is shown in Fig. 3-2-2.

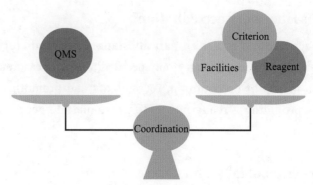

Fig. 3-2-2　Balance of key elements in quality management system (QMS)

2.2　Quality assurance, quality control and proficiency testing

Quality assurance (QA) is process-orientated and ensures the right things being done in the right way. Quality control (QC) is test-orientated and ensures the results being as expected.

Proficiency testing (PT), sometimes referred to as external quality assurance (EQA), provides an independent assessment of the testing methods used and the level of staff competence.

A new aquatic animal disease proficiency testing programme (Fig. 3-2-3) conducted by the Australian Government was assessed against ten priority fish and crustacean diseases from 2018 to 2021.

Fig. 3-2-3　2019 Asia-Pacific laboratory proficiency testing workshop

2.3　Test methods

The national standard and fishery industry standard of China are adopted by the Ministry of Agriculture and Rural Affairs of the People's Republic of China. The international standard is adopted by OIE World Organisation for Animal Health. ISO/IEC 17025 requires the use of appropriate test methods and has requirements for their selection, development and validation to show fitness for purpose. In the veterinary profession, except *Aquatic Manual* of OIE, other standard methods (published in international regional or national standards) or fully validated methods may be preferable to use. Even with the use of standard methods, some in-house evaluation, optimisation and/or validation generally must be done to ensure valid results.

2.4 Facilities and environment

The quality management system should cover all areas of activities affecting the testings done at the laboratory.

As the standard in the field of aquatic animal disease testing usually provides highly sensitive methods for measuring gene amplifications in pathogens, the PCR laboratory can be divided into three or four rooms or areas for reagent preparation, sample preparation, gene amplification, and product analysis (Fig. 3-2-4). The rooms or areas of the PCR laboratory should be independently separated, and there is no air flow through the separated rooms or areas.

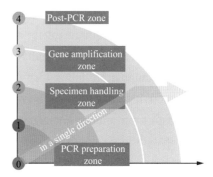

Fig. 3-2-4 Division of PCR laboratory functional areas

2.5 Facilities and environment

The equipment configuration for PCR laboratories is as follows (Table 3-2-1).

Table 3-2-1 The equipment configuration for PCR laboratories

PCR preparation zone	Specimen handling zone	Gene amplification zone	Post-PCR zone
Pipette, refrigerator, balance, low speed centrifuge, vortex mixer, mobile UV lamp	Pipette, table centrifuge (high/low speed), constant temperature devices (water bath or dry bath), refrigerator, vortex mixer, mobile UV lamp	Thermal cycler system for nucleic acid amplification (quantitative fluorescence detection or common)	Pipette, capillary electrophoresis system, transblot system, molecule hybridization chest/oven, water bath, DNA sequencer

3 Element of quality management system and our experience

3.1 Element of quality management system

The element of quality management system is shown as Fig. 3-2-5.

Fig. 3-2-5 Element of quality management system

3.2 Our experience

The experience is that laboratories should have evidence of continual improvement, which is an obligatory requirement for accredited laboratories. The laboratory must be competitive and stay current with the quality and technical management standards, and with methods used to demonstrate laboratory competence and establish and maintain technical validity. Many activities can keep a laboratory progressive, for example, attendance at conferences on diagnostics and quality management, participation in local and international organizations, participation in writing national and international standard, current awareness of publications about diagnostic methods, training programmes, conducting research, participation in cooperative programmes, exchange of procedures, methods, reagents, samples, personnel, and ideas, continual professional development and technical training, management reviews, analysis of customer feedback, root cause analysis of anomalies and implementation of corrective, preventive and improvement actions, and so on.

The process of CNAS laboratory accreditation is generally divided into three steps: application, on-site review and appraisal approval (Fig. 3-2-6). A typical accreditation period is about 6 months (Fig. 3-2-7).

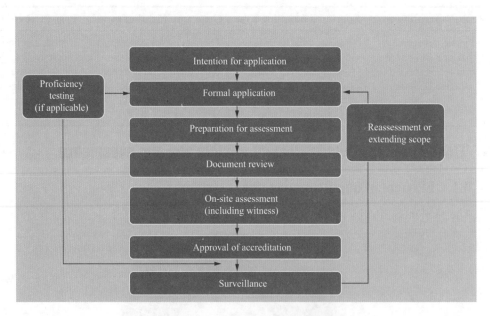

Fig. 3-2-6 Flow chart of accreditation
(Data from https://www.cnas.org.cn/english/accreditation on laboratory/12/717790.shtml)

ISO/IEC 17025 认可证书

Name: Center for Aquatic Animal Disease Diagnosis and Control, Yellow Sea Fisheries Research Institute, Chinese Academy of Fishery Sciences

Address: No.106, Nanjing Road, Qingdao, Shandong, China

Registration No. CNAS L10772

CHINA NATIONAL ACCREDITATION SERVICE FOR CONFORMITY ASSESSMENT

Accreditation Criteria: ISO/IEC 17025:2005 and relevant requirements of CNAS

SCHEDULE OF ACCREDITATION CERTIFICATE

Effective Date: 2019-03-19 Expiry Date: 2024-03-04

SCHEDULE 3 ACCREDITED TESTING SCOPE

№	Test Object	Item/Parameter		Standard or Method	Note	Effective Date
		№	Item/ Parameter			
veterinary						
molecular biological diagnosis						
1	aquatic animal	1	White spot virus	Diagnostic protocols for white spot disease Part2:Nested PCR method GB/T 28630.2-2012		2019-03-19
				Quarantine protocol for white spot disease SN/T 1151.2-2011	Accredited only for 7.1	2019-03-19
				Manual of Diagnostic Tests for Aquatic Animals (OIE, 2017) Chapter 2.2.8 White spot disease 4.3.1.2.4.1.		2019-03-19
		2	Infectious hypodermal and hematopoietic necrosis virus	Polymerase chain reaction(PCR) method for Infectious hypodermal and hematopoietic necrosis virus (IHHNV) GB/T 25878-2010		2019-03-19
				Quarantine protocol for Infectious hypodermal and haematopoietic necrosis SN/T 1673-2013	Accredited only for 5.4	2019-03-19

No. CNAS L10772 第 1 页 共 2 页

The scope of the accreditation in Chinese remains the definitive version.

Fig. 3-2-7 An example of accredited testing scope

References

CNAS. CNAS Introduction [EB/OL]. (2019-07-03) [2020-01-01] https://www.cnas.org.cn/english/introduction/index.shtml.

CNAS. International cooperations & MLA/MRA [EB/OL]. [2020-01-01] https://www.cnas.org.cn/english/intemationalcooperations/index.shtml.

CNAS. Process [EB/OL]. (2012-12-14) [2020-01-01] https://www.cnas.org.cn/english/accreditationonlaboratory/12/717790.shtml.

Rapid detection of aquaculture pathogens

Author: Zhang Qingli
E-mail: zhangql@ysfri.ac.cn

1 Introduction of rapid detection techniques of aquatic pathogens

Aquaculture is one of fastest-growing fields in the developing world. Diseases of aquaculture animals are constant threats to the sustainability and economic viability of aquaculture. Among all the control methods of aquaculture diseases, early diagnosis in time is considered to be the vital approach.

Traditionally, various histological, biochemical and serological tests have been used for aquatic diseases diagnosis. The common diagnosis techniques used in the aquatic diseases control include the electron microscope observation method, the TE staining histological method, the pathological section method, the nucleic acid probe hybridization method, the antibody hybridization method, the PCR detection method and the isothermal amplification method.

Although electron microscopy technology can directly observe the presence of virus particles or the pathogenic bacteria, its operation is complicated, time-consuming, and its accuracy is not high. TE staining and pathological section methods are based on pathological techniques and usually cannot directly detect viruses or pathogenic bacteria. The antibody detection and nucleic acid probe hybridization methods are commonly used to detect the protein or nucleic acid components of the pathogenic agents, and the disadvantages are that their detection sensitivity is low and they cannot be used for the early warning of pathogens infection, especially at the very early stage of the infection. PCR detection method is a widely used technology for the early warning of pathogens, and it can achieve relatively rapid detection of the pathogens under laboratory conditions. However, conventional PCR detection requires expensive PCR instrument, gel electrophoresis and imaging system, this greatly limits the promotion and application of PCR detection method in production. The isothermal amplification methods provide a rapid, sensitive, specific, simple and less expensive procedure for detecting nucleic acid of the pathogenic bacteria, and represent a more prospective early warning detection method of aquatic diseases (Fig. 3-3-1).

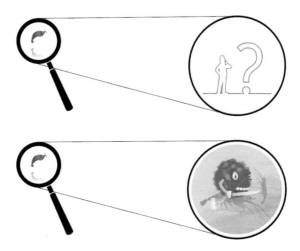

Fig. 3-3-1 Detection techniques help identify the aquatic pathogens

Nowadays, many isothermal nucleic acid amplification techniques have been developed, such as transcription-mediated amplification or self-sustained sequence replication, nucleic acid sequence-based amplification (NASBA), signal-mediated amplification of RNA technology, strand displacement amplification (SDA), rolling circle amplification (RCA), loop-mediated isothermal amplification (LAMP), isothermal multiple displacement amplification (IMDA), helicase-dependent amplification (HDA), single-primer isothermal amplification (SIA) and circular helicase-dependent amplification (cHDA) (Fahy et al., 1991; Sooknanan et al., 1995; Gracias et al., 2010; Hall et al., 2002; Walker et al., 1996; Gadkar et al., 2005; Jin et al., 2006; Mori and Notomi, 2009; Zarogoulidis et al., 2014; Dawei et al., 2015; Myrmel et al., 2017; Yan et al., 2006). Among all these isothermal nucleic acid amplifications techniques, LAMP is the most widely used.

Several methods of rapid detection on the spot for aquatic pathogens has been reported based on the LAMP principles by different research groups in the world (Kono et al., 2004; Savan et al., 2005; Kiatpathomchai et al., 2007; Puthawibool et al., 2009; Yang et al., 2016). Yellow Sea Fisheries Research Institute, Chinese Academy of Fishery Sciences (YSFRI, CAFS) has invented the highly sensitive and rapid detection kits for over twenty aquatic pathogens based on the novel isothermal techniques including LAMP. (Zhang et al., 2009; Zhang, 2019).

2 General information of the highly sensitive and rapid detection kits

The first version of the highly sensitive and rapid detection kits developed by YSFRI is based on the isothermal nucleic acid amplification technique of LAMP. The second version of the kits is based on a novel isothermal nucleic acid amplification technique, which is called as promoter-mediated isothermal amplification (PAMP). By using the kits, it is possible to determine whether the shrimp is carrying 100 virus copies in 60 minutes in the field (Zhang et al., 2018) (Table 3-3-1, Fig. 3-3-2).

Table 3-3-1 Types of highly sensitive and rapid detection kits that can be provided by YSFRI

Pathogen detectable by the kits		Application scope
Shrimp pathogens		
WSSV	IHHNV	
CMNV	SHIV	
MDNV	EHP	
MrNV	IMNV	All the kits are in one specification, of which one kit can be used for four samples detection. The kits can be applied for detection the pathogens of aquaculture animal: selection of healthy broodstock, disease early diagnosis, risk monitoring, epidemiological surveillance, etc.
MBV	PvNV	
YHV	BP	
TSV	HPV	
Vp_{AHPND}	Spiroplasmas	
Fish pathogens		
NNV	SVCV	
GCRV	CCV	
TRBIV	Edwardsiella tarda	
Mollusk viruses		
OsHV-1	AVNV	

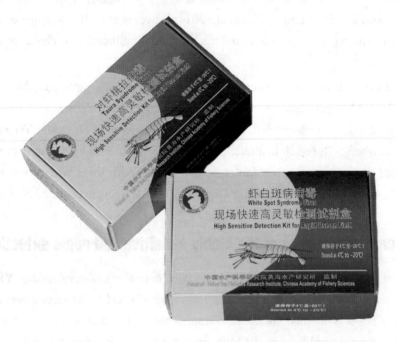

Fig. 3-3-2 The highly sensitive and rapid detection kits invented by YSFRI

3 Principle of the highly sensitive and rapid detection kits

The research team from YSFRI carried out the original innovation and application research of nucleic acid isothermal amplification technology, optimized the LAMP assays, as well as invented the novel technique of PAMP. Through the research and development of aquatic animal pathogenic nucleic acid on-site rapid preparation and isothermal amplification, long-term storage of reagents at room temperature and immobilization of nucleic acid dyes, the research team has broken the restrictions on the large-scale promotion of nucleic acid isothermal amplification technology. By combining the key techniques including nucleic acid on-site rapid preparation and isothermal amplification, highly sensitive and rapid detection kits were developed. The basic principle of the kits is shown in Fig. 3-3-3.

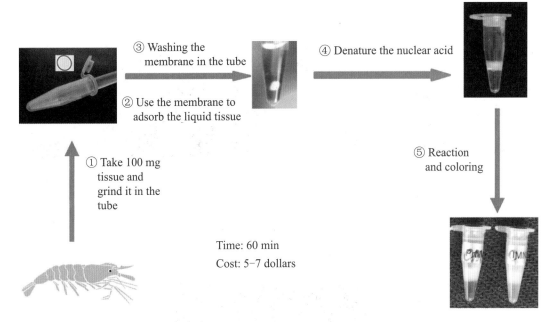

Fig. 3-3-3 The basic principle of the highly sensitive and rapid detection kits from YSFRI

The kits possess the typical advantages of isothermal amplification techniques. The typical characteristics include the following aspects:

A. High sensitivity. The lowest detecting limit of viral detection kits can be as low as 10 copies, and its sensitivity is more than 100 times higher than that of conventional PCR methods.

B. High specificity. The detection primers are designed based on six different regions in the targeted virus conservative gene, and the specificity exceeds that of conventional PCR.

C. The detection time is short. The whole detection takes only 1 hour from nucleic acid preparation to detection completion.

D. The equipment requirements are low and no complicated equipment is needed. With a water bath or a constant temperature metal bath, or even a kettle of boiling water and a thermometer, the aquatic virus detection can be completed by using the kits.

E. Simple operation and obvious results. The entire detection process does not involve complicated instruments or equipment, and can be completed by workers with a little molecular biology basis. the detection results are clear and obvious, and can be judged by direct observation.

F. Safe for people and the environment. No toxic reagents are used in the detection process.

G. Low cost. The cost of the detection technology is much lower than the cheapest PCR detection method available.

The research team in YSFRI have developed more than 20 kinds of rapid and highly sensitive detection kits for aquatic animal pathogens, and promoted to apply more than 12,000 kit sets in 16 provinces in China (Fig. 3-3-4), which provides new technology and product support for the early warning of major and emerging diseases of aquaculture animals.

Fig. 3-3-4 The manufacturing, technical training and practical application of the highly sensitive and rapid detection kits

From 2015 to 2020, YSFRI organized technical training courses of the kits for researchers, farmers, and technicians from countries of ASEAN and Southeast Asia each year (Fig. 3-3-5). YSFRI has supplied diagnostic kits for shrimp pathogens to over 10 countries.

Fig. 3-3-5 Technical training and practical application of the highly sensitive and rapid detection kits in YSFRI in 2018

4 The protocol of detecting pathogens by using the kit

4.1 The components of the kit

The components of the kit are shown in Table 3-3-2.

Table 3-3-2 The components in the kit

Number	Name in English	Quantity	Notes
1	Sampling membrane	4 pieces	Store at −80 ℃ to 4 ℃
2	Membrane for positive control	1 piece	Store at −80 ℃ to 4 ℃
3	Membrane for negative control	1 piece	Store at −80 ℃ to 4 ℃
4	Sample collection tube	4 pieces	Store at −80 ℃ to 4 ℃
5	Buffer A	0.8 mL	Store at −80 ℃ to −20 ℃
6	Washing tube	6 pieces	Store at −80 ℃ to −20 ℃
7	Nucleic acid denature tube	6 pieces	Store at −80 ℃ to −20 ℃
8	CMNV detection tube	6 pieces	Store at −80 ℃ to −20 ℃
9	Grinder (Pestle)	4 pieces	
10	Toothpick	30 pieces	
11	Dropper (Pipette)	1 piece	
12	Manual	1 copy	

4.2 Additional equipment required for the test

A. Thermal protective aid: water-bath, metal bath, PCR machine with lid-heating turn-off or any heating devices with temperature controller.

B. Tools for sampling: scissors, forceps (If no such tools are available, the tissues can be torn off by hand).

C. Standby tools: some toothpicks, disposable gloves.

4.3 Instructions for use

Please read the instruction in Section 4 firstly, and then follow the steps to carry out the test (Figs. 3-3-6 to 3-3-7).

A. Collect about 100 mg tissue from gills, appendages or the larvae body, put it into the sample collection tube, and quickly grind the samples into paste using the grinder. A total of four samples can be managed at one time.

B. Dip the toothpick in the sample paste and then daub the liquid adsorbed by the toothpick onto the sampling membrane until it is fully saturated. The different sampling membranes with number should be wetted by corresponding sample paste. Discard the toothpick. The 0.2 mL

PCR tube containing the sampling membrane is kept in a plastic bag. The sampling membranes are assigned numbers (1-4) on the tube covers.

C. Add 2-3 drops of Buffer A onto the sampling membrane in the 0.2 mL PCR tube, stir the membrane gently for 30 seconds with a new toothpick. Discard the toothpick.

D. Move the sampling membrane into the corresponding numbered washing tube with a new toothpick, and vortex the tube vigorously for 3-4 min. Discard the toothpick.

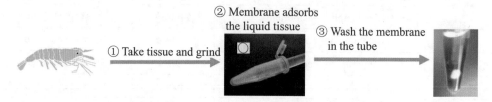

Fig. 3-3-6　The picture displaying the protocol (Step 1-4) of the detecting kit

E. Move the sampling membrane, and the positive and negative control membranes into the corresponding nucleic acid denatured tube with a new toothpick, and incubate the tube at 70 ℃ for 5 min. Then put the tube into cold water or place it at room temperature immediately for 2 min to cool it down.

F. Move the membranes into the corresponding CMNV detection tube with a new toothpick, and then incubate the tube at 63-65 ℃ for 60 min with water bath, metal bath or PCR machine. If PCR machine is used, the lid-heating program must be turned off.

G. Incubate the tube at 95-98 ℃ for 4 min, and then shake the CMNV detection tube immediately (5-6 times within 3 s) to make the dye drop from the lid of the tube and blend with the reaction buffer.

H. Swing the CMNV detection tube downwards immediately to collect all buffer down to the bottom. Observe the color after 1-2 min by naked eye.

Fig. 3-3-7　The picture displaying the protocol (Step 6-8) of the detecting kit

4.4　Read the results

A. Positive control should show green color; negative control should show orange color. If not, the result is invalid.

B. Comparing with positive control and negative control, read the detection result of the sample. Green color is positive, and orange color is negative. The grade of green color represents the amount of CMNV infection.

4.5 Notes

A. The kit must be stored at a low temperature (−80 ℃ to −20 ℃; The lower, the better).

B. To prevent cross contamination between samples, please do not open the CMNV detection tube after amplification (Step 6).

C. Liquid in the CMNV detection tube should be colorless and clear. If the color changes into light red or orange in one CMNV detection tube, please discard the tube.

D. During the test, the toothpick should be changed in every step for every sample. You cannot use the same toothpick to move or contact the sampling membrane, and membrane for positive and negative control.

E. This kit is for research only. It is not intended for any medicine, clinic and food use.

References

Dawei L, Bei L, Hao Z, et al. Disintegration of cruciform and G-quadruplex structures during the course of helicase-dependent amplification (HDA) [J]. Bioorganic & Medicinal Chemistry Letters, 2015, 25(8): 1709−1714.

Fahy E, Kwoh D Y, Gingeras T R. Self-sustained sequence replication (3SR): an isothermal transcription-based amplification system alternative to PCR [J]. PCR methods and applications, 1991, 1(1): 25−33.

Gadkar V, Rillig M C. Suitability of genomic DNA synthesized by strand displacement amplification (SDA) for AFLP analysis: genotyping single spores of arbuscular mycorrhizal (AM) fungi [J]. Journal of Microbiological Methods, 2005, 63(2):157-164.

Gracias K S, Mckillip J L. Nucleic acid sequence-based amplification (NASBA) in molecular bacteriology: A procedural guide [J]. Journal of Rapid Methods & Automation in Microbiology, 2010, 15(3): 295−309.

Hall M J, Wharam S D, Weston A, et al. Use of signal-mediated amplification of RNA technology (SMART) to detect marine cyanophage DNA [J]. Bio. Techniques, 2002, 32(3): 604−606, 608−611.

Jin I' Yasushi S, Tsutomu M. Improvements of rolling circle amplification (RCA) efficiency and accuracy using Thermus thermophilus SSB mutant protein [J]. Nuclc Acids Research, 2006, 34(9): e69.

Kiatpathomchai W, Jareonram W, Jitrapakdee S, et al. Rapid and sensitive detection of Taura syndrome virus by reverse transcription loop-mediated isothermal amplification [J]. J Virol Methods, 2007, 146(1-2): 125−128.

Kono T, Savan R, Sakai M, et al. Detection of white spot syndrome virus in shrimp by loop-mediated isothermal amplification [J]. J Virol Methods, 2004, 115(1): 59−65.

Mori Y, Notomi T. Loop-mediated isothermal amplification (LAMP): a rapid, accurate, and cost-effective diagnostic method for infectious diseases [J]. Journal of Infection & Chemotherapy, 2009, 15(2): 62-69.

Myrmel M, Veslemøy O, Khatri, M. Single primer isothermal amplification (SPIA) combined with next generation sequencing provides complete bovine coronavirus genome coverage and higher sequence depth compared to sequence-independent single primer amplification (SISPA) [J]. PLoS ONE, 2017, 12(11): e0187780-.

Puthawibool T, Senapin S, Kiatpathomchai W, et al. Detection of shrimp infectious myonecrosis virus by reverse transcription loop-mediated isothermal amplification combined with a lateral flow dipstick [J]. J Virol Methods, 2009, 156(1-2): 27-31.

Savan R, Kono T, Itami T, et al. Loop-mediated isothermal amplification: an emerging technology for detection of fish and shellfish pathogens [J]. J Fish Dis, 2005, 28(10): 573-581.

Sooknanan R, Malek L T, Gemen B V. Nucleic acid sequence-based amplification (NASBA) [J]. Molecular Methods for Virus Detection, 1995, 28(28): 261-285.

Walker GT, Nadeau J G, Linn C P, et al. Strand displacement amplification (SDA) and transient-state fluorescence polarization detection of *Mycobacterium tuberculosis* DNA [J]. Clinical Chemistry, 1996, (1): 9-13.

Yan X, Hyun-Jin K, Ann K, et al. Simultaneous amplification and screening of whole plasmids using the T7 bacteriophage replisome [J]. Nuclc Acids Research, 2006(13): e98.

Yang H L, Qiu L, Liu Q, et al. A novel method of real-time reverse-transcription loop-mediated isothermal amplification developed for rapid and quantitative detection of a new genotype (YHV-8) of yellow head virus [J]. Lett Appl Microbiol, 2016, 63(2): 103-110.

Zarogoulidis P, Mairinger F, Vollbrecht C, et al. Isothermal multiple displacement amplification: a methodical approach enhancing molecular routine diagnostics of microcarcinomas and small biopsies [J]. Oncotargets & Therapy, 2014, 7: 1441-1447.

Zhang Q, Shi C, Huang J, et al. Rapid diagnosis of turbot reddish body iridovirus in turbot using the loop-mediated isothermal amplification method [J]. J Virol Methods, 2009, 158(1-2): 18-23.

Zhang Q L. Evidences for cross-species infection in fish of covert mortality nodavirus (CMNV) [C]. 12th Asian Fisheries and Aquaculture Forum, Iloilo Convention Center, Iloilo City, Pilipinas, 8-12 April, 2019.

Acute hepatopancreatic necrosis disease (AHPND) in shrimp

Author: Dong Xuan
E-mail: dongxuan@ysfri.ac.cn

1 Acute hepatopancreatic necrosis disease (AHPND)

1.1 Overview

AHPND refers to infection with strains of *Vibrio* (V_{AHPND}) that contain a pVA1-type plasmid carrying genes encoding homologues of the *Photorhabdus* insect-related (Pir) binary toxin, PirA and PirB (OIE, 2019).

AHPND is also known as early mortality syndrome (EMS).

The economic losses to the global shrimp farming industry from AHPND have been estimated at over 1 billion dollars per year. AHPND has been listed as a notifiable disease of crustaceans by the world organisation for animal health (OIE) in 2016.

1.2 Aetiological agent

Initially, AHPND-causing *V. parahaemolyticus* (Vp_{AHPND}) is regarded as the only known pathogen responsible disease (OIE, 2019).

Recently, it was reported that the bacterial aetiology of AHPND also included *V. harveyi* (Vh_{AHPND}), *V. owensii* (Vo_{AHPND}), *V. campbellii* (Vc_{AHPND}) and *V. punensis* (Dong et al., 2019b).

1.3 Host factors

Susceptible host species: *Penaeus vannamei* and *P. monodon*.

Species with incomplete evidence for susceptibility: *P. chinensis*.

Susceptible stages of the host: mass mortalities often occur within 35 d post stocking with postlarvae or juveniles, and as early as 10 d post stocking (OIE, 2019). Outbreaks were also reported in Philippines occuring as late as 96 d after pond-stocking (OIE, 2019).

1.4 Geographical distribution of AHPND

AHPND was first reported among shrimp farms in Vietnam and China in 2010. The disease was later reported in Malaysia (2011), Thailand (2012), Mexico (2013), Philippines (2014), USA (2017) and Bangladesh (2017) (OIE, 2019).

2 Clinical signs and histopathology

2.1 Clinical signs

The clincal signs of AHPND-affected shrimp:

A. Pale and atrophied hepatopancreas (HP) (Fig. 3-4-1).

B. Empty stomach (ST) and midgut (MG) (Fig. 3-4-1).

C. Mortality from AHPND progresses extremely rapidly, and usually occurs within 30-35 d after stocking.

D. Reduced food consumption.

E. Discoloration of appendages and the body.

F. HP tissue is firmer than normal (NACA, 2014).

Fig. 3-4-1 Clinical signs of AHPND in *P. vannamei*
(Revised from Dong et al., 2017)

2.2 Histopathology

Histopathology of normal hepatopancreas from healthy *P. vannamei* is shown in Fig. 3-4-2a. In the early stage of HP from AHPND-affected shrimp, the HP tubules start degenerate, rounding up and slough into lumen (Fig. 3-4-2b). In the acute phase of HP from AHPND-affected shrimp, acute progressive degeneration of the HP tubules and sloughing of HP tubule epithelial cells into the HP tubules occur (Fig. 3-4-2c). In the terminal phage of HP from AHPND-affected shrimp, extensive haemocyte infiltration and massive bacterial colonization in tubule lumens occur (Fig. 3-4-2d). The only definitive histopathology is massive sloughing occur of HP tubule epithelial cells in the absence of bacteria (Fig. 3-4-3) (NACA, 2014).

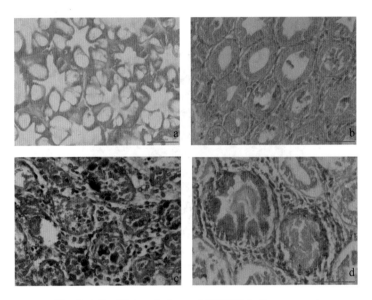

Fig. 3-4-2　Histopathology of AHPND in *P. vannamei*

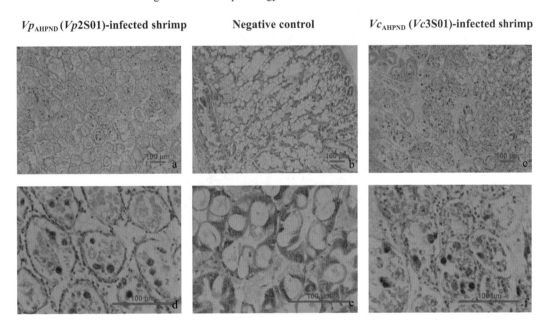

Fig. 3-4-3　Histopathological examination of *Vp*2S01- and *Vc*3S01-infected shrimp

3 Isolation and identification of AHPND-causing *Vibrio*

For AHPND-causing *Vibrio*, bacteria can be isolated from the gut-associated tissues (HP, stomach or gut) with thiosulfate citrate bile salts sucrose (TCBS) plate (Figs. 3-4-4 to 3-4-5). Both *Vp*2S01 and *Vc*3S01 formed green and round colonies on TCBS plate, while AHPND-causing *V. owensii* strain 20170320001 formed yellow colonies (Table 3-4-1). We here suggest that both green and yellow colonies should be tested when using the TCBS plate method.

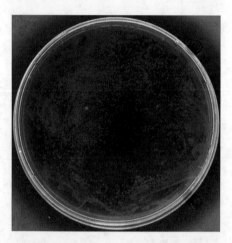

Fig. 3-4-4 Non-sucrose-fermenting bacterial colonies are covered by the green color

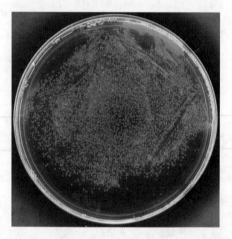

Fig. 3-4-5 Sucrose-fermenting bacterial colonies are covered by the yellow color

Table 3-4-1 Utilization of sucrose of AHPND-causing *Vibrio* species

Name	Species	*pirAB^{vp}*	*scrB*	Sucrose in biology	Color in TCBS
*Vp*2S01	*V. parahaemolyticus*	+	−	No utilization	Green
*Vc*3S01	*V. campbellii*	+	−	No utilization	Green
20170320001	*V. owensii*	+	+	Utilization	Yellow

3.1 Multilocus sequence analysis (MLSA)

MLSA can be used as a tool for identifying *Vibrio* by using the 16S rRNA, *rpoD*, *rctB* and *toxR* (Fig. 3-4-6).

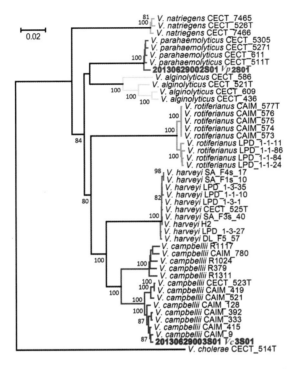

Fig. 3-4-6 MLSA of AHPND-causing *Vibrio* strain *Vp*2S01 and *Vc*3S01

3.2 Isolation and identification of AHPND-causing *Vibrio*

The average nucleotide identity (ANI) analysis using gecome sequence of *Vibrio* is shown in Fig. 3-4-7.

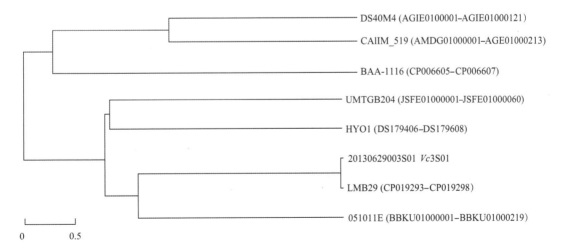

Fig. 3-4-7 ANI analysis using genome sequence of *Vibrio*

4 Molecular diagnostics

4.1 Polymerase chain reaction (PCR)

Several real-time PCR, single-step PCR and nested PCR methods have been standardized (OIE, 2019).

4.1.1 Real-time PCR

A. Primers:

VpPirA-F: 5'-TTG-GAC-TGT-CGA-ACC-AAA-CG-3';

VpPirA-R: 5'-GCA-CCC-CAT-TGG-TAT-TGA-ATG-3';

VpPirA probe: 5'-6FAM-AGA-CAG-CAA-ACA-TAC-ACC-TAT-CAT-CCC-GGA-TAMRA-3'.

B. Target gene: *pirA*.

4.1.2 Single-step PCR

A. Primers:

AP3-F: 5'-ATG-AGT-AAC-AAT-ATA-AAA-CAT-GAA-AC-3';

AP3-R: 5'-GTG-GTA-ATA-GAT-TGT-ACA-GAA-3'.

B. Target gene: a 333 bp fragment from the *pirA* gene in pVA1-type plasmid.

4.1.3 Nested PCR

A. Primers:

AP4-F1: 5'-ATG-AGT-AAC-AAT-ATA-AAA-CAT-GAA-AC-3';

AP4-R1: 5'-ACG-ATT-TCG-ACG-TTC-CCC-AA-3';

AP4-F2: 5'-TTG-AGA-ATA-CGG-GAC-GTG-GG-3';

AP4-R2: 5'- GTT-AGT-CAT-GTG-AGC-ACC-TTC-3'.

B. Target gene: The first-step PCR amplified a 1,269 bp fragment from the *pirAB* genes. The second-step PCR amplified a 230 bp fragment from the *pirAB* genes (Dangtip et al., 2015).

4.2 Loop-mediated isothermal amplification (LAMP)

LAMP method has been developed. This method is easy to perform and require one single temperature, which is ideal for field use.

Chapter 3 Maricultural organism disease control and molecular pathology | 393

5 The horizontal transfer of pVA1-type plasmids

What is the mechanisms underlying the burgeoning number of *Vibrio* species that cause AHPND? Figs. 3-4-8 to 3-4-14 and Table 3-4-2 give some examples and illustrations which can help us understand the mechanisms. One yellow colored *Vibrio* (20160513VC2W) was isolated in the infected group (Figs. 3-4-8 to 3-4-9).

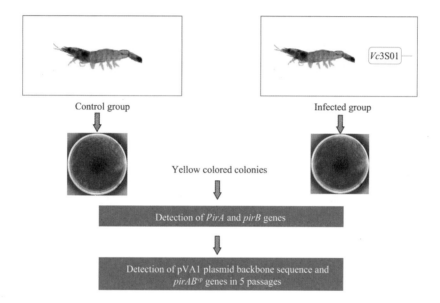

Fig. 3-4-8 The experimental design of horizontal transfer of pVA1-type plasmid

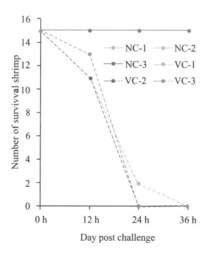

Fig. 3-4-9 Plots of the survival shrimp in each group

Table 3-4-2 Detection of pVA1-type plasmid backbone sequence and $pirAB^{vp}$ genes in 5 passages

Passage number	PCR primers			
	VpPirA-384	VpPirB-392	AP1	AP2
1	+	+	+	+
2	−	−	+	+
3	−	−	+	+
4	−	−	+	+
5	−	−	+	+

Notes: "−" means not detected; "+" means detected.

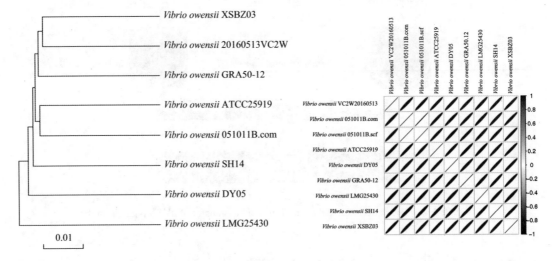

Fig. 3-4-10 ANI analysis revealed that 20160513VC2W strain has the closet relationship with *Vibrio*

As shown in Fig. 3-4-10, the plasmid 3 from *V. owensii* 20160513VC2W displayed a 99% nucleotide identity to the pVA1-type plasmid from *Vc*3S01.

This work provides evidence of the horizontal transfer of pVA1-type plasmids within *Vibrio* species. Horizontal transfer of pVA1-type plasmids within *Vibrio* species can lead to the diversity of AHPND-causing *Vibrio* bacteria. The dissemination of the pVA1-type plasmids in non-parahaemolyticus *Vibrio* also raises the concern of missing detection in industrial settings since the isolation method currently used mainly targets *V. parahaemolyticus* (Figs. 3-4-11 to 3-4-14).

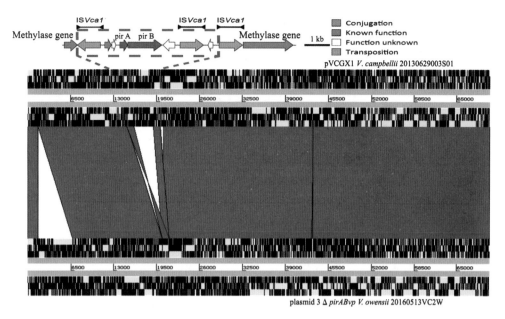

Fig. 3-4-11　Comparative pVA1-type plasmids analysis of AHPND-*Vc*3S01 and *V. owensii* 20160513VC2W
(Dong et al., 2019a)

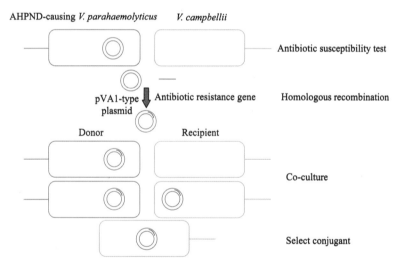

Fig. 3-4-12　The conjugant experiment design

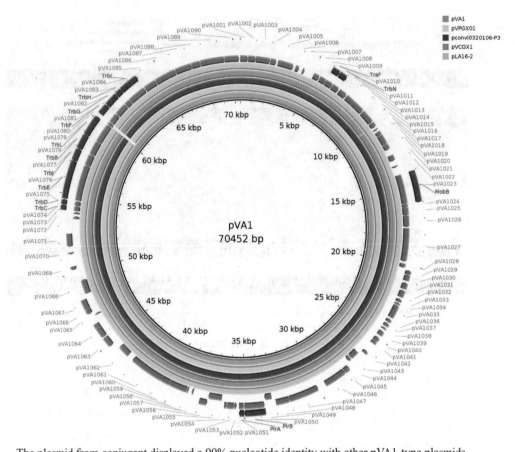

The plasmid from conjugant displayed a 99% nucleotide identity with other pVA1-type plasmids.

Fig. 3-4-13　Comparative sequences analysis of pVA1-type plasmids (Dong et al., 2019b)

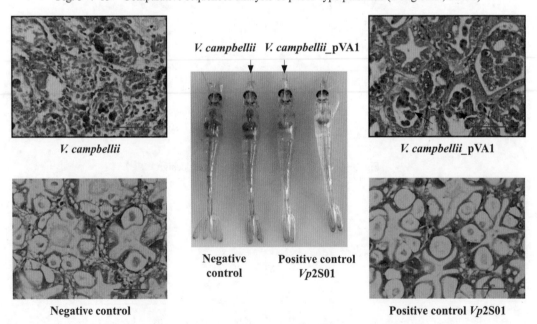

Fig. 3-4-14　Clinical signs and histopathology of AHPND in *Penaeus vannamei* from AHPND-*Vibrio* bacteria challenge studies (Dong et al., 2019b)

References

Dangtip S, Sirikharin R, Sanguanrut P, et al. AP$_4$ method for two-tube nested PCR detection of AHPND isolates of *Vibiro parahaemolyticus* [J]. Aquaculture Reports, 2015, 2: 158-162.

Dong X, Chen J, Song J, et al. Evidence of the horizontal transfer of pVA1-type plasmid from AHPND-causing *Vibrio campbellii* to non-AHPND *V. owensii* [J]. Aquaculture. 2019a, 503: 396-402.

Dong X, Song J, Chen J, et al. Conjugative transfer of the pVA1-type plasmid carrying the *pirAB*^{vp} genes results in the formation of new AHPND-causing *Vibrio* [J]. Frontiers in cellular and infection microbiology, 2019b, 9: 195.

Dong X, Wang H, Xie G, et al. An isolate of *Vibrio campbellii* carrying the *pir*^{VP} gene causes acute hepatopancreatic necrosis disease [J]. Emerging Microbes & Infections, 2017, 6(1): 131.

NACA. Diseases of Crustaceans—Acute Hepatopancreatic Necrosis Syndrome (AHPNS) [Z]. 2014.

OIE. Chapter 2.2.1. Acute Hepatopancreatic Necrosis Disease [M]. In: OIE Manual of Diagnostic Tests for Aquatic Animals. 2019.

Infection with decapod iridescent virus 1 (DIV1), an emerging disease in shrimps

Author: Qiu Liang
E-mail: qiuliang@ysfri.ac.cn

1 Background

1.1 White leg shrimp, *Penaeus vannamei*

White leg shrimp *P. vannamei* is one of the most important crustacean species in worldwide aquaculture, especially for the coastal developing countries. About 98.6% of *P. vannamei* was produced in developing countries in 2014.

1.2 Emerging diseases

Emerging diseases recently occurring in farmed *P. vannamei* have significantly and negatively impacted on shrimp farming industry along with the spread of the species (Fig. 3-5-1).

Fig. 3-5-1 Farmed *P. vannamei*

1.3 Viral metagenomics

Viral metagenomics, which involves viral purification and next generation sequencing, has been proven to be useful for describing novel viruses in new diseases, and has been recognized as an important tool for discovering novel viruses in human and veterinary medicine (Fig. 3-5-2) (Groff et al., 2015).

Chapter 3 Maricultural organism disease control and molecular pathology | 399

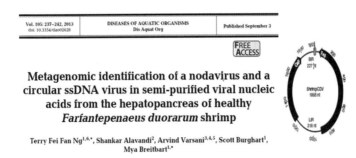

Fig. 3-5-2   Application of viral metagenome in shrimp sample

1.4   Family Iridoviridae

Iridoviridae is a family of large, icosahedral viruses. Infections can be either covert or patent, and they can lead to high levels of mortality among commercially and ecologically vertebrates including important fish and amphibians (Fig. 3-5-3).

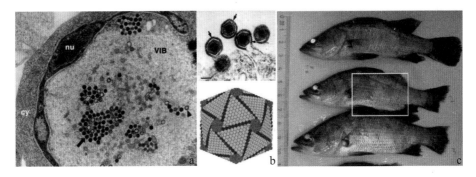

Fig. 3-5-3   Morphology and infection of iridovirids (Jancovich et al., 2012; Groof et al., 2015)

Table 3-5-1 shows the essential feature of iridovirids.

Table 3-5-1   The essential feature of iridovirids

Typical member	Virion	Genome	Taxonomy	Host range
Frog virus 3, genus *Ranavirus*; Infectious spleen and kidney necrosis virus, genus *Megalocytivirus*	150–200 nm (non-enveloped)	Linear, double-stranded circularly permuted DNA, 103–220 kbp	Alphairidovirinae Genus: *Lymphocystivirus* Genus: *Megalocytivirus* Genus: *Ranavirus* Betairidovirinae Genus: *Chloriridovirus* Genus: *Iridovirus*	Ectothermic vertebrates (subfamily Alphairidovirinae); mainly insects and crustaceans (subfamily Betairidovirinae)

2 Disease and infection

2.1 Samples and detection

The samples of *P. vannamei* (No. 20141215) collected from the pond in 2014 with massive die-offs exhibited obvious clinical signs, including empty stomach and guts, pale hepatopancreas and soft shell. The shrimp samples were tested and demonstrated to be free of WSSV, YHV, TSV, IHHNV and Vp_{AHPND} by PCR or RT-PCR methods recommended by the OIE and Flegel & Lo (Table 3-5-2).

Table 3-5-2　Test results of common pathogens

Pathogen detection	Method	Test result
WSSV	nPCR (OIE, 2012)	Negative
IHHNV	PCR (OIE, 2012)	Negative
Vp_{AHPND}	nPCR(Flegel et al., 2013)	Negative
YHV	RT-nPCR (OIE, 2012)	Negative
TSV	RT-PCR OIE (OIE, 2012)	Negative

2.2 Histopathology

Histological examination showed that inclusions and karyopyknosis existed in hematopoietic tissue and hemocytes in gills, hepatopancreas and periopods (Fig. 3-5-4) (Qiu et al., 2017).

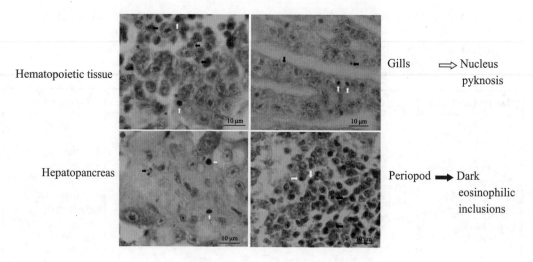

Fig. 3-5-4　Histopathological features of infected shrimp

2.3 Experimental challenge

The shrimp were divided into an injection test, a reverse garvage test and a *per os* test. Each test included a virus challenge group and a control group, and each group included three biological replicates (20 individuals each) (Fig. 3-5-5).

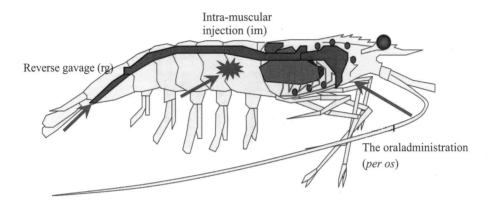

Fig. 3-5-5　Experimental challenge methods

Results of challenge tests showed that cumulative mortality of shrimp in intramuscular injection group (im), anal reverse garvage group (rg) and the oral administration group (*per os*) groups all reached 100% within two weeks of post-infection (Fig. 3-5-6) (Qiu et al., 2017).

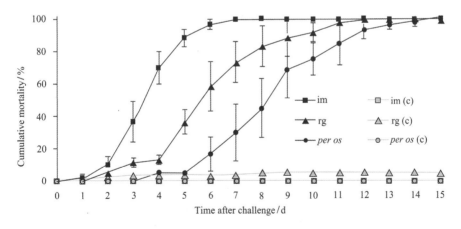

Fig. 3-5-6　Cumulative mortalities of *P. vannamei* in experimental infection (Qiu et al., 2017)

2.4　Gross signs of challenged shrimp

P. vannamei challenged with the viral preparation from the sample 20141215 exhibited the symptoms similar to those in the original individual sample: empty stomach and guts in all diseased shrimp, slight loss of color on the surface and section of hepatopancreas, and soft shell in partially infected shrimp. One third of individuals had slightly reddish body (Fig. 3-5-7) (Qiu et al., 2017).

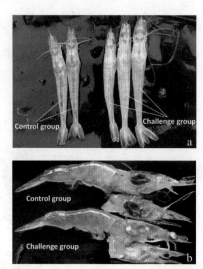

Fig. 3-5-7 Clinical symptoms of *P. vannamei* challenged with decapod iridescent virus 1 (DIV1)

2.5 Transmission electron microscope of diseased shrimp

There were large numbers of virions in the hematopoietic tissue (Fig. 3-5-8a). DIV1 budded and acquired an envelope from the plasma membrane (Fig. 3-5-8b). DIV1 replicated and assembled in hematopoietic cells (Fig. 3-5-8c). The stages of nucleocapsid assembly are indicated with numbers 1–3, and a complete nucleocapsid is indicated with number 4 (Fig. 3-5-8d). The capsids at stage 2 and 3 should have a small opening at one vertex but may not be visible in the picture due to the ultrathin section (Qiu et al., 2019).

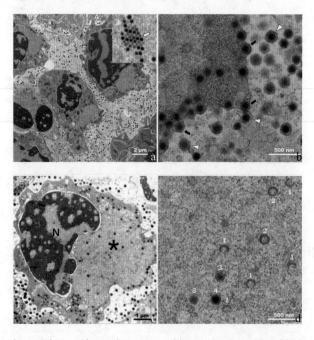

N. nucleus; *. a large electron-lucent virogenic stroma; white arrows. paracrystalline array of viral particles; black arrows. budding virions; white triangles. budded virions that acquired an envelope.

Fig. 3-5-8 Transmission electron microscope of hematopoietic tissue of naturally infected prawn

The crescent-shaped structure curves to mature virion with a dense (Fig. 3-5-9) (Qiu et al., 2019).

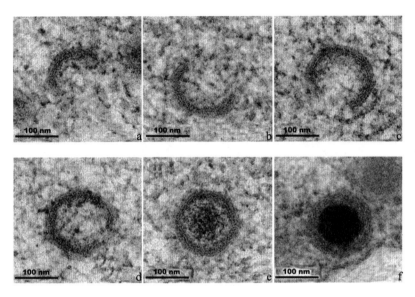

Fig. 3-5-9　The assembling process of nucleocapsid (Qiu et al., 2019)

2.6　*In situ* hybridization

Positive signals of *in situ* hybridization were observed in hematopoietic tissue and hemocytes in gills, hepatopancreas and periopods (Fig. 3-5-10) (Qiu et al., 2017).

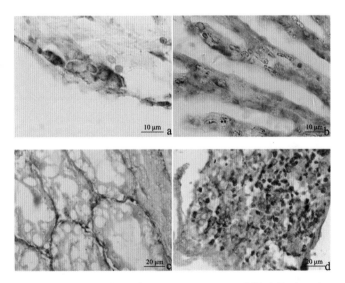

a. Hematopoietic tissue; b. Hepatopancreas; c. Gill; d. Periopod.
Fig. 3-5-10　*In situ* hybridization of infected *P. vannamei*

2.7 *In situ* digoxigenin-labeled loop (ISDL)-mediated isothermal amplification

ISDL results showed that blue signals existed in hematopoietic tissue, hemocytes, some R-cells, and myoepithelial fibers of the hepatopancreas, coelomosac epithelium of antennal gland, and epithelium of ovaries (Fig. 3-5-11) (Qiu et al., 2019).

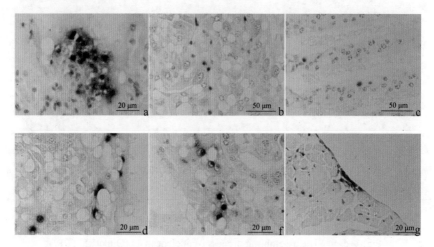

Fig. 3-5-11   ISDL of infected shrimp (Qiu et al., 2019)

2.8 Virus purification

After sucrose density gradient centrifugation, two cloudy bands with slightly blue color showed up in sucrose gradients under the original 40% (mass fraction) sucrose fraction. They were more visible with a laser irradiation (Fig. 3-5-12) (Qiu et al., 2017). Visualization under transmission electron microscope of the negatively stained grid with a drop from the band showed enveloped icosahedral viral particles with diameters of (160.2 ± 7.0) nm $(n = 10)$(v-v) and (142.6 ± 4.0) nm $(n = 10)$ (f-f).

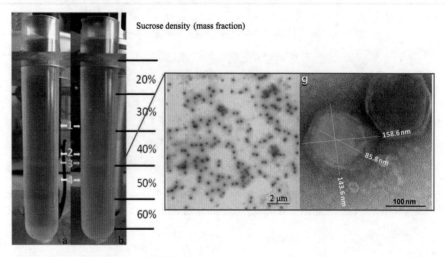

Fig. 3-5-12   Purification of DIV1 using centrifuged sucrose gradients (Qiu et al., 2017)

3 Viral metagenomics

3.1 Experimental procedure

Experimental procedure of viral metagenomics is shown in Fig. 3-5-13.

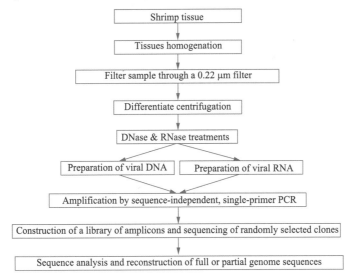

Fig. 3-5-13　Procedure of viral metagenomics

3.2 The proportion of reads

24.9% of clean reads are annotated as viruses and 58.57% of viruses reads are annotated as family Iridoviridae (Fig. 3-5-14).

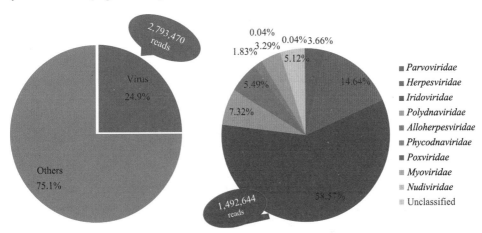

Fig. 3-5-14　The proportation of viral metagenomics data

4 Taxonomy

In March 2019, the Executive Committee of the International Committee on Taxonomy of Viruses (ICTV) approved the proposal made by Chinchar et al. to add a new species of DIV1 in a new genus *Decapodiridovirus*, with SHIV 20141215 and CQIV CN01 as two isolates (Figs. 3-5-15 to 3-5-16) (Qiu et al., 2019).

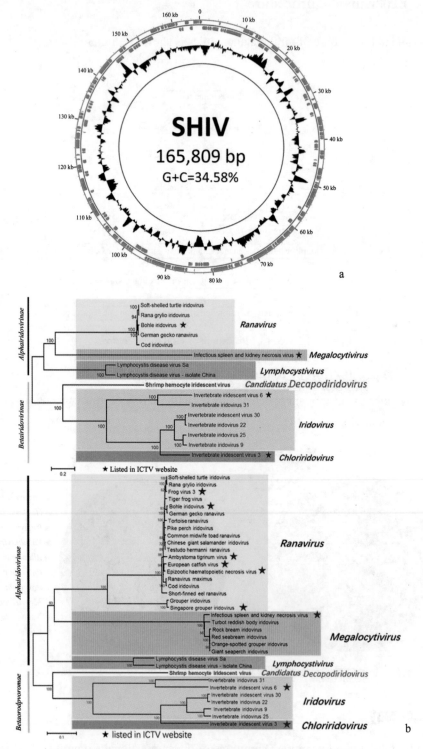

a. Complete genome of isolate SHIV 20141215; b. Phylogenetic tree.

Fig. 3-5-15　Genome analysis

Fig. 3-5-16 Genus division of family Iridoviridae in ICTV website

5 Host Range

Currently known susceptible species infected with DIV1 include *P. vannamei*, *M. rosenbergii*, *Exopalaemon carinicauda*, *M. nipponense*, *Procambarus clarkii* and *C. quadricarinatus* (Fig. 3-5-17).

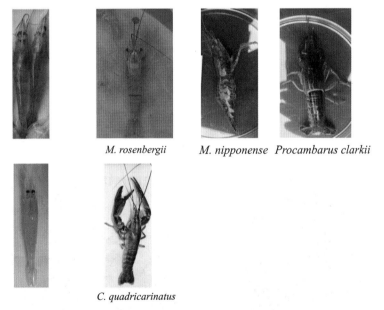

Fig. 3-5-17 Susceptible hosts of DIV1 (confirmed by research)

6　Epidemiology

Target surveillance was carried out in China in 2017 and 2018. It revealed that the virus had been detected in 11 provinces out of the 16 surveyed provinces and caused massive economic losses. Positive rate of samples was 12.2% (221/1,809) (Fig. 3-5-18).

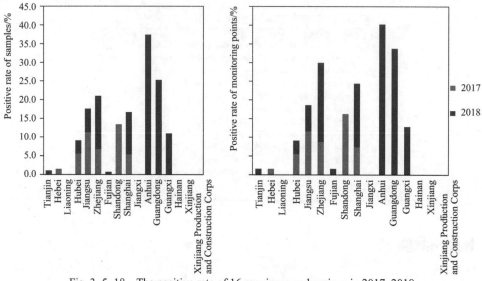

Fig. 3-5-18　The positive rate of 16 provinces and regions in 2017-2018

7　Diagnostic method

7.1　Nested PCR assay

The nested PCR method has been established and validated for DIV1 detection (Fig. 3-5-19) (Qiu et al., 2017).

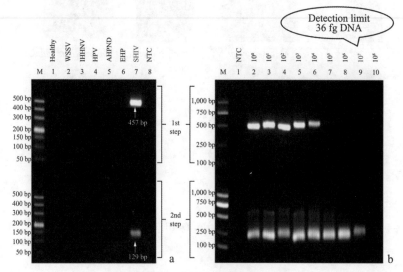

Fig. 3-5-19　A nested PCR for DIV1

7.2 TaqMan probe based PCR

The TaqMan based real-time PCR method has been established and validated for DIV1 detection (Fig. 3-5-20) (Qiu et al., 2018).

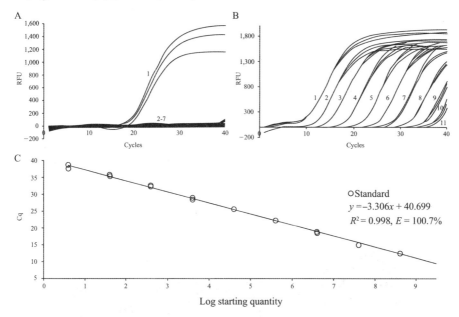

Fig. 3-5-20 A TaqMan real-time PCR for DIV1 (Qiu et al., 2018)

7.3 *In situ* hybridization

The *in situ* hybridization method has been established for DIV1 detection using a 279 bp digoxin labeled probe. The paraffin sections were then subjected to *in situ* hybridization assays according to the protocol of *in situ* hybridization.

7.4 LAMP & *In situ* DIG-labeling-LAMP

A set of specific primers, composed of FIP, BIP, LF, LB, F3 and B3, for LAMP detection of DIV1 was designed to target the gene of the second largest subunit of DNA-directed RNA polymerase II of DIV1 (Fig. 3-5-21). ISDL followed the method by Chen et al., 2019.

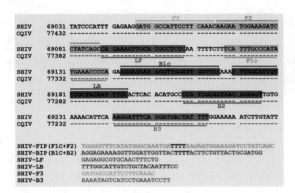

Fig. 3-5-21   Primers of loop-mediated isothermal amplification (LAMP) for DIV1 (Chen et al., 2019)

7.5   High sensitive detection kit for rapid use on field

High sensitive detection kit was established based on the LAMP method for rapid use on field (Fig. 3-5-22).

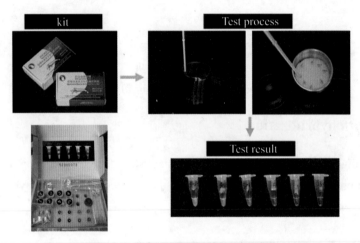

Fig. 3-5-22   Test kit based on LAMP

References

Groof A D, Guelen L, Deijs M, et al. A novel virus causes scale drop disease in *Lates calcarifer* [J]. PLoS Pathogens, 2015, 11(8): e1005074.

Jancovich J K, et al. Family Iridoviridae [M]// King A M Q, Adams M J, Carstens E B, Lefkowitz E J. Virus Taxonomy: Classification and Nomenclature of Viruses. Ninth Report of the International Committee on Taxonomy of Viruses. Amsterdam: Elsevier Academic Press, 2012: 193-210.

Ng T F F, Alavandi S, Varsani A, et al. Metagenomic identification of a nodavirus and a circular ssDNA virus in semi-purified viral nucleic acids from the hepatopancreas of healthy

Farfantepenaeus duorarum shrimp [J]. Diseases of Aquatic Organisms, 2013, 105(3): 237-242.

Qiu L, Chen M M, Wan X Y, et al. Characterization of a new member of Iridoviridae, shrimp hemocyte iridescent virus (SHIV), found in white leg shrimp (*Litopenaeus vannamei*) [J]. Scientific Reports, 2017, 7(1): 11834.

Qiu L, Chen M M, Wan X Y, et al. Detection and quantification of shrimp hemocyte iridescent virus by TaqMan probe based real-time PCR [J]. Journal of Invertebrate Pathology, 2018, 154: 95-101.

Qiu L, Chen M M, Wang R Y, et al. Complete genome sequence of shrimp hemocyte iridescent virus (SHIV) isolated from white leg shrimp, *Litopenaeus vannamei* [J]. Archives of Virology, 2017, 163: 781-785.

Qiu L, Chen X, Zhao R H, et al. Description of a natural infection with decapod iridescent virus 1 in farmed giant freshwater prawn, *Macrobrachium rosenbergii* [J]. Viruses, 2019, 11(4): 354.

Chen X, Qiu L, Wang H, et al. Susceptibility of *Exopalaemon carinicauda* to the infection with shrimp hemocyte iridescent virus (SHIV 20141215), a strain of decapod iridescent virus 1 (DIV1) [J]. Viruses, 2019, 11(4): 387.

Viral nervous necrosis of teleost fish

Author: Shi Chengyin
E-mail: shicy@ysfri.ac.cn

1 Background

1.1 Different names of viral nervous necrosis (VNN)

VNN of teleost fish has different names, for example, viral encephalopathy and retinopathy (VER) and encephalomyelitis. It has emerged as a major constraints on the culture and sea ranching of a number of fish species.

1.2 Short research history of VNN

The research history of VNN is very short (about 30 years). VNN was first described in 1987 and the causative agent of VNN was identified as iridovirus in 1992 (Table 3-6-1) (Munday et al., 2002).

Table 3-6-1 Short research history of VNN

Year	Region	Fish species	Diseases
1987	Australia	Asian seabass	Brain lesions probably associated with nodavirus infection: brief description
1988	Australia	Asian seabass	Detailed description
1988	Caribbean	European seabass	An identical pathological condition
1990	Japan	Japanese parrotfish	Nervous necrosis; non-enveloped, icosahedral virus, 34 nm in diameter
1990	Australia	Asian seabass larvae	Similar virus, 25-30 nm in diameter
1992	Japan	Striped jack larvae	The virus was identified as a new member of the family Nodaviridae

1.3 Significance of VNN

VNN is the major viral disease of marine fish in East Asia and Southeast Asia. More than 64 fish species can be infected by VNN and many of them are mariculture fish. More significantly, VNN can cause high mortalities (nearly 100%) of fish larvae and juveniles (Figs. 3-6-1 to 3-6-2).

Fig. 3-6-1　Infected tongue sole larvae with looping swimming

Fig. 3-6-2　Infected dragon grouper larvae

2　Epidemiology

2.1　Fish species affected by VNN

More than 64 species of fish in 26 families and 14 orders can be naturally affected by VNN. Most of them (19 families) belong to the order Perciformes. Three families are in the order Pleuronectiformes.

The important mariculture fish species affected by VNN and the corresponding epidemic areas are listed below (Table 3-6-2).

Table 3-6-2　The important mariculture fish species affected by VNN and the corresponding epidemic areas

Fish species affected by VNN	Epidemic areas of VNN
Order Perciformes	
Asian seabass (*Lates calcarifer*)	China, Southeast Asia, Australia
Japanese seabass (*Lateolabrax japonicus*)	Japan
Crimson snapper (*Lutjanus erythropterus*)	China
Mangrove red snapper (*Lutjanus argentimaculatus*)	Southeast Asia
Cobia (*Rachycentron canadum*)	China
Dragon grouper (*Epinephelus lanceolatus*)	China, Southeast Asia
Tialipia (*Oreochromis niloticus*)	Mediterranean
European seabass (*Dicentrarchus labrax*)	Mediterranean
Gilthead seabream (*Sparus aurata*)	Mediterranean
Order Pleuronectiformes	
Japanese flounder (*Paralichthys olivaceus*)	Japan
Tongue sole (*Cynoglossus semilaevis*)	China

2.2 Distribution

VNN is widely distributed all around the world, except South America. The main distributing areas include East Asia, Southeast Asia, Oceania, Mediterranean area, United Kingdom, Norway, North America and Caribbean.

2.3 Susceptible stages of the host

The susceptible stages of most affected fish are larvae or juveniles. Losses tend to be very high in these stages. However, significant mortalities have occurred in older fish up to harvest size in recent years. The higher water temperature is, the greater likelihood of disease in older fish occurs (Table 3-6-3) (Munday et al., 2002).

Table 3-6-3 Important clinical features of VNN in larval and juvenile fish

Fish species	Earliest occurrence	Usual onset	Latest occurrence	Usual mortality	Highest mortality
Asian seabass	9 dph	15-18 dph	≥ 24 dph	50%-100% per month	100% in <1 month
European seabass	10 dph	25-40 dph	≥ 12 month	10% per month	—
Redspotted grouper	14 dph (total length: 7-8 mm)	9-10 mm	< 40 mm	80%	Up to 100%
Brownspotted grouper	—	20-50 mm	—	50%-80%	—
Striped jack	1 dph	1-4 dph	< 20 (total length: 8 mm)	100%	—
Japanese parrotfish	Total length: 6-25 mm		< 40 mm	—	Up to 100%
Japanese flounder	35 dph (total length: 17-18 mm)	—	—	100%	—
Turbot	< 21 dph	Total length: 25 mm	50-100 mg	—	Up to 100%

Note: dph, days post-hatch.

2.4 Transmission

2.4.1 Horizontal transmission

NNV can be transmitted horizontally by contaminated feedstock, contaminated water or rearing equipment. As a result, it is very difficult to control the disease. The horizontal transmission of NNV is affected by stocking density, water temperature and virulence of the virus.

The resistance of NNV to environmental conditions is high. The virus particles can tolerate pH 2-9, or keep infectivity in sea water at 15 ℃ for more than a year.

2.4.2 Vertical transmission

NNV can be detected in broodstock gonads. It can spread from broodstock to larvae through eggs or genital fluids.

3 Histopathology

3.1 Clinical signs

In general, the clinical signs relate to the lesions present in the brain and retina (Figs. 3-6-3 to 3-6-6) (Jaramillo et al., 2017). Most infected fish show abnormal movement, *e.g.*, uncoordinated swimming, spiral swimming and darting (Jaramillo et.al., 2017). Swim bladder hyperinflation has been reported in groupers, Asian seabass and European seabass. Larval halibut become paler, whereas groupers usually become darker. Overall, the main outcome is mass mortality, especially for larvae (Munday et.al., 2002).

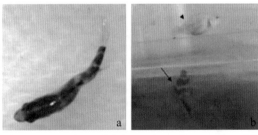

Fig. 3-6-3 Diseased Asian seabass (*Lates calcarifer*)

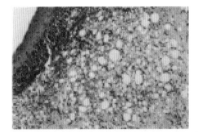

Fig. 3-6-4 Severe vacuolation of the brain

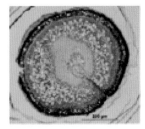

Fig. 3-6-5 Severe vacuolation of the retina

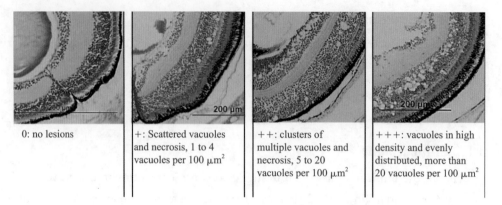

Fig. 3-6-6　Scoring of lesions in brain and retina

3.2　Histopathology: electron microscope

The infected cells of NNV include neurones, astrocytes, oligodendrocytes and microglia (Fig. 3-6-7) (Yoshikoshi et al., 1990).

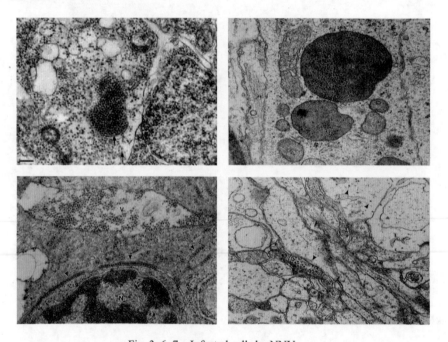

Fig. 3-6-7　Infected cells by NNV

4　Etiology

4.1　Taxonomy of the pathogen

The family of Nodaviridae contains three genera: *Alphanodavirus*, *Betanodavirus* and *Gammanodavirus*. Viruses in genus *Alphanodavirus* are originally isolated from insects. Viruses in genus *Betanodavirus* usually affect the nervous system of fish, leading to behavioural abnormalities and extreme high mortalities. Viruses in genus *Gammanodavirus* (for example, the pathogen of white tail disease which affects the giant freshwater prawn) infect crustaceans. NNVs are classified as species of *Betanodaviruse* (Fig. 3-6-8).

- Atlantic cod nervous necrosis virus
- Atlantic halibut nodavirus
- *Atractoscion nobilis* nervous necrosis virus
- Baramundi nervous necrosis virus
- Barfin flounder nervous necrosis virus
- Chinese catfish nervous necrosis virus
- Cobia nervous necrosis virus
- *Dicentrarchus labrax* encephalitis virus
- Dragon grouper nervous necrosis virus
- *Epinephelus aeneus* encephalitis virus
- *Epinephelus coioides* nervous necrosis virus
- *Epinephelus tauvina* nervous necrosis virus
- European eel nervous necrosis virus
- Firespot snapper nervous necrosis virus
- Guppy nervous necrosis virus
- *Hippoglossus hippoglossus* betanodavirus
- Hump-back grouper nervous necrosis virus

- Japanese flounder nervous necrosis virus
- *Lates calcarifer* encephalitis virus
- Malabaricus nervous necrosis virus
- *Melanogrammus aeglefinus* nervous necrosis virus
- *Mugil cephalus* encephalitis virus
- Redspotted grouper nervous necrosis virus
- Sevenband grouper nervous necrosis virus
- *Solea senegalensis* nervous necrosis virus
- Striped Jack nervous necrosis virus
- Tiger puffer nervous necrosis virus
- Turbot nodavirus
- *Umbrina cirrosa* nodavirus
- White star snapper nervous necrosis virus
- Yellow grouper nervous necrosis virus
- Yellow-wax pompano nervous necrosis virus
- *Zebrias zebra* nervous necrosis virus
- ...

Fig. 3-6-8   Reported species or isolates of NNV

4.2   Virion structure: electron microscope

The virion of NNV is icosahedral, non-enveloped, with the diameter of around 25 nm (20-34 nm). Observed by electronic microscopy, the virions may be membrane bound by endoplasmic reticulum or are free in the cytoplasm and may be present as paracrystalline arrays (Figs. 3-6-9 to 3-6-11) (Tang et al., 2002).

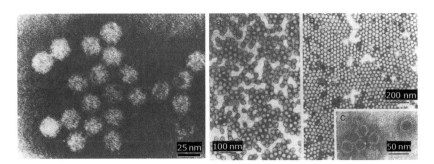

Fig. 3-6-9   Purified NNV particles

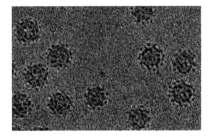

Fig. 3-6-10   CryoEM micrograph of MGNNV

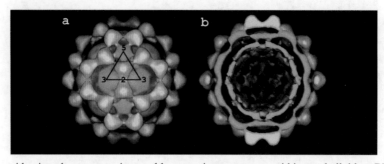

a. outside view; b. cutaway view; gold: protrusions, green: caspid inner shell, blue: RNA.

Fig. 3-6-11　CryoEM maps of MGNNV at 23 Å resolution

4.3　Molecular biology

The genome of NNV is composed of two single stranded positive-sense RNA. The sequence of RNA1 is about 3.1 kb and includes an open reading frame (ORF) encoding a RNA dependent RNA polymerase (RdRp). The sequence of RNA2 is about 1.4 kb and encodes the capsid protein (Cp) (Figs. 3-6-12 to 3-6-14).

(to be continued)

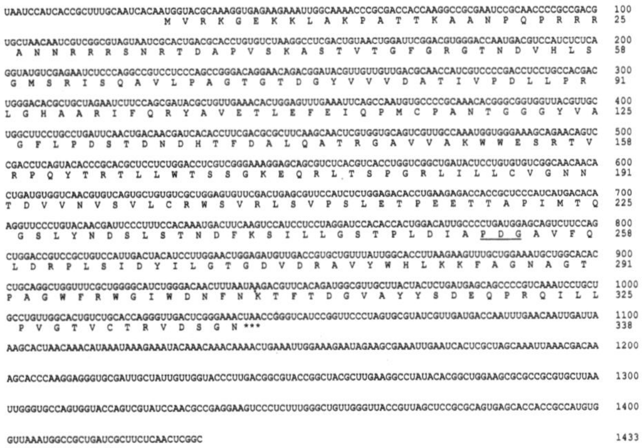

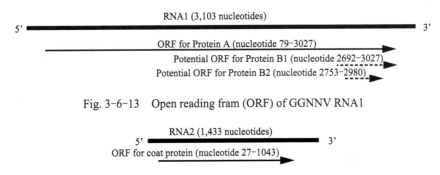

Fig. 3-6-12 RNA1 and RNA2 sequences of GGNNV

Fig. 3-6-13 Open reading fram (ORF) of GGNNV RNA1

Fig. 3-6-14 Open reading fram (ORF) of GGNNV RNA2

4.4 Genotypes (species)

NNV (genus *Betanodavirus*) generally contains four genotypes or species: SJNNV, TPNNV, BFNNV and RGNNV. However, two reassortant viruses, RGNNV/SJNNV and SJNNV/RGNNV, are identified in Mediterranean area (Fig. 3-6-15, Table 3-6-4) (Toffan et al., 2017).

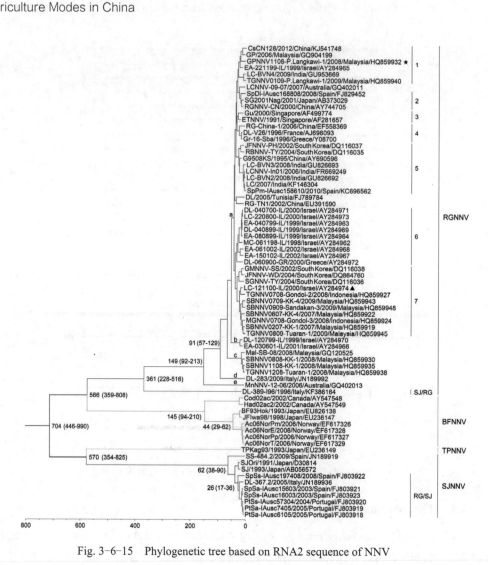

Fig. 3-6-15　Phylogenetic tree based on RNA2 sequence of NNV

Table 3-6-4　Genotypes (species) of NNV

Genotype	Serotype	Host fish	*In vitro* optimal growth temperature/℃
SJNNV	A	Striped jack	20–25
TPNNV	B	Tiger puffer	20
BFNNV	C or B	Cold-water fish: Atlantic halibut, Atlantic cod, flounders, etc.	15–20
RGNNV	C	Warm-water fish: Asian seabass, European seabass, groupers, etc.	25–30
RGNNV/SJNNV (reassortant)	A	Striped jack, Senegalese sole, gilthead seabream, European seabass	25–30
SJNNV/RGNNV (reassortant)	C	European seabass, sevenband grouper	20–25

5 Diagnosis/Detection

5.1 Methods

Four types of technical methods can be used to diagnose VNN (OIE, 2016). The methods include: a. demonstrating characteristic lesions in the brain and/or retina by light microscopy; b. detection of virions, viral antigens or viral RNA by electron microscopy, serology or molecular techniques; c. detection of specific antibodies in sera or body fluids; d. tissue culture of virus (Table 3-6-5).

Table 3-6-5 Rating of detection methods of NNV (OIE, 2016)

Method	Targeted surveillance			Presumptive diagnosis	Confirmatory diagnosis
	Larvae	Juveniles	Adults		
Histopathology	d	d	d	b	d
Histopathology followed by immunostaining	d	d	d	b	d
Transmission electron microscopy	d	d	d	c	d
Isolation in cell culture followed by immunostaining or PCR	a	a	d	a	a
RT-PCR	b	b	b	a	a
RT-PCR followed by sequencing	d	d	d	b	a
Real-time PCR	a	a	a	a	a

The designations used in the table indicate: a, the method is the recommended method for reasons of availability, utility, and diagnostic specificity and sensitivity; b, the method is a standard method with good diagnostic sensitivity and specificity; c, the method is applied in some situations, but cost, accuracy, or other factors severely limits its application; d, the method is presently not recommended for this purpose.

5.2 Molecular techniques

Many molecular methods for diagnosing VNN have been established successively. The methods are indirect ELISA in 1992, RT-PCR in 1994, nested RT-PCR in 1999, real-time RT-PCR in 2004, RT-LAMP in 2009, RT-qPCR in 2010, CPA-LFD in 2015, nanoparticle-based lateral flow biosensor in 2015, molecular-beacon based microfluidic system in 2015, and so on.

5.3 Cross-priming isothermal amplification coupled with lateral flow dipstick (CPA-LFD)

The CPA-LFD method for detection of RGNNV was established in 2015. This method is simple, rapid with high sensitivity and good specificity, and can be widely applied for rapid detection of RGNNV on the spot (Figs. 3-6-16 to 3-6-17).

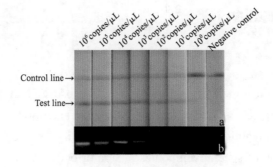

a. CPA-LFD; b. RT-PCR.

Fig. 3-6-16 Sensitivity comparison of CPA-LFD and RT-PCR for detection of NNV

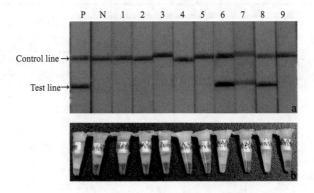

a. CPA-LFD; b. LAMP Kit.

Fig. 3-6-17 Comparison of CPA-LFD and LAMP Kit for detection of NNV

6 Control (Munday et al., 2002)

6.1 Main strategy: exclude virus from premises totally

A. Detect NNV in fingerlings by nested RT-PCR before importing them into a facility.

B. In hatcheries: detect NNV in gonadal tissues of broodfish.

C. Wash eggs with ozone water.

E. The broodfish should not be stressed by too frequent spawning.

6.2 Supplementary measures

A. Clean, dry and disinfect hatching facilities between batches of larvae.

B. Maintenance of biological security between different parts of the hatchery.

C. Apply suitable disinfectants.

D. Lower larvae density to $<10 \text{ L}^{-1}$ in ponds.

6.3 Vaccines (Doan et al., 2017)

6.3.1 Inactivated vaccines

The inactivated vaccines of NNV are listed in Table 3-6-6.

Table 3-6-6 Inactivated vaccines of NNV

Inactivated vaccine	Species and size	Method	Results/RPS
BEI-inactivated HGNNV Formalin-inactivated HGNNV	Orange-spotted grouper (early larval stage about 40 dph. BW: about 0.2 g. TBL: about 2.4 cm)	Injection	79% (BEI) 39% (Formalin)
Formalin-inactivated RGNNV	Sevenband grouper (juvenile about 25.4 g)	Injection	60%
BEI-inactivated HGNNV	Adult orange-spotted grouper (BW about 1.35 kg)	Injection	High efficiency
Formalin-inactivated RGNNV type	Brown-marbled grouper (5 g)	Injection	86%-100%

6.3.2 Recombinant vaccines

The recombinant vaccines of NNV are listed in Table 3-6-7.

Table 3-6-7 Recombinant vaccines of NNV

Recombinant vaccine	Species and size	Method	Results/RPS
Recombinant capsid protein (Artemia-encapsulated recombinant NNV capsid protein)	Orange-spotted grouper (Larvae: 35 dph)	Oral	64.5%
Recombinant capsid protein	Orange-spotted grouper (fry)	Oral	78.3%
Recombinant capsid protein (rT2 vaccine)	Turbot (1-3 g)	Injection	82%
Recombinant capsid protein Vaccine	Turbot (Juvenile: about 2.2 g)	Injection	57%
Recombinant protein vaccine	Seven-band grouper (28 g)	Injection	88%

6.3.3 Virus-like particles (VLPs) vaccines

VLPs vaccines of NNV are listed in Table 3-6-8.

Table 3-6-8 VLP vaccines of NNV

VLPs vaccines	Species and size	Method	Results/RPS
GNNV VLPs	Dragon grouper (20 g), Malabar grouper (20 g)	Injection	Significant efficiency
MGNNV VLPs, SB2 VLPs	European seabass (66 g), European seabass (22 g)	Injection	71.7%-89.4%, 27.4%-88.9%

6.3.4 DNA vaccines

The DNA vaccines of NNV are listed in Table 3-6-9.

Table 3-6-9 DNA vaccines of NNV

DNA vaccines	Species and size	Method	Results/RPS
pFNCPE42	Asian seabass (juvenile)	Injection	77.33%

6.4 Resistant heritability

The heritability estimates of resistance to viral diseases in farmed fish species are listed in Table 3-6-10.

Table 3-6-10 Heritability estimates of resistance to viral diseases in farmed fish species

Pathogen	Species (host)	Heritability: h^2 ($\pm$ SE)	
		Binary traits	Time until death
Viral nervous necrosis viruses (VNNV)	Atlantic cod (*Gadus morhua*)	$h^2 = 0.75 \ (\pm 0.11)$	
Viral hemorrhagic septicemia virus (VHSV)	Rainbow trout (*Oncorhynchus mykiss*)	$h^2 = 0.57$	$h^2 = 0.11 \ (\pm 0.10)$
Infectious salmon anaemia virus (ISAV)	Atlantic salmon (*Salmo salar*)	$h^2 = 0.40 \ (\pm 0.04)$	
Infectious pancreatic necrosis virus (IPNV)	Atlantic salmon (*Salmo salar*)	$h^2 = 0.43$	$h^2 = 0.16$
Salmon pancreases disease virus (SPDV)	Atlantic salmon (*Salmo salar*)		$h^2 = 0.21 \ (\pm 0.005)$
Koi herpesvirus (KHV)	Common carp (*Cyprinus carpio*)	$h^2 = 0.79 \ (\pm 0.14)$	

6.5 Different transmission routes of NNV and possible prevention modes

NNV can spread via both vertical and horizontal transmissions. The transmission routes and possible prevention modes of NNV are shown in Fig. 3-6-18 (Doan et al., 2017). The possible prevention modes are as follows: a. vaccination; b. serological diagnostic (ELISA) to screen and eliminate seropositive individuals; c. direct diagnostic (RT-qPCR) to screen and eliminate positive individuals or germplasm; d. ozone/UV/bleach treated water; e. strict control of feed input to avoid NNV infected trash fish; f. unique equipment kit for each tank/pond/cage and adapted decontamination of equipment after use; g. biosecurity measures during all production cycles; h. ozone treatment of artemia before feeding.

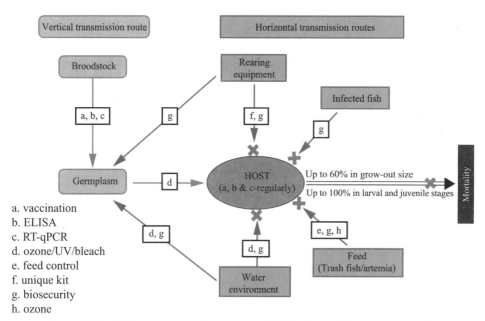

Fig. 3-6-18 Different transmission routes of NNV and possible prevention modes

References

Doan Q K, Vandeputte M, Chatain B, et al. Viral encephalopathy and retinopathy in aquaculture: a review [J]. Journal of Fish Diseases, 2017, 40(5): 717–742.

Jaramillo D, Hick P, Whittington R J. Age dependency of nervous necrosis virus infection in barramundi *Lates calcarifer* (Bloch) [J]. Journal of Fish Diseases, 2017, 40(8): 1089–1101.

Munday B L, Kwang J, Moody N. Betanodavirus infections of teleost fish: A review [J]. Journal of Fish Diseases, 2002, 25(3): 127–142.

OIE. Chapter 2.3.12. Viral encephalopathy and retinopathy [M]. In: OIE Manual of Diagnostic Tests for Aquatic Animals, 7th edn, Paris: OIE, 2016: 444.

Tang L, Lin C S, Krishna N K, et al. Virus-like particles of a fish nodavirus display a capsid subunit domain organization different from that of insect nodaviruses [J]. Journal of Virology, 2002, 76(12): 6370–6375.

Toffan A, Pascoli F, Pretto T, et al. Viral nervous necrosis in gilthead sea bream (*Sparus aurata*) caused by reassortant betanodavirus RGNNV/SJNNV: An emerging threat for Mediterranean aquaculture [J]. Scientific Reports, 2017, 7: 46755.

Yoshikoshi K, Inoue K. Viral nervous necrosis in hatchery-reared larvae and juveniles of Japanese parrotfish, *Oplegnathus fasciatus* (Temminck & Schlegel) [J]. Journal of Fish Diseases, 1990, 13(1): 69–77.

Viral covert mortality disease and its pathogen

Author: Zhang Qingli
E-mail: Zhangql@ysfri.ac.cn

1 Viral covert mortality disease

An emerging disease, called covert mortality from disease (CMD), became widely pandemic, and has attacked the shrimp farming industries in China since 2009 (Fig. 3-7-1). At ponds level, the moribund shrimp suffering disease of CMD hided under the bottom of deep water rather than swimming to the surface, so it was called "covert mortality" by the local farmers. The farming shrimp suffering CMD showed continuous low-level mortalities, so the CMD was also called running mortality syndrome (RMS). The clinical signs of shrimp individuals attacked by the CMD included hepatopancreatic atrophy and necrosis, empty stomach and guts, slow growth, soft shell, in many cases, abdominal muscle whitening and necrosis (Fig. 3-7-2) (Zhang et al., 2016). The CMD or RMS was more common in the farming white leg shrimp *Penaeus vannamei*, than in the other farming shrimp species. Recently, a new RNA virus, covert mortality nodavirus (CMNV), was proved to be the infectious agent of the CMD (Zhang et al., 2014). Therefore, the CMD were renamed as viral covert mortality disease (VCMD) academically to specify and emphasize the viral cause (Zhang et al., 2017).

Fig. 3-7-1 Mass mortality of farming *P. vannamei* caused by the viral covert mortality disease (VCMD) at ponds level

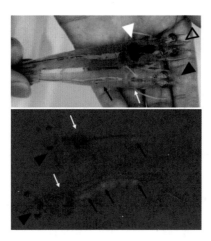

Black arrows show whitening muscle of abdominal segment. White arrows indicate hepatopancreas atrophy and color fading. The framed triangles showed the healthy shrimp and the black triangles indicate the shrimps suffering CMD.

Fig. 3-7-2 Clinical signs of *P. vannamei* suffering from covert mortality disease (CMD) (Zhang et al., 2014)

VCMD usually occurs on 30-80 d post stocking, with cumulative mortality up to 80%. In some serious cases, it occurs on 10-20 d of stocking post larvae into grow-out ponds. Cases of asymptomatic infection detecting with CMNV kits were also found in farms.

The mortality caused by VCMD would be accelerated by the sudden change of ponds environment including the high pond water temperature above 28 ℃, or the sudden decrease of salinity because of the rain, or the sharp increase of temperature, ammonium nitrogen and pH value.

The VCMD was found to emerge in white leg shrimp prior to 2009 and caused substantial shrimp mortality in China. Besides China, the prevalence of VCMD (RMS) was also found in farming *P. vannamei* from several countries in the Southeast Asia and South Asia, including Thailand, India, Malaysia, Vietnam and Indonesia (Siripong et al., 2016; Zhang et al., 2017; Pooljun et al., 2016; Varela, 2018).

2 Pathogen—CMNV

Histopathological examination of shrimp suffering the VCMD revealed coagulative necrosis of striated muscle similar to typical histopathology features of infectious myonecrosis virus (IMNV), *P. vannamei* nodavirus (PvNV) and *Macrobrachium rosenbergii* nodavirus (MrNV). However, shrimp of VCMD was tested to be negative for IMNV, MrNV and PvNV by RT-PCR (Zhang et al., 2014). A novel nodavirus (tentatively named as CMNV) was proved to be the pathogen of the VCMD of shrimp. CMNV was a spherical virus that showed icosahedral characters, and its diameter was about 32.1 nm ± 5.5 nm (Fig. 3-7-3).

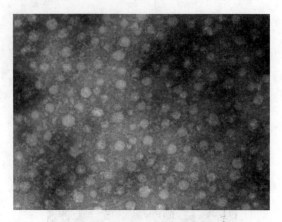

Fig. 3-7-3 Transmission electron micrographs of the negative stained CMNV virions

CMNV was a new viral species of in the genus *Alphanodavirus*. The genomic RNA-1 and RNA-2 of CMNV were characterized recently (Fig. 3-7-4) (Xu et al., 2020). RNA-1 is 3,228 bp in length, and it contains two putative open reading frames (ORFs), one encoding the RNA dependent RNA polymerase (RdRp) with a length of 1,043 amino acids and another encoding the protein B2 with a length of 132 amino acids. RNA-2 is 1,448 bp in length and it encodes a capsid protein of 437 amino acids. It is found that the CMNV subgenomic RNA-3, originating from RNA-1, does not encode the protein B1. The lack of protein B1 shows the difference between CMNV and the other alphanodaviruses (except BoV). That is, the current data analysis shows that the RNA-1 of CMNV, as well as the RNA-1 of BoV in *Alphanodavirus*, lacks the ORF encoding protein B1. CMNV shared the highest similarity of 51.78% for RdRp with the other known nodaviruses (Xu et al., 2020).

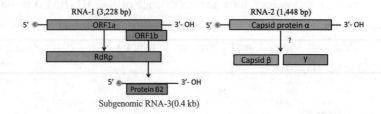

Fig. 3-7-4 Genomic structure and the deduced proteins in CMNV genomic RNA-1 and RNA-2 (Xu et al., 2020)

The pathogenicity of CMNV was not as high as the white spot syndrome virus (WSSV) to the *P. vannamei*. In a previous study, the artificial challenge test showed that cumulative mortality of shrimp in the *per os* infection was 84.85% ± 2.14% by Day 10 post-injection. In contrast, there was no mortality of shrimp in the blank group injected with PBS (Fig. 3-7-5) (Zhang et al., 2014). Artificial infection by CMNV injection would cause more mortality according to this research.

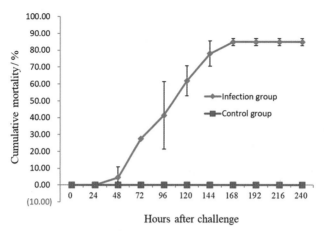

The shrimp individuals in the infection group were feed with the diseased shrimp and the shrimp individuals in the control group were injected with PBS buffer.

Fig. 3-7-5 Cumulative mortality curves of *P. vannamei* in the experimental infection of CMNV (Zhang et al., 2014)

It was found that CMNV would cause obvious hepatopancreatic necrosis in the VCMD cases of farming *P. vannamei*, *P. japonicus* and *P. chinensis* (Zhang et al., 2014; Zhang et al., 2017). The typical pathological changes presented in the abdominal muscle was muscle fragmentation tending towards coagulative, hyaline degeneration, muscular lysis and myonecrosis. Multifocal myonecrosis in the striated muscle, as well as the hemocytic infiltration and karyopyknosis of haemocytes could be observed in the necropsied muscles (Figs. 3-7-6a and b) (Zhang et al., 2014). *In situ* hybridization (ISH) of the muscle with the CMNV probe showed that the strong CMNV probe signals occurred at the region of the muscle necrosis (Fig. 3-7-6c).

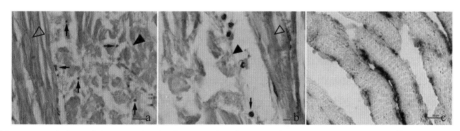

a. and b. H&E staining of abdominal muscle necrosis (Zhang et al., 2014); c. ISH photos of abdominal muscle. Note the dark purple color of the hybridization of the CMNV probes. Karyopyknotic nuclei were marked by the black arrows and the light purple intranuclear inclusions were indicated by the white arrows. The black triangles showed the muscle that showed fragmentation tending towards coagulative and dissolving necrosis.

Fig. 3-7-6 Histopathological and ISH photos of abdominal muscle of *P. vannamei* suffering from viral covert mortality disease

3 Transmission

3.1 Vertical transmission

Vertical transmission of CMNV in *Exopalaemon carinicauda* was reported several years ago (Liu et al., 2017). The ovarian and testis tissues in the artificial infected parental *E. carinicauda* were determined to be CMNV-positive in the molecular detection of CMNV. Fertilized eggs were also proved to be CMNV-positive of Reverse transcript nest PCR(RT-nPCR) whether the fertilized eggs coming from the CMNV-positive female broodstock mating with the CMNV-negative male broodstock, or the eggs originating from the CMNV-negative female broodstock mating with the CMNV-positive male. CMNV were shown to be present in the oocytes of ovarian, spermatocytes of testis, fertilized eggs and the cell of nauplii by the ISH assays. CMNV virions were observed in oogoniums, oocytes, spermatocytes, fertilized eggs and nauplii in the TEM analysis (Liu et al., 2017). The results of the molecular, histopathology and TEM assays suggested that CMNV could transmit vertically via sperm and oocyte in *E. carinicauda*, which indicated that we should pay more attentions to the risk of vertical transmission of CMNV in the other crustaceans.

3.2 Horizontal transmission

The horizontal transmission of CMNV in the farming ponds was well studied in the past several years. It had been found that many organisms co-inhabiting with farming shrimp in the ponds could be infected by the CMNV, and these co-inhabiting organisms were deduced to be acting as the vectors of CMNV spread in farming shrimp. Eleven species of invertebrates inhabiting shrimp ponds were detected RT-nPCR or RT-LAMP positive, including brine shrimp *Artemia sinica*, barnacle *Balanus* sp., rotifer *Brachionus urceus*, amphipod *Corophium sinense*, Pacific oyster *Crassostrea gigas*, hermit crab *Diogenes edwardsii*, common clam *Meretrix lusoria*, ghost crab *Ocypode cordimundus*, hyperiid amphipod *Parathemisto gaudichaudi*, fiddler crab *Tubuca arcuata*, and an unidentified gammarid amphipod (Liu et al., 2018). Five wild crustacean species, including *C. sinense, D. edwardsii, O. cordimanus, P. gaudichalldi* and *T. arcuata*, were also tested to be ISH positive of CMNV. These five species might act as reservoir hosts of CMNV in horizontal transmission (Liu et al., 2018). In addition, CMNV was proved to be capable of naturally crossing the species barrier and infecting fish species (Zhang, et al., 2018; Zhang et al., 2019; Wang, et al., 2019a; Wang et al., 2019b.).

4 Prevalence

A continuous epidemiological survey of pathogens of farming shrimp was conducted in China in the past ten years.

The CMNV epidemic in China first originated in Guangxi and Hainan provinces in 2002–2003, and then gradually spread to Guangdong and Fujian provinces. With the wide application of shrimp post larvae with CMV positive produced by companies in Fujian Province, the virus spread to northern coastal provinces and cities, and even inland Xinjiang, Henan and other places. The investigation results revealed that CMNV was one of the most important viral diseases in farming shrimp species in the coastal areas in China in 2017–2019 (Fig. 3-7-7). Among the major cultured shrimp species in eleven coastal provinces and cities in China from 2013 to 2019, the annual detection rates of CMNV in farming crustaceans were 45.93% (130/283), 27.91% (84/301), 20.85% (54/259), 26.8% (68/254), 16.3% (63/387), 29.15% (58/199) and 15.9% (55/345). CMNV showed a national prevalence, wide range of hosts with a high prevalence (Zhang et al., 2017; Li et al., 2018; Li et al., 2019).

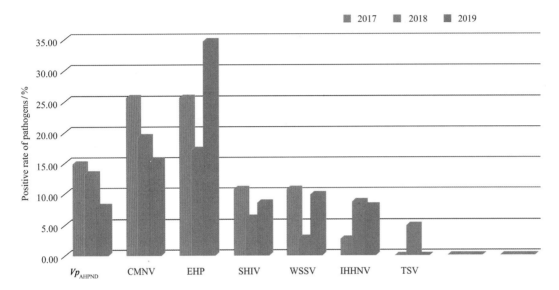

The abbreviations were represented as following, *Vibrio parahaemolyticus* (Vp_{AHPND}) for the acute hepatopancreatic necrosis disease, enterocytozoon hepatopenaei (EHP), shrimp hemocyte iridescent virus (SHIV), white spot syndrome virus (WSSV), infectious hypodermal and hematopoietic necrosis virus (IHHNV), Tora syndrome virus (TSV), yellow head virus (YHV), and infectious myonecrosis virus (IMNV).

Fig. 3-7-7 Prevalence of CMNV in the major farming crustaceans in the coastal areas in China in 2017–2019

5 Epidemic area

In the process of tracing the origin of CMNV, we found that a high proportion of CMNV positive was detected in samples of *Penaeus monodon* with slow growth syndrome in Thailand. In addition, Professor Flegel from Thailand conducted an epidemiological investigation of 200 ponds in 2014 and found that the positive detection rate of CMNV in sick shrimp was as high as 43%. In 2018, our laboratory found that shrimp samples from Malaysia, Indonesia and India were detected to be CMNV positive. Meanwhile, shrimp samples collected from Thailand were found positive for CMNV by RT-PCR and ISH by other research groups (Pooljun et al., 2016; Thitamadee et al., 2016). In addition, CMNV positive samples were also reported in the other major shrimp farming areas including the countries from the South Asia, the middle-South America (Varela, 2016, 2018; Li et al., 2019).

6 Molecular diagnostics

Up to now, several detection methods of CMNV based on the RdRp gene sequence has been established for the different diagnostic purposes. The reported molecular diagnostic methods included the reverse transcript nest PCR (RT-nPCR) (Zhang, et al., 2014), the reverse transcript loop-mediated isothermal amplification (RT-LAMP) (Zhang et al., 2017b), the TaqMan real time RT-PCR (Pooljun et al., 2016) and a TaqMan real time RT-PCR with high compatibility (Li et al., 2018).

Here will enclose a recommended protocol for CMNV molecular detection based on RT-nPCR.

According to internationally recognized requirement of viral detection criterion included in the CNAS-CL62 (2016) as "Application note for detection and calibration laboratory capability approval criteria in the field of gene amplification detection", the experimental operation rules for detecting CMNV using reverse transcription nested PCR were formulated in this protocol (Zhang et al., 2014; Zhang et al., 2018).

6.1 Prepared works before starting

6.1.1 Laboratory space layout

The overall layout and various parts' arrangements of the laboratory should consider to reduce the potential contamination of samples and the harm to personnel. In principle, separate working areas should be provided, including (but not limited to) PCR mixture preparation area, nucleic acid extraction area, nucleic acid amplification area and amplification product analysis area.

6.1.2 Instrument and equipment

Refrigerator, pipette (Measuring range: 2 μL, 20 μL, 200 μL, 1,000 μL), clean bench, tissue grinder, benchtop centrifuge, metal bath, microwave oven, PCR instrument, nucleic acid electrophoresis analysis system (Electrophoresis apparatus, electrophoresis tank) and Gel imager.

6.1.3 Consumables and reagents

Grinding beads, disposable tweezers, disposable grinding rod, 1.5 mL RNase-free centrifuge tube, agarose, 50 mL measuring cylinder, 50 mL or 100 mL Erlenmeyer, tips (10 μL, 200 μL, 1,000 μL), 95% ethanol, TAE running buffer, sterile double distilled water and nucleic acid dye.

6.1.4 Precautions

Nucleic acid dyes should be stored at 2−8 ℃ in the dark; this reagent is a low toxicity reagent and disposable gloves must be worn when handling.

Before starting, personnel should confirm that the reagents in the kits are in a dissolved state, and reagent crystals due to low temperature should to be avoid, which will affect the working concentration.

6.2 Steps in detailed

6.2.1 Sample preparation

This step should be completed in the nucleic acid extraction area.

Sample selection: Select VCMD susceptible target organs (hepatopancreas, appendage). For samples with complex components such as feces, special attention should be paid to the quality of RNA extraction. In this protocol, quality control is performed by amplification of the intrinsic genes.

Sample homogenization: Take approximately the amount of tissue equivalent to 1/5 of the volume of 1.5 mL EP tube, place it into a 1.5 mL nuclease-free EP tube, then add 700 μL of absolute ethanol and 3−4 grinding beads, and grind it for 1 min at 1,500 r/min with a tissue grinder, until the sample is ground to a slurry. Take 60 μL (30−100 mg) of the homogenate slurry into a new 1.5 mL nuclease-free EP tube and centrifuge it 12,000 r/min for 1 min, discarding the supernatant, and then dry the precipitation at room temperature for 10−20 min to allow the residual ethanol to fully evaporate.

Note 1: Zirconia or yttrium oxide grinding beads of Beijing Sigma Data Technology Co., Ltd., Qingdao Aikebao Biotechnology Co., Ltd, and Boao Yijie (Beijing) Technology Co., Ltd., are recommended with a specification of about 3 mm.

Note 2: You can choose other homogenate method of tissue according to the equipment in the laboratory.

6.2.2 Extraction of total RNA from tissues

This step should be completed in the nucleic acid extraction area.

A. Add 0.5–0.75 mL of TRIzol™ reagent to the above prepared sample tube, mix well and shake it at room temperature for 5 min, centrifugate it at 10,000 r/min for 5 min at 4 ℃, and then take the supernatant to a new RNase-free 1.5 mL EP tube.

B. Add chloroform (volume about 1/5 of supernatant) into the EP tube, shake vigorously for 15 s, and then stand for 5 min at room temperature.

C. Centrifuge at 12,000 r/min for 15 min at 4 ℃.

D. Pipette the upper aqueous phase carefully and transfer it to a new RNase-free 1.5 mL EP tube.

E. Add isopropanol (volume equal to the upper aqueous phase) into the new 1.5 mL EP tube, mix the mixture thoroughly by upside down the tube for several times, let it stand at room temperature for 10 min.

F. Centrifuge at 12,000 r/min for 10 min at 4 ℃, and then discard the supernatant.

G. Add 1 mL of RNase-free 75% ethanol, wash the precipitation upside down, centrifugate it at 7,000 r/min for 5 min at 4 ℃, discard the supernatant carefully.

H. Dry the precipitation at room temperature.

I. Add 20–100 μL of RNase-free water to dissolve the RNA precipitation and immediately use it for RT-PCR or store it at –80 ℃ until use. Before use, adjust the RNA template concentration to 100–200 ng/μL.

Note: Total RNA extraction can be carried out by commercial kits. Recommended kit: ① RNAiso Plus (Brand: TaKaRa; Item number: 9108Q); ② TaKaRa MiniBEST Universal RNA Extraction Kit (Brand: TaKaRa; Item number: 9767); ③ TRIzol™ Reagent (Brand: ThermoFishier Scientific; Item number: 15596026).

6.2.3 RT-nPCR

Reverse transcription

Prepare the reverse transcription primer premix in the PCR system configuration area. Divide the premix in 5 μL per tube so that each reaction tube contains 4 μL of RNase-free water and 1 μL of 10 mmol/L primer Noda-R1 (Table 3-7-1) and store the sub-packaging premix at −20 ℃. Before reverse transcription, 1 μL of the sample RNA to be tested should be added to the above mentioned sub-packaging premix, and then pre-denatured the premix containing sample RNA at 70 ℃ for 5 min, immediately place the tube containing the premix into the ice-water mixture for 2 min.

Reverse transcription should be completed in the nucleic acid extraction area. Prepare each reverse transcription reaction premix 4 μL by mixing the following reagents: 2 μL 5× M-MLV reverse transcriptase buffer, 0.5 μL dNTP (10 mmol/L), 0.25 μL RNase inhibitor (40 U/μL), 0.5 μL M-MLV reverse transcriptase (200 U/μL), 0.75 μL of RNase-free water. Mix the premix well by pipetting or vortaxing, add the premix into the tube, slightly centrifugate it for collecting the liquid to the bottom of the tube, and then place it in metal bath at 42 ℃ for 1 h. After denaturation at 70 ℃ for 5 min, a cDNA template can be obtained.

When performing reverse transcription of total tissue RNA of the sample, a positive control, a negative control, and a blank control should be set. The positive control uses the CMNV positive shrimp tissue RNA as template, the negative control uses the CMNV negative shrimp tissue RNA as template, and the blank control uses the RNase-free water as template.

Meanwhile, the reverse transcription of the internal reference gene should be performed on the sample tissue total RNA, positive control, negative control and blank control to monitor the quality of the extracted nucleic acid, the process of reverse transcription and the possibility of contamination.

RT-PCR reaction

The reaction mixture of RT-PCR (Tables 3-7-2 to 3-7-3) should be prepared in the PCR system configuration area. It is recommended to prepare large volume premixes of enzyme-free for dispensing. Specific operations: Prepare large volume of premixes not including the Taq DNA polymerase, and store it at -20 ℃. Before the starting of test, the Taq DNA polymerase-free premix is thawed firstly, and then dispensed in 0.2 mL EP tube. The corresponding volume of Taq DNA polymerase is added in proportion into the 0.2 mL EP tube for test use.

Table 3-7-1　Primer sequences and amplified fragments

Primer name	Sequence (5'-3')	Length of amplicon	Effect
Noda-F1	AAATACGGCGATGACG	619 bp	CMNV first round
Noda-R1	ACGAAGTGCCCACAGAC		
CMNV-F	TCGCGTATTCGTGGAT	413 bp	CMNV second round
CMNV-R	TAGGGTCAAAAGGTGTAGT		
Internal reference-F	GCCTGAGAAACGGCTACCA	180 bp	Decapod internal control
Internal reference-R	CAGACTTGCCCTCCACTCG		

Table 3-7-2　First round of RT-nPCR

Ingredient	Volume/μL
10×Ex Taq Buffer	2.5
MgCl$_2$ (25 mmol/L)	2
dNTP (2.5 mmol/L each)	2
Noda-F1 (10 μmol/L)	1
Noda-R1 (10 μmol/L)	1
Ex Taq enzyme (5 U/μL)	0.1
Sterile double distilled water	15.4
cDNA template	1
Total capacity	25

Table 3-7-3 Second round of RT-nPCR

Ingredient	Volume/μL
10×Ex Taq Buffer	2.5
$MgCl_2$ (25 mmol/L)	2
dNTP (2.5 mmol/L each)	2
CMNV-F (10 μmol/L)	1
CMNV-R (10 μmol/L)	1
Ex Taq enzyme (5 U/μL)	0.1
Sterile double distilled water	15.4
Template: first round PCR product	1
Total capacity	25

For example, when preparing nine sample PCR amplification systems, the method of calculation is (23.9 μL Taq DNA polymerase-free premix + 0.1 μL Ex Taq)×10 = 239 μL Taq DNA polymerase-free premix + 1 μL Ex Taq enzyme.

PCR amplification should be completed in the nucleic acid amplification area; Electrophoretic analysis should be performed in the amplification product analysis area.

Amplification is performed on a PCR equipment, and the amplification procedure is as follows:

A. The first round of amplification procedure of CMNV: denaturation at 94 ℃ for 3 min; 94 ℃ for 20 s, 50 ℃ for 20 s, 72 ℃ for 40 s, 30 cycles; extension at 72 ℃ for 7 min. The second round of amplification procedure: denaturation at 94 ℃ for 3 min; 94 ℃ for 20 s, 50 ℃ for 20 s, 72 ℃ for 30 s, 30 cycles; extension at 72 ℃ for 7 min.

B. Internal reference gene amplification program: denaturation at 94 ℃ for 3 min; 94 ℃ for 45 s, 60 ℃ for 45 s, 40 cycles; extension at 60 ℃ for 7 min.

C. Prepare a 1% agarose gel and add the nucleic acid dye GeneFinder into the gel at a ratio of 1 : 10,000. For example, add 0.7 g of agarose to a buffer of 70 mL, heat the mixture and melt it by microwave, and then mix it with 7 μL nucleic acid dye GeneFinder.

D. Prepare the loading buffer according to this ratio of GeneFinder: 6 × Loading buffer = 1 : 9; mix 5 μL of the PCR product with 1 μL of the loading buffer, drop the mixture into agar gel wells, and then electrophorese it in running buffer at the voltage of 5 V/cm. Finally observe the gel by using a gel imager.

Notes: GeneFinder can be replaced with other commercial nucleic acid dyes with the same nucleic acid staining effect, such as GoldView, GeneGreen and GelRed.

6.2.4 Result analysis

Prerequisites: The first round and second round PCR of the positive control produced

619 bp/413 bp bonds in the gel. The internal reference PCR of the positive control produced 180 bp band in the gel. Meanwhile, the first round and second round PCR of the negative control did not produce 619 bp/413 bp bonds in the gel and the internal reference PCR of the negative control produced 180 bp bonds. The blank control showed no any bands (except the primer dimer band). The experimental results were credible.

The sample which produced the 619 bp band in the first round and 413 bp bands in the second round should be determined as strongly CMNV positive. If the 413 bp band appeared only in the second round PCR of one sample, the sample should be determined as weakly CMNV positive. Otherwise, the sample should be determined as CMNV negative. For confirmation of the specific amplification, the PCR amplicon of 619 bp/413 bp should be sequenced and compared with the reference sequence (see Appendix A).

Appendix A

AAATACGGCGATGACGGCTTGAGCCACAACCGAGTCAAACCTTTCATCAACAA
AGTAGCGAACGCACTAGGCCTATCGACTAAATATGAGAAATTTAATCCAGAAATAGG
AATATCATTTCTTGCTCGCGTATTCGTGGATCCTTTTAACACAAACACTTCCATTACG
GATCCCCTTCGCTGTTTACGCAAAATCCATCTCACCGCCCGGAACCCTTCCGTCCCT
TTAGCAGACGCATGTTGCGATCGAGTTGAAGGCTATCTCGTGACAGATGCCCTTACA
CCATTGGTGGGTGATTACTGTCGCTCTATGATTCGATTGTATGGTGGAGCTGCGTCAA
GTCAATTGGTAAGACTGAAGCGCAAAACCAGCAATTCCGAGAAGCCCAATTGGTTG
ACGAATGATGGTTCCTGGCCCCAAAATGCCGCCGATAAAGATGCTATGTTCAACGTC
CTTTGTGCCCGCACTCAAATTCATCCCGAGACGGTAAATAGCCTTATCGAACGCCTG
GCCAACATAACTACACCTTTTGACCCTATAATTACTGAGTTTAATGATGCCGCAAGTA
ACAGTAATACCATCGGCGTTGATGGTCCAGTAGGGTCTGTGGGCACTTCGT

Note: ① The GenBank accession number of the above sequence is KM112247.1 ② The positions of the outer primer and the inner primer are indicated by a single underline and a double underline respectively.

REFERENCES

Li X P, Wan X Y, Huang J, et al. Molecular epidemiological survey on covert mortality nodavirus (CMNV) in cultured crustacean in China in 2016–2017 [J]. Progress in Fishery Sciences ,2019, 40(2): 65–73 (In Chinese).

Li X P, Wan X Y, Xu T T, et al. Development and validation of a TaqMan RT-qPCR for the detection of convert mortality nodavirus (CMNV) [J]. J Virol Methods, 2018, 262: 65–71.

Liu S, Li J T, Tian Y, et al. Experimental vertical transmission of covert mortality nodavirus (CMNV) in *Exopalaemon carinicauda* [J]. J Gen Virol, 2017, 98(4): 652–661.

Liu S, Wang X H, Xu T T, et al. Vectors and reservoir hosts of covert mortality nodavirus (CMNV) in shrimp ponds [J]. J Invertebr Pathol, 2018, 154: 29-36.

Pooljun C, Direkbusarakom S, Chotipuntu P, et al. Development of a TaqMan real-time RT-PCR assay for detection of covert mortality nodavirus (CMNV) in penaeid shrimp [J]. Aquaculture, 2016, 464: 445–450.

Siripong T. Review of current disease threats for cultivated penaeid shrimp in Asia [J]. Aquaculture, 2016, 452: 69–87.

Thitamadee S, Prachumwat, A, Srisala J, et al. Review of current disease threats for cultivated penaeid shrimp in Asia [J]. Aquaculture, 2016, 452: 69–87.

Varela A. Nodavirus of covert mortality (CMNV) in marine shrimp culture. Technical note. Rev. Scientific Repertoire. State University at a Distance, 2016, 19 (1): 33-40 (in Spanish).

Varela A. Pathologies of the hepatopancreas in marine shrimp cultivated in America and its differential diagnosis by histopathology. AquaTIC Mag, 2018, 50: 13-30 (in Spanish).

Wang C, Liu S, Li X P, et al. Infection of covert mortality nodavirus in Japanese flounder reveals host jump of the emerging *Alphanodavirus* [J]. J Gen Virol, 2019b, 100(2): 166-175.

Wang C, Wang X H, Liu S, et al. Preliminary study on the natural infection of *Carassius auratus* with covert mortality nodavirus (CMNV) [J]. Progress in Fishery Sciences, 2019a, 40(2): 25-32.

Xu T T, Liu S, Li X P, et al. Genomic characterization of covert mortality nodavirus from farming shrimp: Evidence for a new species within the family *Nodaviridae* [J]. Virus Res, 2020, 286: 198092.

Zhang Q L. Evidences for cross-species infection in fish of covert mortality nodavirus (CMNV) [C]. 12th Asian Fisheries and Aquaculture Forum, Iloilo Convention Center, Iloilo City, Pilipinas, 2019.

Zhang Q L, Liu Q, Liu S, et al. A new nodavirus is associated with covert mortality disease of shrimp [J]. J Gen Virol, 2014, 95: 2700–2709.

Zhang Q L, Liu S, Li J, et al. Evidence for cross-species transmission of covert mortality Nodavirus to new host of *Mugilogobius abei* [J]. Front Microbiol, 2018, 9: 1447.

Zhang Q L, Liu S, Yang H L, et al. Reverse transcription loop-mediated isothermal amplification for rapid and quantitative assay of covert mortality nodavirus in shrimp [J]. J Invertebr Pathol, 2017b, 150: 130–135.

Zhang Q L, Xu T T, Liu S, et al. Prevalence and distribution of covert mortality nodavirus (CMNV) in cultured crustacean [J]. Virus Res, 2017, 2(233): 113-119.

Threats of concern to mollusk aquaculture in China

Author: Bai Changming
E-mail: baicm@ysfri.ac.cn

1 Main threats and diseases in mollusk aquaculture in China

The main threats to mollusk aquaculture in China come from three fields: bad weather, including sudden drop of salinity caused by heavy rainfall and elimination of aquaculture installations etc.; pollution, including harmful toxins associated with red tide and chemical pollution; and overintensive aquaculture. Overintensive aquaculture has caused multiple troubles to mollusk aquaculture industry in China, e.g. slow growth and increasing disease occurrences of cultured animals, low quality of the products, and finally the low profit of the industry (Fig. 3-8-1).

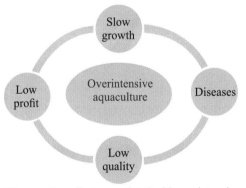

Fig. 3-8-1 The negative effects associated with overintensive aquaculture

1.1 China: the biggest mollusk producer

The mollusk production has increased steadily and quickly since 1990s benefited from the introduction of intensive aquaculture techniques. China is undoubtedly the most biggest mollusk producer around the world in recent decades (Fig. 3-8-2).

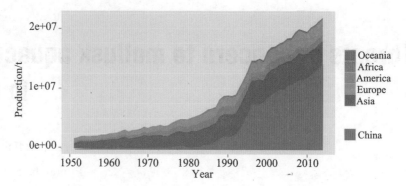

Fig. 3-8-2 Mollusk production of different continents and China

In 2018, the production of maricultured mollusks was about 14.4 million tonnes in China, accounting for 71.1% of China's total mariculture production, and 83.9% of the world's maricultured mollusk production. The cultured mollusks consist of 8 categories and 48 species in China. Their production percentages in 2018 are shown in Fig. 3-8-3.

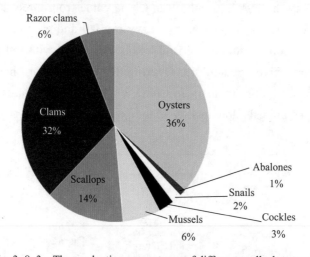

Fig. 3-8-3 The production percentages of different mollusk categories

1.2 Threats associated with overintensive aquaculture

China has the coastline with a length of about 18,000 km comprising by different climate zone, all suitable for mollusk aquaculture. It is still a challenge for the industry to cultivated over 14 million tonnes mollusks per year, although various intensive and multi-trophic aquaculture modes have been introduced. Overintensive aquaculture that exceed the carrying capacity of given sea areas is common to discover along China's coast. Overintensive aquaculture could increase aquaculture production and bring economic profit to this industry. While the operation will bring more troubles over the long period, which includes but not limited to slow growth, increasing disease occurrences, and thus low output efficiency and economic profit. The weakness and problems of intensive aquaculture were aggravated by the organization patterns of the mollusk cultivation industry. Small family producers with no or weak associations were the main subjects of mollusk production in China. Therefore, it is susceptible to an economic problem of overconsumption that called "the tragedy of the commons". Now the governors and farmers have notified the negative effect of overintensive aquaculture, and policies and measures have been adopted to resolve the problem (Song et al., 2020).

1.3 Important mollusk pathogens in China

Increasing disease outbreaks are one of the main drawbacks associated with intensive mollusk aquaculture (Song et al., 2020). Furthermore, with the purpose of promoting production and quality of mollusks, exchange of economically important mollusk among different geographic areas was common, which has resulted in the rapid spreading of devastating pathogens (Barbosa et al., 2015). Severe outbreaks of infectious disease have resulted in significant economic loss and even the collapse of entire industries in China (Fig. 3-8-4). In recent decades, different herpesviruses have emerged both in bivalve and gastropod mollusks, and caused significant economic destruction in many countries around the world (Arzul et al., 2017), including China.

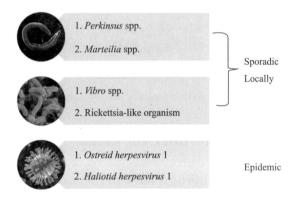

Fig. 3-8-4 The main pathogens faced by mollusks in China

2 Herpesvirus infecting bivalves

Herpesviruses comprise a large number of species infect a range of invertebrate and vertebrate animals. Malacoherpesviruses refer to a group of herpesviruses infecting invertebrates, which were formally classified into one family (Malacoherpesviridae) within the order Herpesvirales. The structure of malacoherpesviruses is similar to that of vertebrate herpesviruses. A linear, double-stranded DNA is packed in an icosahedral capsid, which is covered by a proteinaceous tegument and an outer lipid envelope. The diameters of capsids determined by transmission electron microscopy (TEM) varied (from 71 nm to 111 nm) among different studies (Renault et al., 2004). The herpesviruses infecting bivalves were then named as Ostreid herpesvirus 1 (OsHV-1), and allocated Malacoherpesviridae family (Davison et al., 2017).

2.1 Spread of herpes-like virus in bivalves

Herpes-like virus infection in bivalves was firstly reported in the 1970s, and then spread around the world (Fig. 3-8-5). This was also the first herpesviral infection in mollusks and invertebrates.

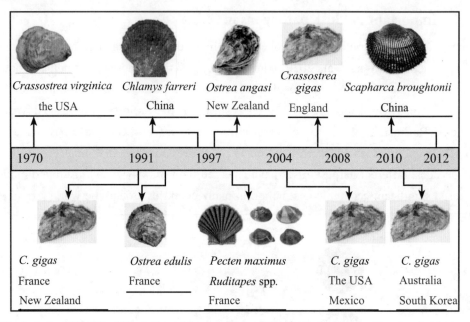

Fig. 3-8-5 A simplified illustration of the emergency and spreading of herpes-like virus in bivalves

2.2 Bivalve herpes-like viral disease in China

2.2.1 Bivalves infected by herpes-like virus reported in China

In China, the first case of herpes-like virus disease was reported in cultured *Chlamys farreri* in 1997 (Song et al., 2001; Wang et al., 2004a). Mortalities always occurred when water temperature raised to 23 ℃. The production of *C. farreri* dropped from about 600,000 t in 1996 to less than 200,000 t in 1998 in China. The virus was firstly named as acute viral necrosis virus (AVNV) (Wang et al., 2004a). Subsequently, herpes-like virus infection and associated mortalities were reported firstly in *C. gigas* larvae in hatcheries in 2009, in adult *Scapharca broughtonii* in 2012 (Bai et al., 2016), and in adult *Scapharca subcrenata* in 2019. High mortalities associated with OsHV-1 infections often outbroke during larval and juvenile stages, or in adult bivalves when they were taken away from their initial habitats and maintained in extremely high densities (Bai et al., 2016).

Temperature plays a key role in driving the rapid virus replication and disease outbreak associated with OsHV-1 infection (Arzul et al., 2017). OsHV-1 disease always occurred in the warm season when the water temperature was increasing and exceeded a certain temperature threshold, which has made the increasing water temperature like a trigger of mortalities associated with OsHV-1 infection (Bai et al., 2016). However, the temperature thresholds associated with bivalve herpes-like viral disease may vary according to the host species, variants and geographic position (Arzul et al., 2017; Xin et al., 2020).

2.2.2 Geographic distribution of herpes-like virus

According to our literature review, OsHV-1 infection has been reported in 13 bivalve species distributed over 17 countries around the world. *Crassostrea gigas* was the mostly affected species, infections in which have been reported in all the 17 countries. The repeated mortalities of mollusks associated with OsHV-1 infection have constituted a serious issue in this sector in many countries (Arzul et al., 2017). Epidemiological studies indicated that OsHV-1 was widely distributed along the coastline in China. 21.1% of the sampled 1599 samples in 27 aquaculture sites during 2001−2013 were tested positive by PCR in China (Bai et al., 2015). The positive samples comprised by seven bivalve species and distributed in 14 sites (Bai et al., 2015).

3 Herpesvirus infecting abalone

The herpesviruses infecting abalone species were phylogenetically close related to OsHV-1 (Savin et al., 2010), and were latterly named as Haliotid herpesvirus 1 (HaHV-1), under a newly created genus *Aurivirus*, which together with the genus *Ostreavirus* comprised the family Malacoherpesviridae. HaHV-1 infection was characterized by acute mortality process and necrotizing ganglioneuritis, therefore, was also termed abalone viral ganglioneuritis (AVG) (Corbeil et al., 2020). The known four abalone species affected by herpesvirus infection around the world are shown in Fig. 3-8-6.

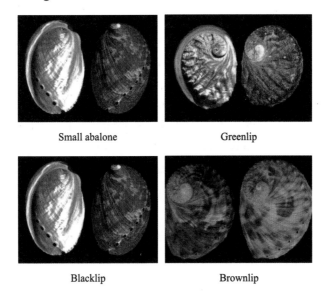

Fig. 3-8-6 Morphological characteristics of the four abalone species

3.1 History of abalone aquaculture in China

The abalone aquaculture was initiated in the 1980s, and the annual production has increased consistently since then (Fig. 3-8-7). It developed rapidly in the 1990s, and spread along the coasts from north to south. The two main cultivated abalone species were the Pacific abalone (*Haliotis discus hannai*) and the small abalone (*Haliotis diversicolor supertexta*), which varied according to different regions and different period of time (Wu et al., 2016). Farming of both these two species has suffered devastating disease losses during the mid-1990s and early 2000s, respectively. Viral etiology and a *Vibrio* species were clinically detected in the northern and southern regions of China (Wu et al., 2016).

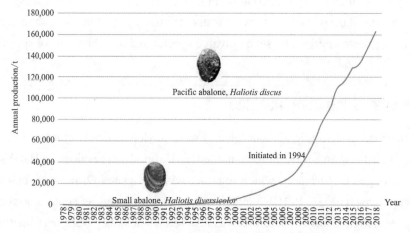

Fig. 3-8-7 The development and annual production of abalone in China

3.2 Abalone herpes-like viral disease in China

The first report of HaHV-1 infection in gastropod mollusks could be traced back to 1999, which caused the collapse of small abalone, *Haliotis diversicolor supertexta*, cultivation industry in Chinese mainland (Bai et al., 2019; Wang et al., 2004b). The small abalone has been widely cultivated in southern China in 1990s, the production of which has accounted for 65% of China's total abalone production. While the serious HaHV-1 infection has ruined the whole industry. The production of small abalone only accounted for less than 5% of the total abalone production in China (Wu et al., 2016). Similar herpesvirus infections were subsequently identified in Taiwan in 2003 (Chang et al., 2005), and in Australia in 2005 (Corbeil et al., 2020). The affected abalone populations in Taiwan were cultivated *H. diversicolor supertexta*, the same as that in Chinese mainland. In Australia, more species were infected by the herpesvirus infection, which including greenlip abalone (*Haliotis laevigata*), blacklip abalone (*Haliotis rubra*), brownlip abalone (*Haliotis conicopora*) and the hybrid of greenlip and blacklip abalone (Corbeil et al., 2016).

The occurrences of HaHV-1 were also closely associated with tempreture. However,

HaHV-1 disease always occurred in the cold season of southern Chinese mainland and Taiwan when the temperature is below 20 ℃ (Wang et al., 2004b). These results definitely indicated the divergence of the biology of the two viruses and their responses to temperature. While in Southern Australia, some cases of HaHV-1 infection did occur in summer in December 2005 and January 2006 (Corbeil et al., 2020). There is no exact temperature range associated with HaHV-1 infection in Australia, but experimental infections were always arranged in the range of 15-18 ℃ (Corbeil et al., 2016).

References

Arzul I, Corbeil S, Morga B, et al. Viruses infecting marine molluscs [J]. J Invertebr Pathol, 2017, 147: 118-135.

Barbosa S V, Renault T, Travers M A, et al. Mass mortality in bivalves and the intricate case of the Pacific oyster, *Crassostrea gigas* [J]. J Invertebr Pathol, 2019, 131: 2-10.

Bai C M, Gao W, Wang C, et al. Identification and characterization of Ostreid herpesvirus *1* associated with massive mortalities of *Scapharca broughtonii* broodstocks in China [J]. Dis Aquat Organ, 2016, 118(1): 65-75.

Bai C M, Li Y N, Chang P H, et al. Susceptibility of two abalone species, *Haliotis diversicolor supertexta* and *Haliotis discus hannai*, to *Haliotid herpesvirus 1* infection [J]. J Invertebr Pathol, 2019, 160: 26-32.

Bai C M, Wang C M, Xia J Y, et al. Emerging and endemic types of Ostreid herpesvirus 1 were detected in bivalves in China [J]. Journal of Invertebrate Pathology, 2015, 124: 98-106.

Chang P H, Kuo S T, Lai S H, et al. Herpes-like virus infection causing mortality of cultured abalone *Haliotis diversicolor supertexta* in Taiwan [J]. Dis Aquat Organ, 2005, 65(1): 23-27.

Corbeil S. Abalone viral ganglioneuritis. Pathogens [J], 2020, 9(9): 720.

Corbeil S, Williams L M, McColl K-A, et al. Australian abalone (*Haliotis laevigata*, *H. rubra* and *H. conicopora*) are susceptible to infection by multiple abalone herpesvirus genotypes [J]. Dis Aquat Organ, 2016, 119(2): 101-106.

Davison A J. Journal of general virology-introduction to 'ICTV virus taxonomy profiles' [J]. Journal of General Virology, 2017, 98(1): 1.

Savin K W, Cocks B G, Wong F, et al. A neurotropic herpesvirus infecting the gastropod, abalone, shares ancestry with oyster herpesvirus and a herpesvirus associated with the amphioxus genome [J]. Virology Journal, 2010, 7(308): 1-9.

Song L S. An early warning system for diseases during mollusc mariculture: exploration and utilization [J]. Journal of Dalian Ocean University, 2020, 35(1): 1-9.

Song W B, Wang C M, Wang X H, et al. New research progress on massive mortality of cultured scallop *Chlamys farreri* [J]. Marine Sciences, 2001, 25(12): 23-26.

Renault T, Novoa B. Viruses infecting bivalve molluscs [J]. Aquatic Living Resources, 2004, 17(4): 397-409.

Wang C M, Wang X, Ai H, et al. The viral pathogen of massive mortality in *Chlamys farreri* [J]. Journal of Fisheries of China, 2004a, 28(5): 547-553.

Wang J, Guo Z, Feng J, et al. Virus infection in cultured abalone, *Haliotis diversicolor* Reeve in Guangdong Province, China [J]. J Shellfish Res, 2004b, 23(4): 1163-1168.

Wu F C, Zhang G F. Pacific abalone farming in China: Recent innovations and challenges [J]. J Shellfish Res, 2016, 35(3): 703-710.

Xin L S, Wei Z X, Bai C M, et al. Influence of temperature on the pathogenicity of *Ostreid herpesvirus* 1 in ark clam, *Scapharca broughtonii* [J]. J Invertebr Pathol, 2020, 169: 107.

Chapter 4
Fishery environment and bio-remediation

Environmental monitoring for marine aquaculture

Author: Jiang Tao
E-mail: jiangtao@ysfri.ac.cn

1 Background

Sea food (Fig. 4-1-1) contains high protein and is delicious. However, natural resources cannot satisfy the consumption.

Fig. 4-1-1 Several favorite sea food species in China

Aquaculture has the potential to play a major role in feeding human being in the future. The rapid expansion of aquaculture activities in coastal marine areas during the past decades has induced a general concern for the impact on critical environmental variables. It has been well known that aquaculture (especially fish and shellfish aquaculture) has a negative impact on the marine environment and ecology (Fig. 4-1-2). Sedimentation of a large amount of organic matter will cause depletion of oxygen and produce hypoxia in the bottom seawater.

Aquaculture activity can release excessive nutrients into the surrounding seawater, which may cause harmful algae blooms (HABs) events.

Fig. 4-1-2 Impacts of aquaculture on the environment

Suspended bivalves aquaculture is a double-edged sword (Fig. 4-1-3)!

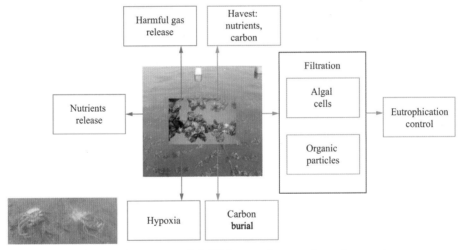

Fig. 4-1-3 Impacts of shellfish aquaculture on the environment

Impacts of shellfish aquaculture on the environment include positive and negative effects. Positive effects include sea food (carbon) harvest, eutrophication control and carbon bury. Negative effects include emission of harmful gas, nutrients release and hypoxia.

Environmental monitoring for marine aquaculture is very important (Fig. 4-1-4). Location, time and capacity should be considered in the layout of aquaculture. It is important in:

A. Disaster prevention, such as harmful algal blooms.

B. Providing suggestions for regulation and management of aquaculture activity.

C. Improving the food quality of cultured species.

Fig. 4-1-4 Field investigation in the aquaculture zone

2 Sampling strategy

2.1 Site selection and sampling design

A. Sampling and monitoring to detect the impacts of aquaculture on marine environment and ecology require very careful consideration (Fig. 4-1-5).

B. Several related issues need to be considered, such as the purpose of sampling, the scale and time-course of the problem, the available resources, and the duration of sampling.

C. Environmental consequences of aquaculture are site-specific, depending on various factors such as regional temperature, local hydrodynamics, geochemical sediment features, benthic shear, sediment texture and composition, water depth, nutrient loading, species reared, feed type and feeding techniques.

Fig. 4-1-5 Intensive aquaculture in the nearshore area

D. Make clear the location and area of aquaculture zone, species cultured, density, amount and feeding method of cultured animals (Fig. 4-1-6).

E. Collect history data of the aquaculture zone including time, cultured species and density, and total/unit production amount.

F. Draw the spatial distribution of aquaculture zone.

Fig. 4-1-6 Fish and shellfish aquaculture in the Daya Bay, China

G. In general, the density of sampling should be as great as possible. However, the following factors should be considered: aims of the investigation, capacity of laboratory services, data storage, and handling and processing capabilities.

H. Sea current data such as water speed and direction on the whole water column (Fig. 4-1-7) are very important.

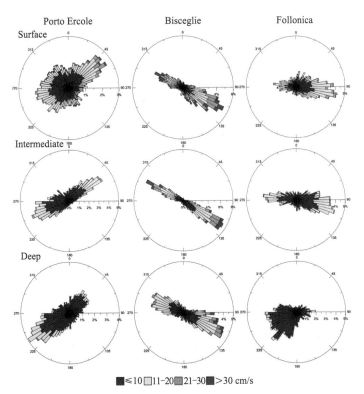

Fig. 4-1-7 Sea current data (sampling sites and current speed and direction) in three aquaculture farm in Italy (Tomassetti et al., 2016)

I. Pollutants and suspended particles are easily dispersed along the marine current (Fig. 4-1-8).

J. Dispersion rises as current velocities increase.

K. Down-flux of suspended particles around cages are mainly confined to 200 m in coastal areas.

L. Kutti et al. examined an off-coast farm located at 230 m water depth and found increased rates of sedimentation at distances up to 900 m away from the farm, suggesting that deep-water farms can induce enrichment of sediments over large areas.

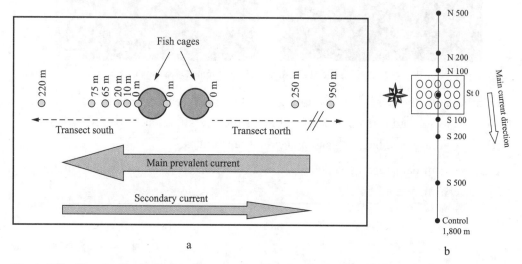

Fig. 4-1-8 The sampling design in adjacent areas of fish farm in Garrucha, Spain (a) (Borja et al., 2009) and along the Cálida Coast (b) (Aguado-Giménez et al., 2007)

2.2 Environmental variables

To measure a variable, you need to estimate of the magnitude of the variable for all elements in the population of interest. As for seawater and sediments (Fig. 4-1-9), environmental variables are quite different.

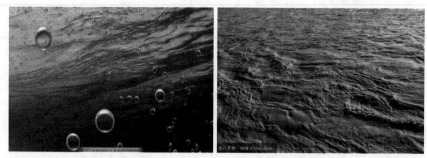

Fig. 4-1-9 Seawater and sediments

2.2.1 Environmental parameters for seawater

Fish and shellfish farming results in the release of large amounts of dissolved nutrients to the surrounding water. Chemical parameters include DIN (NO_3^-, NO_2^-, NH_4^+), PO_4^{3-}, SiO_3^{2-}. Physical parameters include dissolved oxygen (DO), temperature, salinity, pH, suspended solids, etc. Biological parameters include phytoplankton, especially harmful algae, zooplankton, nekton, etc.

2.2.2 Environmental parameters for sediments

Physical parameters, including the grain size distribution of the sediments, and E_h (electrode potential), are qualitative metric of the intensity of reduction conditions, which can reflect organic enrichment. Chemical parameters for organic enrichment include organic matter (OM), total organic carbon (TOC), total nitrogen (TN), C:N, total phosphorus (TP), etc. Secondary chemical hazardous materials including ammonia, sulphides and methane, are the main toxic compounds derived from the anaerobic mineralization of organic matter.

3 Sampling method

The dissolved constituents of seawater are grouped into two categories, namely the major or conservative components and the trace components. The conservative components are not influenced significantly by biological processes (*e.g.*, salinity). The trace components, in addition to being affected by physical processes, are influenced by the biological processes of uptake, excretion and biodegradation (*e.g.*, nutrients). During the field work, the marine analytical chemist must be aware that the components dissolved or dispersed in the sea vary from place to place, with depth and time (season) because of the physical and biogeochemical processes (Fig. 4-1-10).

Fig. 4-1-10 Sample collection in the field investigation

Sampling techniques include some typical equipment for the collection of discrete seawater samples at various depths as well as pumping systems for continuous sampling. Discrete seawater samples are most popular. Depth levels depend on the demand of chemists, hydrographers, biologists and scientists from other disciplines.

The biological activity in seawater does not stop with sample collection, since bacteria and micro- and nano-plankton continue to digest and excrete materials. The concentrations of nutrients and other bioactive elements are liable to change due to the activity of microorganisms naturally present in seawater. Therefore, as a general rule, samples should be kept away from light and analysed within a few hours after collection. If it is difficult to determine the elements in a short time, the following two approaches to preservation are outlined: refrigeration (Fig. 4-1-11) and poisoning.

Fig. 4-1-11　Equipment for cryopreservation

3.1　Sample storage for the determination of nutrients

Freezing (to −20 ℃) is the method chosen by many workers when nutrient samples have to be stored for several weeks or even months. When refrigeration of samples is not possible, addition of poisonous chemicals is another option. The purpose of this procedure is to kill those species which are responsible for consumption of nutrients. Of the various agents investigated, only the three have been found widespread applicated: acidificated (mostly with sulphuric acid), chloroform and mercury chloride.

3.2　Sample storage for the determination of trace elements

Trace elements in seawater are extremely dilute analytes. In most cases, however, long-term storage is necessary. The proper treating procedure for storage can ensure the accuracy of determined elements such as Cd, Co, Cu, Fe, Mn, Ni, Pb or Zn for a period of up to 2 years (or possibly longer). The method is acidifying water samples to pH 1.5-2.0 with ultraclean acid (approximately 1 μL of HNO_3 per milliliter of seawater sample).

4　Determination method

4.1　Salinity, dissolved oxygen, pH

During multi-disciplinary cruises, standard hydrographic or hydrochemical variables such as salinity, temperature, oxygen and nutrient concentrations are commonly recorded directly by CTD sensors (Fig. 4-1-12).

Fig. 4-1-12　Multi-parameter water quality analyzer: determining the environmental factors and collecting seawater during field investigation on the board (a) and different sensors (b, c)

4.2 Nutrients

There are three major groups of analytical methods for the determination of nutrients:

— manual methods, in which each sample is treated individually and manually for each variable;

— automated methods, which are generally automated versions of the manual methods;

— sensors, which provide a physical signal representing the analyte concentration on contact with the seawater sample, preferably without prior chemical treatment.

4.3 Trace elements

The analytical techniques most often used for the determination of dissolved and particulate trace elements in seawater are:

— electrothermal atomic absorption spectrometry (ETAAS);

— mass spectrometric methods, such as inductively coupled plasma mass spectrometry, (ICP-*MS*);

— inductively coupled plasma atomic emission spectrometry (ICP-AES);

4.4 Chlorophll *a* (Chl *a*): total biomass of phytoplankton

Spectrophotometric method is used to determine Chl *a*.

4.5 Phytoplankton pigments: Chemtax

Pigment composition has been used to assess the contributions of various algal species to phytoplankton communities.

HPLC has become the method of choice for the accurate quantitative determination not only of chlorophylls but also of carotenoids (Fig. 4-1-13).

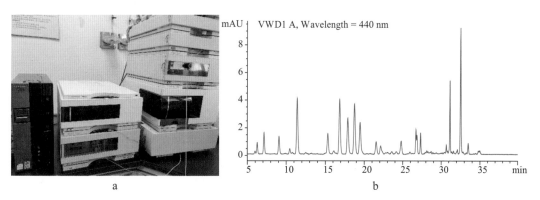

Fig. 4-1-13 HPLC instrument (a) and chromatographic figure (b)

5 Impacts of fish and shellfish farming on marine environment

5.1 Environmental impacts of fish and shellfish farming on surrounding seawater

Water quality around coastal fish/shellfish farms is affected by the release of dissolved and particulate inorganic and organic nutrients.

Higher concentrations of nutrients in seawater in fish farm than at control site have been reported by numerous studies (*e.g.*, Mantzavrakos et al., 2007). For example, concentrations of ammonium-nitrogen, phosphate and suspended solids in a Greek fish farm were proved to be obviously higher than the surrounding sea areas (Fig. 4-1-14) (Mantzavrakos et al., 2007).

Assessments of water quality are, however, often confounded by the methods used, addressing static parameters such as concentrations of nutrients and Chl *a* rather than productivity and nutrient uptake.

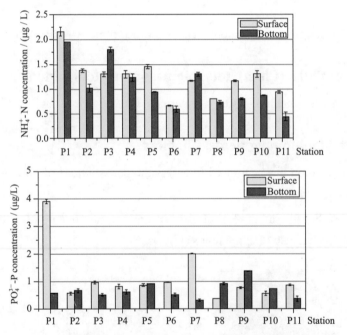

Fig. 4-1-14 Ammonium and phosphate concontration in a Greek fish farm

5.1.1 Long-term effects of fish farming: an example of a gulf in Italy

Chl *a* concentration decreased sharply after N and P loading reduction from land. Chl *a* concentration showed a marked increase from 2001 (fish farming starting) onwards. In the same period, Chl *a* concentrations measured inside and outside the gulf have significantly diverged (Sarà et al., 2011).

As all the other possible causes can be ruled out, aquaculture remains the sole explanation for the observed situation (Fig. 4-1-15).

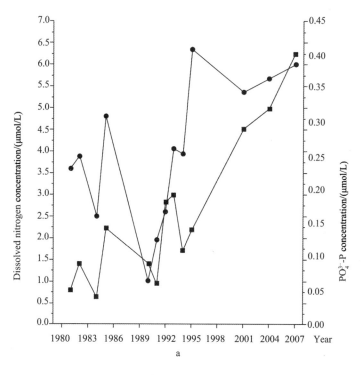

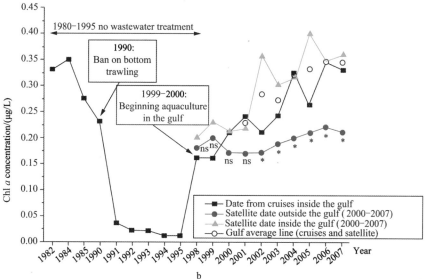

Fig. 4-1-15 Long-term changes of concentration of nutrients (a) and Chl *a* (b) in a fish farm (sarà et al., 2016)

5.1.2 Environmental impacts of shellfish farming on surrounding seawater

Fig. 4-1-16 shows an example in Daya Bay, South China Sea. During the oyster culture period, the average concentration of total Chl *a* (sum of size-fractionated Chl *a*) within the farming area was approximately 60% lower than that at the reference site. Growth rate of shellfish slowed down.

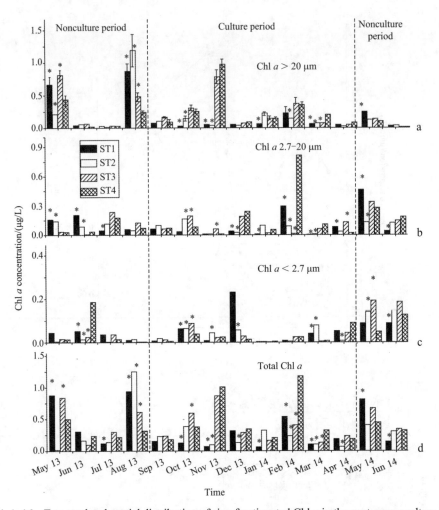

Fig. 4-1-16 Temporal and spatial distribution of size-fractionated Chl *a* in the oyster aquaculture farm

5.2 Environmental impacts of fish/shellfish farming on sediment environment

Benthic impacts are of primary concern in mariculture, particular under eutrophic conditions where accumulation of organic matter in the sediments may result in anoxia and loss of secondary production and biodiversity (Hargrave et al., 2008). Only a small proportion of the carbon supplied to the fish via the feed is retrieved through harvest, whereas a considerable amount reaches the seabed in the form of uneaten food and faeces. This may cause organic enrichment of the sediments. The most evident effects of organic enrichment on bottom sediments are the progressive transformation of the substrate into a flocculent anoxic environment (Fig. 4-1-17). The average sulphide, TOC, OM and TN content of surface sediments at the fish site was considerably higher than that at the oyster and control sites (Fig. 4-1-18). Similarly, the levels of these chemical parameters at the oyster site were significantly higher than those at the control site. Phytopigments also can be indicators for the sediments environment.

Fig. 4-1-17 Impacts of fish farming on the bottom creature

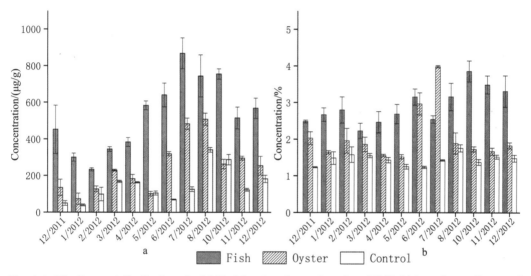

Fig. 4-1-18 Seasonal distribution of sulphide (a) and total organic carbon (TOC) (b) in surface sediments of Daya Bay

5.2.1 Benthic impacts of floating cage farming

The theoretical basis of macrobenthic assemblages dynamics is the secondary succession which describes spatial-temporal changes in the macrofauna composition related to a disturbance (Fig. 4-1-19) (Tomassetti et al., 2016). As sediments become enriched and deoxygenation ensues, sulfate metabolism becomes the major metabolic pathway driving the shifts in macrofaunal communities. Hypoxia and sulfide toxicity inhibit the persistence of most sensitive taxa and the benthic species shift from sensitive species toward tolerant generalists characterized by high abundance and high biomass.

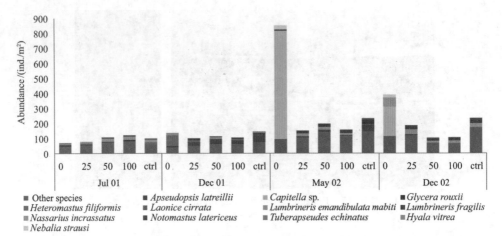

Fig. 4-1-19　Spatial-temporal changes in the macrofauna composition under a fish farm (Tomassetti et al., 2016)

5.2.2　Benthic impacts of floating cage farming on microbial groups

In cage sediments, organic enrichment and the consequent modification of the characteristics of the benthic environment result in an increase in aerobic heterotrophic bacteria and *Vibrio* density, indicating that they are efficient colonizers of organic-rich sediments (Fig. 4-1-20) (Rosa et al., 2004).

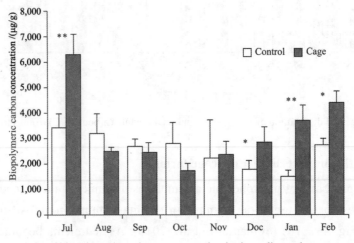

Fig. 4-1-20　Comparison of biopolymeric carbon concentration in the sediment between cage and control site

6　Management and control

The fallowing period is a management measure in aquaculture, when the production is paused for a few months even one year, to reduce the impact on the benthic environment. Fallowing is used in coastal aquaculture with good results, where the regeneration of the benthic communities is stimulated by rapid-colonized benthic fauna.

An example in Norway was shown in Fig. 4-1-21.

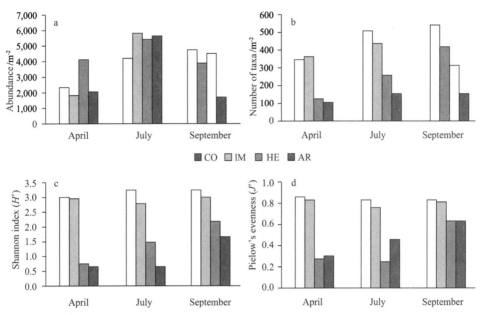

Fig. 4-1-21 Abundance (a), taxa number (b), Shannon index (c) and Pielow's evenness (d) revealing recovery of macrofauna after the fallowing period (Zhulay et al., 2015)

The fallowing initiated macrofauna recovery at farm stations. Significant changes in taxa composition occurred only after 6 months.

Fish welfare in offshore farms is expected to be improved due to higher water quality and less influence from terrestrial run-off and coastal activities. Waste products from farming are rapidly diluted, reducing the local environmental effects and increasing the carrying capacity of the farming sites.

IMTA may reduce the environmental impacts directly for the uptake of dissolved nutrients by primary producers (*e.g.*, macroalgae) and uptake of particulate nutrients by suspension feeders (*e.g.*, mussels), and through harvesting removing the nutrients from the location.

Monitoring HABs:

— HAB events are characterized by the proliferation and occasional dominance of some toxic or harmful algae.

— In some cases, these microscopic cells increase in abundance until their pigments discolor the water, hence the term "red tide" is commonly used.

— When HABs occur, with some dominated algal species releasing toxins and other compounds into the water, marine fauna can be killed.

Bloom control is the most challenging and controversial aspect of HAB management. Removal of HAB cells from the water by dispersing clay over the water surface seems to be an effective and environmental friendly method. In Korea, where a fish-farming industry worth hundreds of millions of dollars is threatened by HABs, this control strategy makes sense economically and socially so it has been progressed (Anderson, 2009). Modified clays can significantly improve the efficiency of the removal algae blooms in the seawater. This method has been widely used in some sea areas in China for tourism reason.

References

Aguado-Giménez F, Marín A, Montoya S, et al. Comparison between some procedures for monitoring offshore cage culture in Western Mediterranean Sea: sampling methods and impact indicators in soft substrate [J]. Aquaculture, 2007, 271(1–4): 365–370.

Anderson D M. Approaches to monitoring, control and management of harmful algal blooms (HABs) [J]. Ocean & Coastal Management, 2009, 52(7): 342–347.

Borja A, Rodríguez J G, Black K. Assessing the suitability of a range of benthic indices in the evaluation of environmental impact of fin and shellfish aquaculture located in sites across Europe [J]. Aquaculture, 2009, 293(3): 231–240.

Jiang T, Yu Z, Qi Z. Effects of intensive mariculture on the sediment environment as revealed by phytoplankton pigments in a semi-enclosed bay, South China Sea [J]. Aquaculture Research, 2016, 48(4): 1923–1935.

Jiang T, Chen F, Yu Z. Size-dependent depletion and community disturbance of phytoplankton under intensive oyster mariculture based on HPLC pigment analysis in Daya Bay, South China Sea [J]. Environmental Pollution, 2016, 219: 804–814.

Klaoudatos S D, Klaoudatos D S, Smith J. Assessment of site specific benthic impact of floating cage farming in the eastern Hios Island, Eastern Aegean Sea, Greece [J]. Journal of Experimental Marine Biology & Ecology, 2006, 338(1): 96–111.

Kutti T, Ervik A, Hansen P K. Effects of organic effluents from a salmon farm on a fjord system. I. Vertical export and dispersal processes [J]. Aquaculture, 2007, 262(2–4): 370–381.

Mantzavrakosa E, Kornaros M, Lyberatos G, et al. Impacts of a marine fish farm in Argolikos Gulf (Greece) on the water column and the sediment [J]. Desalination, 2007, 210: 110–124.

Rosa T, Mirto S, Mazzola A, et al. Benthic microbial indicators of fish farm impact in a coastal area of the Tyrrhenian Sea [J]. Aquaculture, 2004, 230: 153–167.

Sarà G, Martire M L, Sanfilippo M, et al. Impacts of marine aquaculture at large spatial scales: Evidences from N and P catchment loading and phytoplankton biomass [J]. Marine Environmental Research, 2011, 71(5): 317–324.

Tomassetti P, Gennaro P, Lattanzi L. Benthic community response to sediment organic enrichment by Mediterranean fish farms: Case studies [J]. Aquaculture, 2016, 450: 262–272.

Zhulay I, Reiss K, Reiss H. Effects of aquaculture fallowing on the recovery of macrofauna communities [J]. Marine Pollution Bulletin, 2015, 97: 381–390.

Treatment of aquaculture wastewater with constructed wetlands

Author: Cui Zhengguo
E-mail: cuizg@ysfri.ac.cn

1 Introduction of constructed wetlands

1.1 Definition

A constructed wetland is an artificial wetland to treat municipal or industrial wastewater, greywater or stormwater runoff. It may also be designed for land reclamation after mining, or as a mitigation step for natural areas lost to land development (Wang, 2007).

Constructed wetlands are engineered systems that use natural functions vegetation, soil (substrates) and organisms (microbes) to treat wastewater (Fig. 4-2-1).

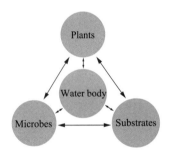

Fig. 4-2-1 Composition of constructed wetland

1.2 General advantages and disadvantages of constructed wetlands

1.2.1 advantages

A. Site location flexibility (compared to natural wetlands).

B. Simple operation and maintenance.

C. Can be integrated attractively into landscaping.

D. Economical and ecological advantages.

1.2.2 Disadvantages/Challenges

A. Mosquitoes (in free water surface systems).

B. Start-up problems.

C. Space requirement.

D. Variable performance possible.

E. Designs still largely empirical (to date).

1.3 Classification of constructed wetlands (Zhang, 2011)

1.3.1 Based on water flow characteristics (Fig. 4-2-2)

A. Free water surface flow (abbreviated as FWS or SF).

B. Subsurface flow (abbreviated as SSF).

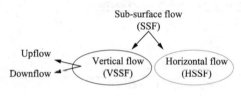

Fig. 4-2-2 Types of SSF

1.3.2 Based on plant species characteristics

A. Floating plants (*e.g.*, *Lemna*, *Nymphaea*).

B. Submerged plants (*e.g.*, *Elodea*).

C. Emergent plants (*e.g.*, *Phragmites*, *Papyrus*).

1.4 Applications–types of wastewater

The constructed wetlands have been used in many types of wastewater treatment (Fig. 4-2-3).

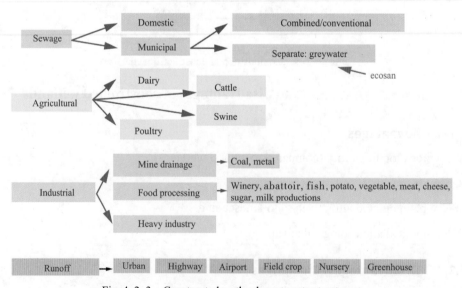

Fig. 4-2-3 Constructed wetland wastewater treatment

2 Construction of wetlands

2.1 Compartments in wetlands

Treatment is the result of complex interactions between all these compartments (Fig. 4-2-4).

A. Plants.

B. Sediment/gravel (substrates).

C. Bacteria growing in biofilms.

D. Root zone/pore water.

E. Litter/detritus.

F. Water.

G. Air.

H. Roots.

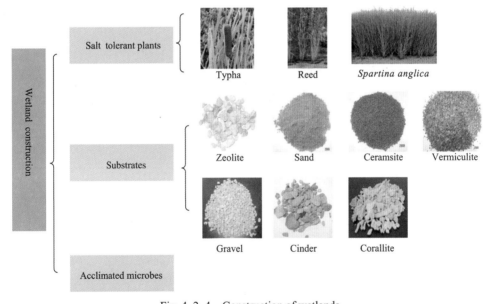

Fig. 4-2-4　Construction of wetlands

2.2 Compartments selection of plants

When choosing aquatic plants, the following factors should be considered (Fig. 4-2-5) (Tao, et al., 2019).

Fig. 4-2-5　Different constructed wetland plants

2.2.1　Selection principle of plants

The selection principle of plants was shown in Fig. 4-2-6.

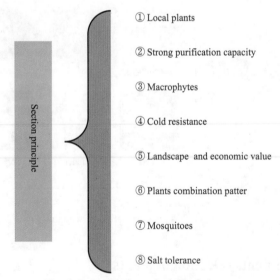

Fig. 4-2-6　Section principle of plants

2.2.2　Other factors

Other factors such as wetland bed, hydraulic retention time (HRT), hydraulic load, season/temperature, DO, pH and nutrients also should be considered (Table 4-2-1).

Table 4-2-1 Rule of thumb area requirements for different wetlands

Type of constructed wetlands	Design area requirement (m²/PE)
Type 1: FWS	5-10
Type 2: HSSF	3-5
Type 3: VSSF	2-3
Type 4: Hybrids (horizontal-vertical, or H-V)	2.5-3

Notes: 1 PE = 1 Person Equivalent = 60 g /(cap/d) (in terms of BOD) = 120 L/(cap/d) (in the Netherlands, valid for mixed domestic wastewater and lower values for greywater); H-V is better nitrogen removal (denitrification) than VSSF).

2.3 Construction of wetlands

An integrated vertical-flow constructed wetland was built in the laboratory (Fig. 4-2-7).

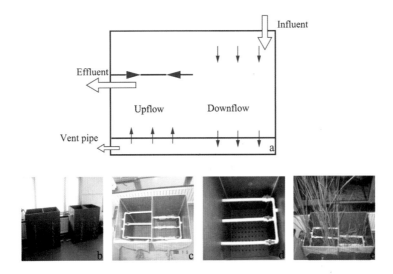

Fig. 4-2-7 An integrated vertical-flow constructed wetland (a) and physical pictures of each part of constructed wetland (b-e).

3 Removal effects and mechanism of constructed wetlands

3.1 Removal effects of constructed wetlands

The removal effects of constructed wetlands on nutrients were shown in Figs. 4-2-8 to 4-2-10 (Wu et al., 2016).

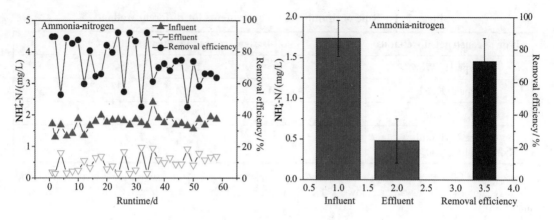

Fig. 4-2-8　The concentrations and removal efficiency of ammonia-nitrogen

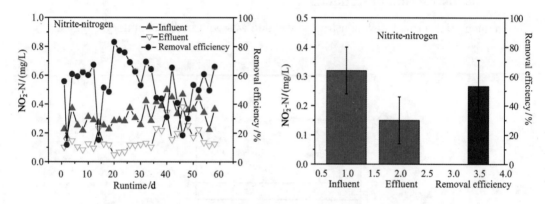

Fig. 4-2-9　The concentrations and removal efficiency of nitrite-nitrogen

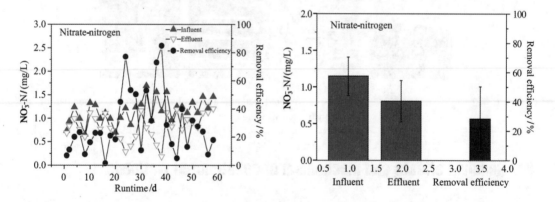

Fig. 4-2-10　The concentrations and removal efficiency of nitrate-nitrogen

Removal effects of the pollutants in the integrated vertical-flow were constructed (Figs. 4-2-11 to 4-2-17).

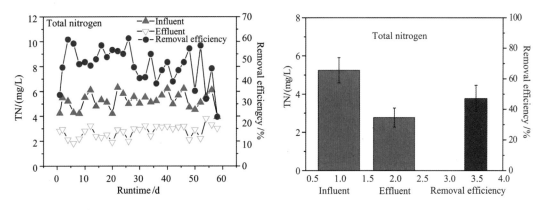

Fig. 4-2-11　The concentrations and removal efficiency of total nitrogen

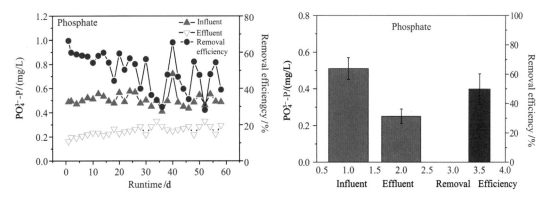

Fig. 4-2-12　The concentrations and removal efficiency of phosphate

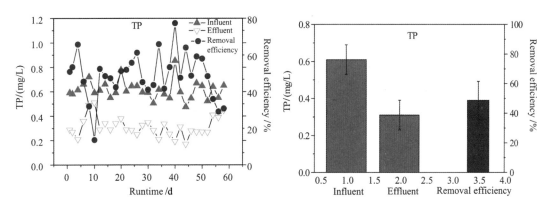

Fig. 4-2-13　The concentrations and removal efficiency of TP

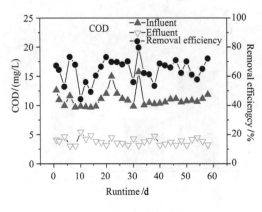
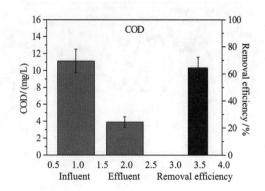

Fig. 4-2-14　The concentrations and removal efficiency of COD

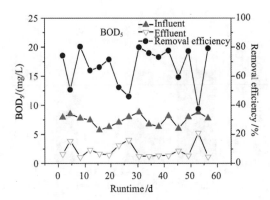
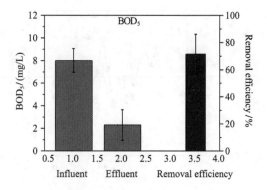

Fig. 4-2-15　The concentrations and removal efficiency of BOD_5

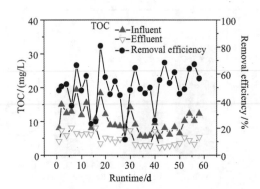
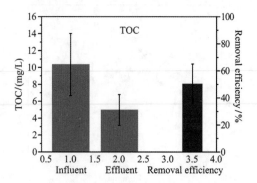

Fig. 4-2-16　The concentrations and removal efficiency of TOC

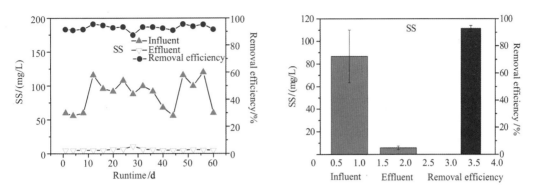

Fig. 4-2-17　The concentrations and removal efficiency of SS

3.2　Removal mechanism of Constructed wetlands

Removal effects of the pollutants in the integrated vertical-flow were constructed (Zhao et al., 2006) (Fig. 4-2-18, Table 4-2-2).

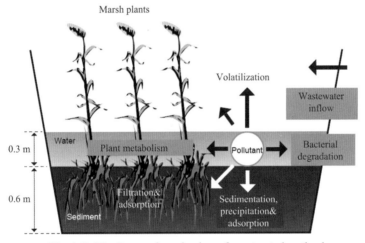

Fig. 4-2-18　Removal mechanism of constructed wetlands

Table 4-2-2　Removal mechanism of constructed wetlands

Mechanism or interreaction	N	P	COD	BOD	DOC	SS	Heavy metal	Pathogenic microbes
Physical sedimentation	+	+	+	+	○	+++	+	+
Substrate filtration	○	+++	○	++	○	+++	○	+++
Substrate adsorption	○	+++	○	○	○	++	○	○
Coprecipitation	○	+++	○	○	○	○	+++	○
Chemical adsorption	○	+	++	○	○	○	+++	○
Chemical decomposition	○	○	+++	+	+	○	○	+++
Microbial metabolism	+++	+++	+++	+++	+++	+++	○	○
Plant metabolism	+	+	++	○	○	○	○	++

(*to be continued*)

Mechanism or interreaction	N	P	COD	BOD	DOC	SS	Heavy metal	Pathogenic microbes
Plant adsorption	++	++	++	○	○	○	++	○
Death	○	○	○	○	○	○	○	+++
Volatilization	+	○	○	○	+++	○	○	○

Notes: +++, most important role; ++, major role; +, minor role; ○, little or no role.

Removal mechanism of nitrogen was shown in Fig. 4-2-19.

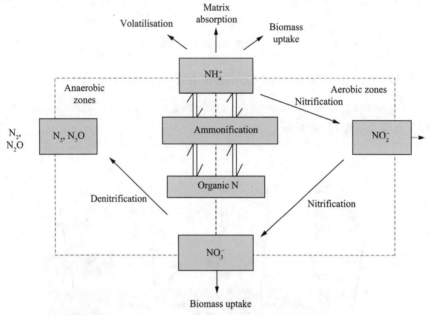

Fig. 4-2-19 Removal mechanism of nitrogen

4 Application of constructed wetlands

The constructed wetland was designed for marine aquaculture wastewater treatment (Figs. 4-2-20 to 4-2-22).

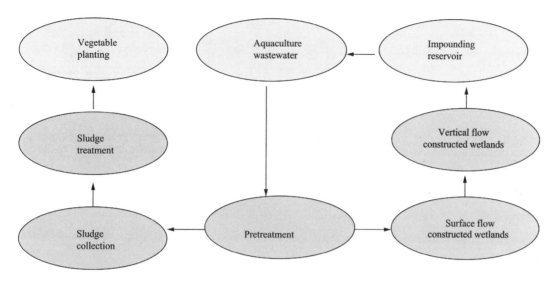

Fig. 4-2-20　Recycling and reuse of marine aquaculture wastewater with constructed wetlands

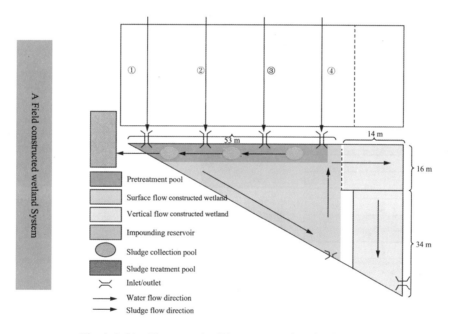

Fig. 4-2-21　Planar graph of the constructed wetland system

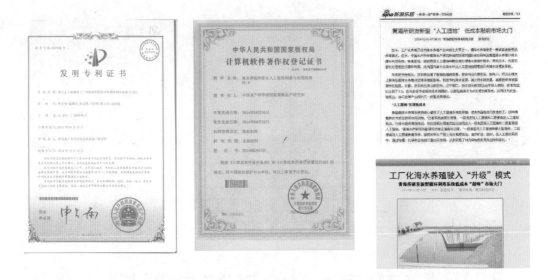

Fig. 4-2-22 Application of the constructed wetland

5 Construction of wetlands

The process of construction of wetlands was shown in Fig. 4-2-23.

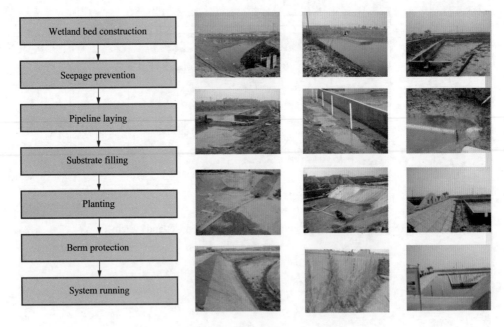

Fig. 4-2-23 Construction of wetlands

References

Tao Z K, Tao M N, Wang Y, et al. Selection and Application of constructed wetland plants [J]. Hubei Agricultural Science, 2019, 58(1): 44-48 (in Chinese).

Wang D. Application of constructed wetland in sewage treatment [J]. Science and Technology Information (Academic Research), 2007(23): 292-293 (in Chinese).

Wu Z B, Cheng S P, He F, et al. Study on the design and purification function of vertical flow constructed wetland [J]. Journal of Applied Ecology, 2016(6): 715-718.

Zhang Q. Construction and application of constructed wetland [J]. Wetland science, 2011, 9(4): 373-379 (in Chinese).

Zhao L F, Zhu W, Zhao J, et al. Mechanism of denitrification of constructed wetlands with low carbon nitrogen ratio pollution in river water [J]. Journal of Environmental Science, 2006, 26(11): 1821-1827 (in Chinese).

Chapter 5
Food engineering and nutrition

Seafood products processing

Author: Cao Rong

E-mail: caorong@ysfri.ac.cn

1 Basic concepts

1.1 Seafood

Seafood refers to food which is processed from aquatic organisms that live in the sea. Seafood has nutritional, functional and economic values.

1.2 Seafood products processing

Seafood products processing refers to the technical processes that take marine animals and plants as materials and apply a variety of mechanical, physical, chemical and microbiological methods.

2 Overview

2.1 Achievements of aquatic products processing in China (Zhao et al., 2016)

2.1.1 Gross output

The total output of aquatic products in China has ranked the first of the world for more than 20 years (Fig. 5-1-1), which accounts for 60%–70% of the world's total aquaculture production.

Fig. 5-1-1　Fishing and inshore cultivation in China

2.1.2 Development of aquatic products processing industry

In the past 20 years, China's aquatic products processing industry has achieved remarkable results (Table 5-1-1, Fig. 5-1-2).

Table 5-1-1 Number of enterprises, processing capacity, total output of processed products in China (1999 *vs* 2018)

Year	Number of enterprises	Processing capacity/t	Total amount of processed products/t
1999	6,443	11,271,317	6,242,131
2018	9,336	28,921,556	21,568,505

Fig. 5-1-2 Typical seafood production lines in China

2.1.3 Composition of aquatic processing products

There are many varieties of aquatic products in China, mainly frozen goods (Fig. 5-1-3).

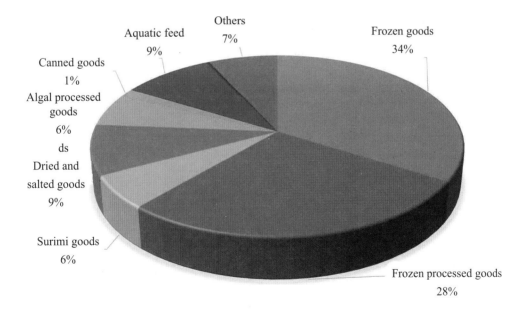

Fig. 5-1-3 The composition of aquatic processing products in China

2.2 Problems

2.2.1 The fishery's available resources (wild, natural, non-farmed) have declined significantly

Because of the overfishing and environmental deterioration, China's fishery biological resources have declined significantly.

2.2.2 The conversion rate of scientific achievements to practical production is still low

The conversion rate of scientific achievements in aquatic products processing is less than 30% in China, while in developed countries the rate is generally more than 70%.

2.2.3 Low-technology content in the products and low education level of labor force

Aquatic processing products are mainly primarily processed. The cultural literacy of the workers is generally at a low level.

2.2.4 The current product categories cannot meet consumer's demand sufficiently

More convenient, ready-to-eat, or accurate nutrition products are welcomed by consumers.

2.2.5 The level of mechanization needs to be improved

Many plants mainly rely on manual operation. There is still a big gap in the level of automation (Fig. 5-1-4).

Fig. 5-1-4 Typical seafood production lines in China

2.2.6 Energy conservation and emission reduction

Aquatic products must be stored and transported under low temperature. Large amount of water is used, and large amount of wastes is generated. Therefore, energy conservation and emission reduction remains to be a difficult task.

2.2.7 The quality and safety assurance system needs to be improved

Quality and safety issue is one of the most important problems (Fig. 5-1-5).

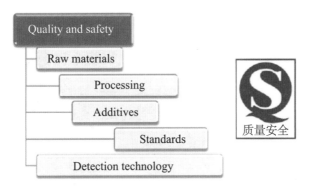

Fig. 5-1-5 Main contents of quality and safety assurance system

2.3 Development strategy

Strategies are needed for better development in the future (Fig. 5-1-6) (Yang et al., 2016).

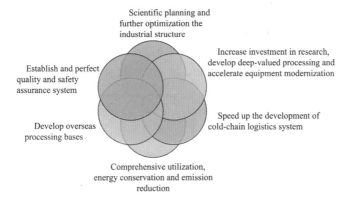

Fig. 5-1-6 Main contents of development strategies

2.3.1 Build modernized cold-chain logistics system for fresh aquatic products

In China, the loss rate of fresh aquatic products during logistics is about 25%-30%, while that in developed countries is under 5% (Fig. 5-1-7).

Fig. 5-1-7 Example of transportation of live fish

2.3.2 Use modern biotechnology in functional products

Biotechnology is necessary in the development of functional products (Fig. 5-1-8).

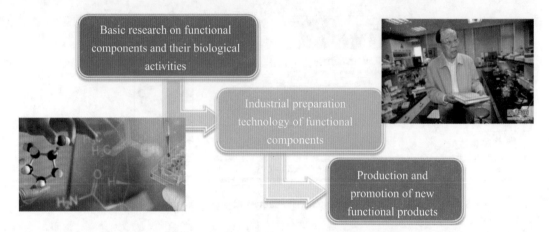

Fig. 5-1-8 Main contents of application of biotechnology

2.3.3 Promote nutritional, refined, convenient and ready-to-eat products

The proportion of convenience foods in the food industry is about 20% in the United States, 15% in Japan and 3% in China.

2.3.4 Upgrade equipment applied in aquatic products processing

Equipments with specific functions, high production capacity and low energy consumption are needed (Fig. 5-1-9).

Fig. 5-1-9 An example of automated production line

2.3.5 Develop green processing technology and comprehensive utilization

China is advocating green development. By-products with significant economic and social benefits should be utilized.

ચેChapter 5 Food engineering and nutrition

3 Raw seafood materials

3.1 Variety of seafood materials

Aquatic food raw materials refer to the species that live in the sea and inland waters (Fig. 5-1-10). The coastline of China is more than 18,000 kilometers long. The marine species are abundant (Fig. 5-1-11).

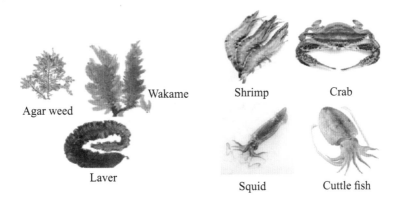

Fig. 5-1-10 Some kinds of plant, crustacean and cephalopod materials

Fig. 5-1-11 Some kinds of fish and shellfish

3.2 Characteristics of seafood materials

3.2.1 Perishable

Aquatic raw materials are generally rich in nutrients and contain very high moisture content. Aquatic products can be easily invaded by bacteria. For example, a large number of bacteria can attach to the body surface and gills of fishes. Moreover, the fish body is easy to be damaged by mechanical actions due to their thin skins and vulnerable scales. Therefore, fishes are perishable even under ice condition. In addition, the proteases of aquatic raw materials are more active than terrestrial biota. These proteases may cause postmortem changes significantly. After death, aquatic materials are very easy to be spoiled (Fig. 5-1-12).

Fig. 5-1-12　Perished shrimp with obvious black spots

3.2.2 Instability of output

The stable supply of raw materials is one of the most important factors for aquatic products processing. However, output of seafood materials is instable. It is almost impossible to guarantee a steady seafood material supply throughout the year due to season, fishery policy, sea conditions, climate, environmental ecology and other factors (Fig. 5-1-13).

Fig. 5-1-13　Material supply is unstable throughout a year

3.2.3 Variability of composition

The general chemical composition in muscles of fish, shrimp and shellfish is as follows: Moisture 70%–85%; protein 16%–22%; fat 6.5%–20%; ash 1%–2%; carbohydrate <1%. The species, individual, age and sex, fishing grounds, season and other factors are related. The content of crude protein and ash in fish and shrimp meat varies slightly, while the content of moisture and fat varies greatly.

3.2.4 Bioactive compounds

Marine organisms can produce many compounds with special structures and wonderful physiological functions in the process of biological evolution and competition (Fig. 5-1-14).

A. Functional peptides;

B. Polyunsaturated fatty acids (DHA, EPA);

C. Polysaccharides;

…

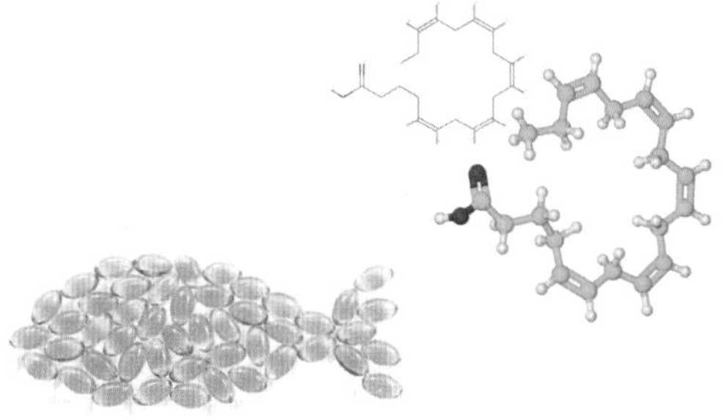

Fig. 5-1-14   Chemical structure of DHA and its product

4 Freshness-keeping

4.1 Basic theory (Rastogi et al., 2007; Torrieri et al., 2006)

4.1.1 Freshness-keeping

Use physical or chemical methods to delay spoilage or inhibit the activity of microorganisms, so as to maintain the freshness, quality and commodity value of the seafood materials.

4.1.2 Factors that should be considered

Many factors which lead to the quality deterioration of seafood materials should be considered, such as bacteria growth and reproduction, enzyme activity, protein degradation and fat oxidation.

4.1.3 Freshness-keeping methods

The most common technique used for freshness keeping is low temperature preservation, in addition to chemical or biochemical preservatives, modified atmosphere packaging, non-thermal technology (irradiation, electron beam, high pressure processing), etc.

Cold-chain logistics refers to a systematic project in which the foods are always under a specific low temperature in the process of production, storage, transportation and sales, in order to guarantee a good quality before consumption (Fig. 5-1-15).

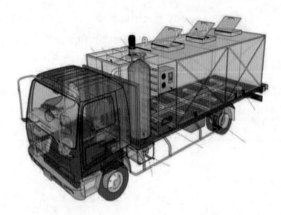

Fig. 5-1-15 A model of cold-chain transporter

4.2 An example: Cold-chain for Antarctic krill

Antarctic krill is quite different from other aquatic materials. A better cold-chain system is essential to guarantee a stable quality of Antarctic krill materials (Fig. 5-1-16).

Fig. 5-1-16 The operation of Antarctic krill

4.2.1 Background

Pelagic fishery has become an important global fishery. The exploitation and utilization of Antarctic krill is of strategic significance. However, the transportation of Antarctic krill requires a much longer time and distance. The quality of Antarctic krill is more prone to degradation. Therefore, cold-chain is necessary in the transportation of Antarctic krill.

4.2.2 Antarctic krill cold-chain logistics system (Fig. 5-1-17)

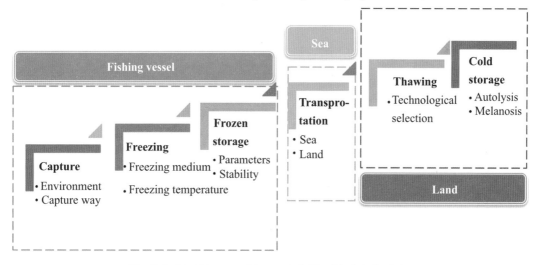

Fig. 5-1-17 Diagram of Antarctic krill cold-chain logistics

4.2.3 Key point 1: How does the quality of Antarctic krill change after catching?

Problem

A. The ship is not qualified for scientific experiments (Fig. 5-1-18).

B. Sensory evaluation alone is insufficient and lack of accuracy.

C. Non-destructive preservation and transportation of Antarctic krill materials is impossible.

Fig. 5-1-18 China's Antarctic krill fishing vessel

Solution

Non-protein nitrogen (NPN) is suitable for the analysis on the Antarctic krill protein changes (Fig. 5-1-19).

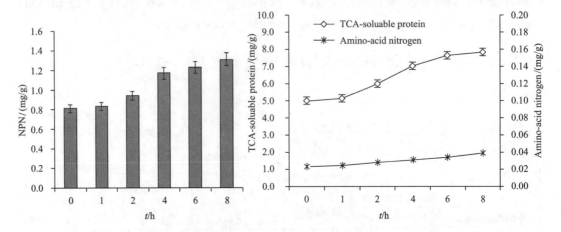

Fig. 5-1-19 Changes in NPN and its related compounds of Antarct kill after captured on the fishing vessel

A. Autolysis of Antarctic krill protein with time.

Faster processing after capture can guarantee better quality.

B. Effect of ultraviolet irradiation on protein autolysis.

Ultraviolet irradiation had no significant effect on autolysis of Antarctic krill during 8 h after capture (Fig. 5-1-20).

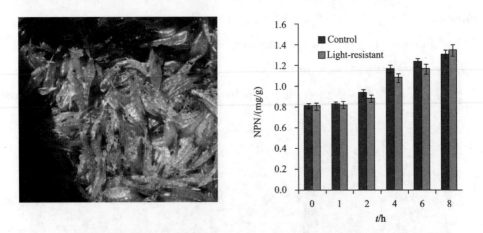

Fig. 5-1-20 Effect of light on NPN content of Antartic krill after captured on the fishing vessel

C. Effect of extrusion damage on protein autolysis.

The dragnet technology for Antarctic krill fishing needs to be further improved (Figs. 5-1-21 to 5-1-22).

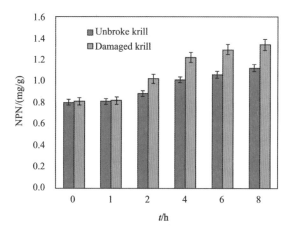

Fig. 5-1-21 Effect of trawling on autolysis process of Antarctic krill protein

Fig. 5-1-22 Image of Antarctic krill fishing

4.2.4 Key point 2: How to choose Antarctic krill thawing method?

In practical production, the choice of thawing method of Antarctic krill should take the type of product into full consideration (Fig. 5-1-23). Water thawing method is suitable for general food processing. Air thawing or water thawing is suitable for fishmeal production. Low temperature air thawing method is suitable for krill oil extraction (Fig. 5-1-24).

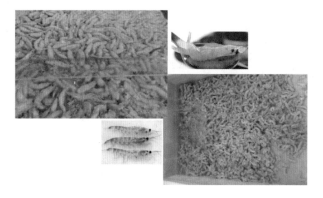

Fig. 5-1-23 Raw Antarctic krill materials

Fig. 5-1-24　Example of Antarctic krill products

4.2.5　Key point 3: How to control melanosis?

Melanosis (black spots forming) is a common problem in crustaceans (Fig. 5-1-25). The blackening process in Antarctic krill is much more rapid and severe.

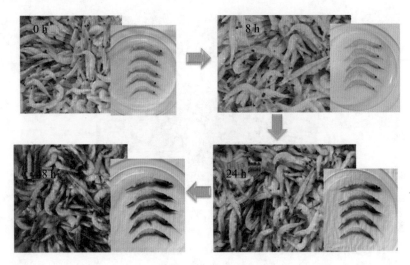

Fig. 5-1-25　Melanosis progress of Antarctic krill

An anti-melanosis preservative with 4-HR, sodium phytate, VC and EDTA has been developed. The efficiency of this preservative has been proven in practice (Fig. 5-1-26).

Fig. 5-1-26　Effect of biopreservative on the control of melanosis

5 Processing technology

A variety of aquatic products processing technologies are represented, including drying, pickling, smoking, surimi processing and some innovative technologies (Fig. 5-1-27) (Li et al., 2009).

Fig. 5-1-27 "One knife, two salt, three days in the sun"

5.1 Freezing

Aquatic products are rich in protein, fat and other nutrients, which are easy to change under high temperature. Low-temperature processing technology can better maintain the original quality of aquatic products (Fig. 5-1-28).

Fig. 5-1-28 Example of freezing products and equipments

Key points of freezing technology:

A. In order to produce frozen products with high quality, rapid and deep temperature freezing method must be applied.

B. Pass through the maximum ice crystal formation temperature interval (0-5 ℃) as quickly as possible.

C. The center temperature of frozen goods should be below -18 ℃.

5.2 Drying

Drying technology is the process of dehydrating the raw materials of aquatic products under natural or artificial conditions (Fig. 5-1-29).

Advantages of drying products:

A. Convenient for preservation.

B. Space saving.

C. Easy to transport.

D. Special flavor.

Fig. 5-1-29　Example of drying products

5.3　Smoking

Smoking processing generally includes raw material treatment, salt, desalination air drying and fumigation. According to the different processing technology, it can be divided into cold fumigation, warm fumigation, hot fumigation, liquid fumigation and electric fumigation.

Fumigation smog plays an important role in the formation of color and aroma of products (Fig. 5-1-30). The most important components in fumigation smog are phenols, aldehydes, ketones, alcohols, organic acids, esters and hydrocarbons.

Fig. 5-1-30　Smoked fish fillets

5.4 Sousing

Sousing refers to the process that salt, vinegar, sugar, wine grains, spices and other materials are used to souse fish and other aquatic products. The equipment for sousing process is simple and easy to operate. Sousing is able to process large amount of raw materials in a short time and the products have unique flavor (Fig. 5-1-31).

Sousing consists of two stages: salting and maturing. Salting is the process that salt permeates food. When some water is removed from the food, decrease in water activity will be caused. Maturing is the process that the tissues become soft and the amino acid contents increase due to the action of microorganism and tissue enzymes. Special flavors are formed at this stage.

Fig. 5-1-31　Example of sousing products

5.5 Canning

Canning is the process putting materials into tin cans (Fig. 5-1-32), glass bottles, soft packaging and other containers through exhausting, sealing and thermization.

Fig. 5-1-32　Example of canning products

5.6 Surimi products

Surimi: sticky meat paste made from fish meat by rolling and chopping with 2%-3% salt added. Surimi products are formed by adding some additives to the meat paste and heating to solidify (Fig. 5-1-33).

Fig. 5-1-33 Example of surimi products

5.7 Example of marine collagen peptide production

Marine collagen products (Fig. 5-1-34) are welcomed by consumers.

Fig. 5-1-34 Examples of marine collagen products

5.7.1 Market analysis

Collagen (peptides) (Fig. 5-1-35) can be used in functional food, health care products, cosmetics and other fields. The world market has reached 100 billion yuan RMB and the demand increases by 20% every year. The marine collagen market prospect is broad.

Fig. 5-1-35 Collagen

5.7.2 Profile of the technology

The production process of marine collagen products contains many steps, among which enzymatic hydrolysis and membrane filtration are the key steps (Fig. 5-1-36).

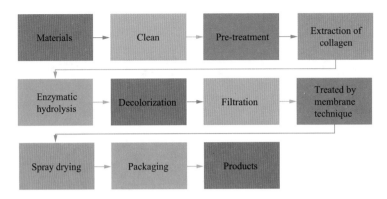

Fig. 5-1-36   Production process of marine collagen products

5.7.3   Key points of the technology

A. Extraction of collagen.

B. Peptides (enzyme) preparation .

C. Decolorization and deodorization.

D. Concentration, desalination and sterilization.

The membrane treatment system (Fig. 5-1-37) is made up of micro-filtration (MF), ultra-filtration (UF) and nano-filtration (NF).

Fig. 5-1-37   Membrane treatment equipments

REFERENCES

Li J, Lu H, Zhu J, et al. Aquatic products processing industry in China: Challenges and outlook [J]. Trends in Food Science & Technology, 2009, 20 (2): 73-77.

Rastogi N K, Raghavarao K S M S, Balasubramaniam V M, et al. Opportunities and challenges in high pressure processing of foods [J]. Critical Reviews in Food Science and Nutrition, 2007, 47 (1): 69-112.

Torrieri E, Cavella S, Villani F, et al. Influence of modified atmosphere packaging on the chilled shelf life of gutted farmed bass (*Dicentrarchus labrax*) [J]. Journal of Food Engineering, 2006, 77 (4): 1078-1086.

Yang Z, Li S, Chen B, et al. China's aquatic product processing industry: Policy evolution and economic performance [J]. Trends in Food Science & Technology, 2016, 58: 149-154.

Zhao W, Shen H. A statistical analysis of China's fisheries in the 12th five-year period [J]. Aquaculture and Fisheries, 2016, 1: 41-49.

Best management practices for fish feed and feeding

Author: Xu Houguo
E-mail: xuhg@ysfri.ac.cn

1 Introduction

1.1 Without efficient feed, there is no efficient aquaculture

A. Feed accounts for 70% of the cost in aquaculture (Fig. 5-2-1).

B. Aquafeed is relevant to human food safety.

C. Feed impacts environment via output/input ratio.

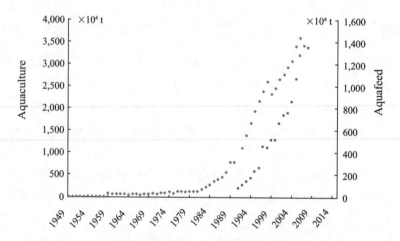

Fig. 5-2-1 Production of aquaculture and aquafeed from 1949 to 2014

1.2 Components of the best feed management

What does best feed management include? The components are as follows (Fig. 5-2-2).

A. Suitable feed type.

B. Proper ingredients.

C. Good formulation.

D. Efficient feeding strategy.

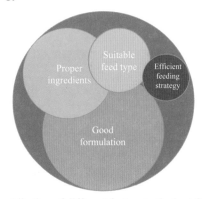

Fig. 5-2-2　Contribution of different factors to the best feed management

2　Feed type and selection

Fig. 5-2-3 shows the feed type and selection.

A. Vegetarian feed.

B. Home-made feed based on trash fish.

C. Pelleted feed.

D. Extruded feed.

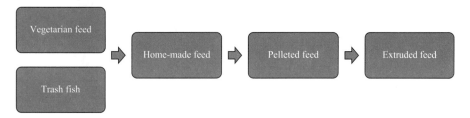

Fig. 5-2-3　Development of aquafeed

2.1　Vegetarian feed

Vegetarian feed, such as wheat bran, rice bran, weed, soy dregs, flour and peanut cakes, are suitable for freshwater fish (Fig. 5-2-4).

Fig. 5-2-4　Vegetarian feed

2.2 Home-made feed based on trash fish

Trash fish, fishing by-catch or small fish are suitable for marine fish (Fig. 5-2-5).

Fig. 5-2-5 Trash fish

Trash fish are firstly baked and ground. Then, vegetarian ingredients, vitamins, oils, binders and other additives are added. The mixture is extruded with simple machines and puffed up to produce dry pellets (Fig. 5-2-6). Some are fed freshly.

Fig. 5-2-6 Process of home-made feed making

The use of trash fish is criticized for the following negative effects:

A. Bad feed efficiency, resource waste.

B. Environmental pollution.

In China, 400-500 million tons of wild trash fish and 30 million tons of feed ingredients are directly fed to semi intensive aquaculture each year.

The emissions of nitrogen and phosphorus by using trash fish as feed are 4-5 times higher than by using pellet feed. The environment impact must be considered (Fig. 5-2-7).

Fig. 5-2-7 Fish mortality caused by environmental problems

2.3 Pellet feed

Ingredients are mixed together according to the nutritional requirements of fish. The pellets are extruded by a pellet machine (Figs. 5-2-8 to 5-2-9). Pellet feed is suitable for both freshwater and marine species.

Fig. 5-2-8 Pellet feed   Fig. 5-2-9 Pellet machine

2.4 Extruded feed

High pressure, friction and high temperature in the extrusion process lead to high gelatinization, sterilization and possibility of modulation of the floatability of feed (Fig. 5-2-10).

The utilization efficiency of extruded feed is higher. Floating feed on the water surface allows observation of fish feeding behaviour and controling of appropriate feeding rate.

Feed waste due to overfeeding can be avoided, which is beneficial to the environment.

In China, the proportion of extruded feed is increasing rapidly, especially in mariculture.

Fig. 5-2-10　Extruding machine

3　Feed ingredient selection —Criteria for ingredient selection

Important factors in ingredients selection are as follows (Fig. 5-2-11).

A. Nutritive (nutritional and anti-nutritional factors).

B. Cost-effective (cost less).

C. High availability.

D. Proper physical properties.

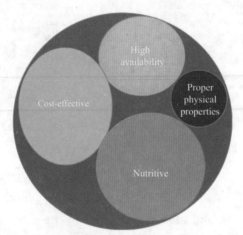

Fig. 5-2-11　Important factors in ingredients selection

3.1　Protein source

3.1.1　Fish meal and fish by-products

Fish meal is the most important protein ingredient used in the diets of fish. Fish meal contains high levels of essential amino acids (Fig. 5-2-12). The major wild fish used to produce fish meals include anchovy, capelin, menhaden, sand eel, herring, sardine and whitefish (Fig. 5-2-13).

The waste products of shrimp and crab processing contain about 40% and 32% protein, respectively. They are also the good sources of trace elements and carotenoid pigments.

Fish protein hydrolysates (Fig. 5-2-14) based on stickwater and fish by-products have high utilization efficiency, which provide recycled use of precious marine protein.

Fig. 5-2-12　Fish meal　　　　Fig. 5-2-13　Wild fish used to　　　Fig. 5-2-14　Dried
　　　　　　　　　　　　　　　　　produce fish meal　　　　　　　fish hydrolysate

3.1.2　Terminal animal by-products

These products are dried terrestrial animal tissues (Figs. 5-2-15 to 5-2-16), excluding hair, hooves, horn, hide trimmings, manure and stomach contents. The protein content of meat meal is about 51%, while that of meat and bone meal is close to 50%. Fat levels in these products average 9.1%-9.7%.

Both ingredients have relatively high ash content, about 27% for meat meal and 31% for meat and bone meal, respectively.

Fig. 5-2-15　Animal by-products　　　　Fig. 5-2-16　Meat and bone meal

Blood meal is a dry product made from clean, fresh animal blood, excluding all extraneous materials (Figs. 5-2-17 to 5-2-18). The most common blood meal is produced by spray-drying after an initial low-temperature vacuum evaporation, which reduces the moisture content to about 70%. Blood meal has minimum protein content of 85% and lysine content of 9%-11%, as well as lysine availability of over 80%.

Fig. 5-2-17　Fresh blood

Fig. 5-2-18　Blood meal

Poultry by-product meal is made from waste generated from poultry processing, excluding feathers and the intestine contents (Figs. 5-2-19 to 5-2-20). Regular poultry by-product meal contains about 58% protein and 13% fat. The ash content should be controlled to be lower than 16%, of which no more than 4% can be acid-insoluble ash.

Fig. 5-2-19　Chicken

Fig. 5-2-20　Poultry meal

Feather meal is made from poultry feathers, which have been hydrolyzed under pressure in the presence of $Ca(OH)_2$ and dried (Figs. 5-2-21 to 5-2-22). It has a protein content of 80%–85%, and no less than 75% of the protein must be digestible by the pepsin digestibility method. Its use in fish feeds is restricted, however, due to its low protein digestibility by fish (52.4%–70.5%).

Fig. 5-2-21　Feather

Fig. 5-2-22　Feather meal

3.1.3 Plant protein sources

Soybean meal is the most commonly used protein source. Dehulling soybean meal contains 48% protein, while defatted soybean meal contains 44% protein (Figs. 5-2-23 to 5-2-24). Soybeans contain several anti-nutritional factors such as the trypsin inhibitors and phytic acid. Full-fat soybean meal is obtained by processing raw soybeans from which no oil has been removed. Its use is restricted in catfish feeds because of the high oil content.

Fig. 5-2-23 Soybean　　　　　　　　　　Fig. 5-2-24 Soybean meal

Canola meal has a protein content of 36% (Figs. 5-2-25 to 5-2-26). Like most oil seeds, it contains anti-nutritional factors. When it is used in fish diet, the content should be lower than 30%.

Fig. 5-2-25 Oilseed rape　　　　　　　　Fig. 5-2-26 Canola meal

3.1.4 Other plant proteins

Plant proteins can be derived from grains. Both corn and wheat gluten meals are by-products of starch production with high protein content (>60% and >75% protein, respectively) (Figs. 5-2-27 to 5-2-28).

Brewery and distiller's by-products are made from the remaining residue after beer and liquor production (Fig. 5-2-29). Brewer's dried grains contain 27% protein and 13% fiber. Brewer's dried yeast contains about 44% protein and only 3% fiber.

Fig. 5-2-27 Wheat gluten meal   Fig. 5-2-28 Corn gluten meal   Fig. 5-2-29 Distillers dried grains

Peanut meal and sunflower meal are shown in Fig. 5-2-30.

Fig. 5-2-30 Peanut meal and sunflower meal

3.1.5 Other new protein sources

The unconventional feedstuffs include single-cell proteins derived from yeast or bacteria, worm meal, insect meal, and so on (Fig. 5-2-31). They have not yet reached the level of availability or acceptance that allows them to be used routinely in fish feed, but someday they may.

Fig. 5-2-31 Materials for algae and insect meals

3.2 Energy or lipid sources

3.2.1 Why lipid nutrition specially important to fish?

A. Unique sources of long chain polyunsaturated fatty acids (LC-PUFA) (Fig. 5-2-32).

Fig. 5-2-32　Fish oil for human consumption

B. LC-PUFA accumulation (Fig. 5-2-33).

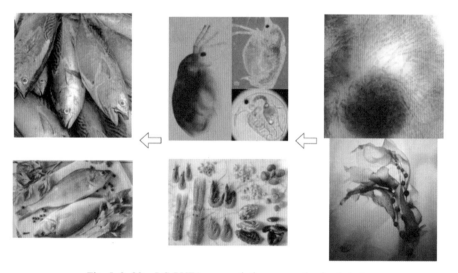

Fig. 5-2-33　LC-PUFA accumulation across the food chain

C. Lipid as an important energy source (Fig. 5-2-34).

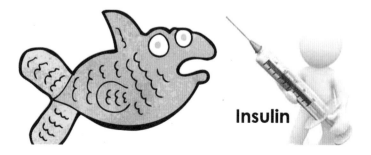

Fig. 5-2-34　Fish are natural diabetics

D. Determine flesh quality (texture and LC-PUFA contents) plastically (Fig. 5-2-35).

Fig. 5-2-35   Lipid content is important for fillet quality (*e.g.*, salmon and tiger puffer)

3.2.2   General guideline of fatty acid necessity in fish

The general pathways of fatty acid biosynthesis in fish can be found in Fig. 5-2-36.

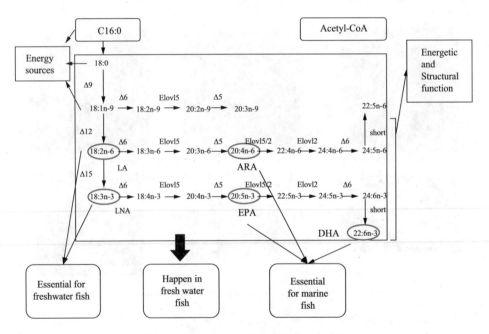

Fig. 5-2-36   Fatty acid metabolism in fish

3.2.3   Lipid sources in fish feeds

Fish oil is the most commonly used lipid source in aquafeed (Fig. 5-2-37).

A. Rich in LC-PUFA (DHA, EPA, ARA).

B. Essential for most marine carnivorous fish.

C. High price.

Soybean oil is the most important vegetable oil used in aquafeed (Fig. 5-2-38).

A. Rich in linoleic acid (C18:2n-6).

B. Having the highest production.

C. Most widely used for both marine and freshwater fish.

Features of sunflower oil (Fig. 5-2-39):

A. Rich in linoleic acid (C18:2n-6).

B. High levels of sunflower oil inducing impairment on fish health.

Fig. 5-2-37　Fish oil　　　Fig. 5-2-38　Soybean oil　　　Fig. 5-2-39　Sunflower oil

Features of rapeseed oil (Canola oil) (Fig. 5-2-40):

A. Low in erucic acid and glucosinolate.

B. Rich in oleic acid (C18:1n-9).

C. High production.

D. Used for both marine and freshwater fish.

Features of linseed oil (Fig. 5-2-41):

A. Rich in linolenic acid (C18:3n-3).

B. Good utilization efficiency in marine fish.

C. Low production and high price.

Fig. 5-2-40　Rapeseed oil　　　Fig. 5-2-41　Linseed oil

Features of palm oil (Fig. 5-2-42):

A. Rich in palmitic acid (C16:0), stearic acid (C18:0) and oleic acid (C18:1n-9).

B. Highly produced in Southeast Asia.

Fig. 5-2-42　Palm and palm oil

There are other oils used in aquafeed, such as:

A. Olive oil: rich in oleic acid (C18:1n-9), high price (Fig. 5-2-43).

B. Animal fat: poultry oil, lard, beef tallow; rich in stearic acid (C18:0) and palmitic acid (C16:0) (Fig. 5-2-44).

Fig. 5-2-43　Olive oil　　　　　　　　　Fig. 5-2-44　Animal fat

3.2.4　Criteria for lipid source selection

A. Fatty acid requirements of specific fish species, *e.g.*, tilapias like linoleic acid but puffers hate n-6 fatty acids (Fig. 5-2-45).

B. Price.

C. Availability.

Fig. 5-2-45　Guidelines of lipid source selection

3.2.5 Lipid requirement in fish

A. The lipid requirement for common carp, which utilizes carbohydrate better, is only 3%–6% of the diet (Fig. 5-2-46).

B. The lipid requirement of marine fish such as Japanese seabass is 10%–15% of the diet.

C. The lipid requirement of fatty fish such as Atlantic salmon is as high as 35% of the diet (Fig. 5-2-47).

Fig. 5-2-46 Fish with low lipid requirement   Fig. 5-2-47 Fillet of fish with high lipid requirement

Lipid types

Both neutral lipids and polar lipids (mainly phospholipids), which function differently, are required in fish feeds (Fig. 5-2-48).

Larvae can not synthesize enough phospholipids while juveniles can. Even though fish juveniles can synthesize phospholipid themselves, soy lecithin is commonly added in feed practice.

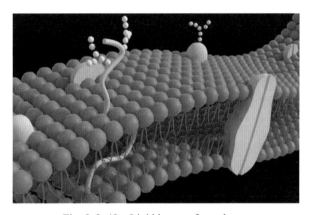

Fig. 5-2-48 Lipid layers of membrane

Some other aspects regarding lipid nutrition

A. Oxidized oils greatly impair fish health (liver function, intestinal inflammation, lipid over-accumulation).

B. Dietary lipid has protein sparing effects (energy balance).

C. Lipid nutrition affects flesh flavor.

D. Lipid digestibility is high in fish.

E. Lipid distribution pattern, which is diverse among fish species, affects the lipid requirement.

F. Different fish species have different fatty acid preference in lipid utilization.

G. High-lipid diets are commonly used in fish farming, which may have adverse effects on fish health.

3.2.6 Conclusion

The characteristics of lipid nutrition in fish are as follows.

A. Lipid nutrition greatly affects the flesh quality.

B. Lipid requirement varies drastically among different fish species.

C. Lipid nutrition is complicated in terms of fatty acids.

D. Mixed oils are commonly used.

E. Attention should be paid to the adverse effects of high dietary fat.

F. The precious marine LC-PUFA resources should be used efficiently.

3.3 Other nutritive ingredients

3.3.1 Vitamin

A. Lipid soluble: Vitamin A, Vitamin D, Vitamin E, Vitamin K.

B. Water soluble: B family, C, H, folic acid, biotin, choline, inositol.

3.3.2 Mineral

A. Macro: Ca, P, Mg, Na, K, Cl, S.

B. Micro: Fe, Cu, Mn, Zn, Co, I, Se.

Usually these nutritive ingredients are added as commercial premix.

3.4 Nonnutritive feed additives

3.4.1 Pellet binders

Fish feeds need to be strong enough to withstand normal handling and shipping without disintegrating. Fish feeds also must be water-stable. Some binders are by-products of cereal grains or plants and also provide nutrients to the diet (Figs. 5-2-49 to 5-2-50). Bentonite is a natural clay mainly consisting of tri-layered aluminum silicate (Fig. 5-2-51).

Fig. 5-2-49　Wheat meal　　Fig. 5-2-50　Precooked potato starch　　Fig. 5-2-51　Bentonite

3.4.2　Pigment

A. Xanthophylls are found in plants, such as corn (Fig. 5-2-52).

B. Carotenoid pigments are used in the diet of crustaceans.

C. Astaxanthin has been used in the diet of salmon.

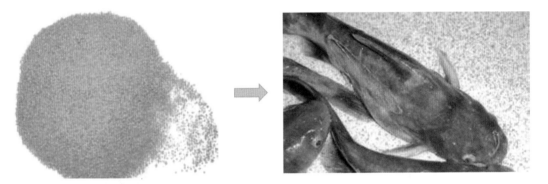

Fig. 5-2-52　Pigment enrichment in yellow catfish with corn gluten meal

3.4.3　Immunostimulants

A. β-glucans: cell walls of yeast.

B. Other artificial or natural polysaccharides.

C. Chinese herbs such as *Pilea pumila* (Fig. 5-2-53).

Fig. 5-2-53　*Pilea pumila*

3.4.4　Probiotics and prebiotics

Live bacteria (probiotics) (Fig. 5-2-54) or substances boosting beneficial intestinal bacteria (prebiotics) affect fish growth by affecting the microbiota in the gut of the animal. Obviously, probiotics must be added to feeds after pelleting.

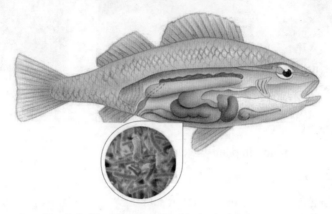

Fig. 5-2-54　Planting of prebiotics in fish intestine

3.4.5　Exogenous enzymes

Exogenous enzymes can enhance the digestion of feed components for the fish that either cannot digest or cannot digest efficiently. Enzyme supplements are typically sprayed on feeds after pelleting (Figs. 5-2-55 to 5-2-56).

Fig. 5-2-55　Amylase　　　　　Fig. 5-2-56　Phytase

There are some other nonnutritive feed additives such as antioxidant, attractants (Fig. 5-2-57) and mould inhibitors.

4　Feed formulation

A good feed formulation is based on the precise nutrient requirements of target fish (Fig. 5-2-58).

There are various software available about how to make an efficient feed formulation based on nutrient requirements of fish.

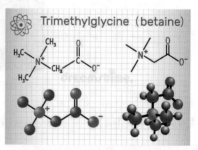

Fig. 5-2-57　Betaine (attractant)

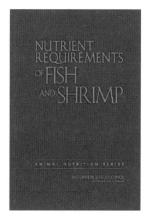 

Fig. 5-2-58 Valuable books about nutrient requirements of fish

5 Feeding strategy management

There are several factors contributing to a good feeding strategy.

A. Feeding regime including "how often" "when" "how much" and "where" is dependent on fish species and the environment (Fig. 5-2-59).

B. Feedrate and feeding frequency should be dynamic based on fish size.

C. Temperature < 10℃, no feeding; 14-15℃, 1 time per day; 18-20℃, 1-2 times per day; 20-30℃, 3-4 times per day.

D. Hand feeding (Fig. 5-2-60): operator can observe feeding behavior; labor cost is high; feed cost increased by 8%.

E. Automatic feeding (Fig. 5-2-61): labor cost is reducing; the quantity of feed used is knowable; less observation of the fish.

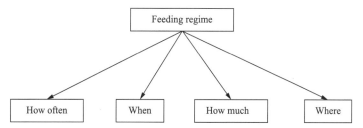

Fig. 5-2-59 Factors of feeding regime

Fig. 5-2-60 Hand feeding

Fig. 5-2-61 Automatic feeding

The general guidelines of feeding practice are as follows.

A. Feeding technique should be based on the principle of "four fixed" (fixed quality, fixed amount, fixed time and fixed location) (Fig. 5-2-62).

B. Feeding should be based on fish species, growth stage, fish behavior, season, climate, water temperature and water quality.

Fig. 5-2-62　A fixed feeding location example

6　How to get optimal feed for sustainable aquaculture?

In order to get optimal feed for sustainable aquaculture, proper ingredients (quality control, cost), adequate research on nutrient requirements, advanced feed processing technology (extrusion) and good feeding management are needed.

References

Sargent J R, Tocher D R, Bell J G. The lipids [M]//Halver J E. Fish Nutrition. San Diego, CA: Academic Press, 2002: 181-257.

Tocher D R. Metabolism and functions of lipids and fatty acids in teleost fish [J]. Reviews in Fisheries Science, 2003, 11: 107-184.

Tocher D R. Omega-3 long-chain polyunsaturated fatty acids and aquaculture in perspective [J]. Aquaculture, 2015, 449: 94-107.

Tocher D R, Betancor M B, Sprague M, et al. Omega-3 long-chain polyunsaturated fatty acids, EPA and DHA: bridging the gap between supply and demand [J]. Nutrients, 2019, 11: 89.

Turchini G M, Torstensen B E, Ng W K. Fish oil replacement in finfish nutrition [J]. Reviews in Aquaculture, 2009, 1: 10-57.

Live food technology of marine fish larvae

Author: Yu Hong
E-mail: yuhong@ysfri.ac.cn

1 Introduction

Successful and economical production of marine fish juveniles is highly multidisciplinary, requiring high-level larval and live-food technology.

The live-food development includes zootechnical, nutritional and microbial aspects. There is a well-established international live food production technology for *Brachionus* and *Artemia*. This technology has been modified for the culture of most marine fish species (Fig. 5-3-1).

Marine copepods are believed to be important as natural food for many larval fish species. Both nutritional and microbial aspects are covered in researches, and marine copepods serve as nutritional references.

a. *Epinephelus lanceolatus*; b. *Plectropomus leopardu*;
c. *Cromileptes altivelis*; d. *Plectropomus laevis*;
e. *Epinephelus akaara*; f. *Epinephelus septemfasciatu*.
Fig. 5-3-1 Marine fish species of breeding

The methods from the normal procedures are used worldwide on some points: The production and highly unsaturated n-3 fatty acids (n-3 HUFA) enrichment of rotifers can readily be done simultaneously.

The focus on *Artemia* will primarily involve specific problems of n-3 HUFA enrichment and stability.

1.1 Live food species

Rotifers and brine shrimp nauplii are two common live-food organisms for early life stages of marine finfish in hatcheries. The most commonly used species within these groups are *Brachionus plicatilis* and *Artemia franciscana*. There is an increasing use of the smaller rotifer species *Brachionus rotundiformis* in marine fish larviculture.

Copepods provide additional desirable characteristics such as size and nutritional value to finfish larvae and, until now, have played a supplemental role in larval rearing. Nauplii of some copepod species have been successfully used to raise fish species which cannot use rotifers as primary feed.

Microalgae are important components of fish larval diets, either directly or indirectly as food for *Brachionus* and *Artemia*. The addition of microalgae to the water during early first feeding is normally termed as the "green-water technique".

1.2 Cultivation systems

1.2.1 Larval survival and growth enhancement

A. Food size.

B. Fatty acid balance and content.

C. Amino acid balance.

D. Nutrient density of foods given to marine fish larvae.

1.2.2 First feeding

Extra care is necessary during first feeding and weaning to avoid high mortality.

The possible technical arrangement includes water supply and pre-treatment, tank design, feeding system and control functions. Standard filter units (*e.g.*, sand filters) will not retain zooplankton individuals as small as 20-30 μm (*e.g.*, many ciliates). Cylindrical-conical shaped tanks allow efficient precipitation and easy removal of organic wastes from the bottom during intensive feeding and high animal densities.

B. plicatilis and *Artemia* have their optimum for growth in sea water (*e.g.*, temperature and salinity is 25-30 ℃ and 10-15 respectively). Adequate aeration of cultures is probably among the main problems of producing rotifers in large and intensive units (>2 m^3) and of *n*-3 HUFA enrichment of *Artemia*. Commercial large-scale production of live feed requires the installation of partly automated feeding systems, enrichment systems, and systems for the cooling, storage and concentration of cultures.

2 Feeding regime components

The live food species of marine fish larvae are illustrated in Fig. 5-3-2.

Chapter 5 Food engineering and nutrition | 517

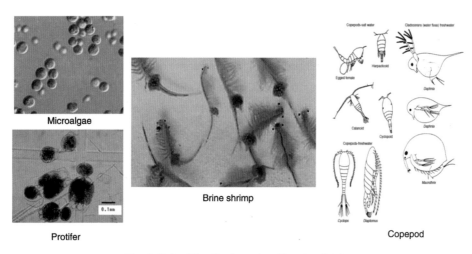

Fig. 5-3-2 Live food species of marine fish larvae

2.1 Microalgae

Microalgae are not strictly necessary for all species during larval feeding, but there is overwhelming documentary evidence that microalgae enhance the production yields and quality of many species. The two species of microalgae that have commonly been used in the the first feeding of marine fish larvae are *Isochrysis galbana* (Prymnesiophyceae) and *Tetraselmis* sp. (Prasinophyceae).

2.2 Rotifers

Rorifer is a major food component for most marine species of fish larvae. The species *B. plicatilis* (the so-called L-strain) is available in many sizes (lorica length 130-340 mm, mean 240 mm), as is the smaller *B. rotundiformis* (the so-called S-strain, lorica length 100-210 mm, mean 160 mm). *B. plicatilis* has been used most frequently for all cold-water species, while *B. rotundiformis* is introduced to temperate and tropical-water species these days. The nutritional value (*e.g.*, energy, protein, lipid essential fatty acids, and micro-nutrients such as vitamins and minerals) totally depends on the cultivation method, feed composition and the optional postharvest treatment of the culture between cultivation and feeding (*e.g.*, short-term enrichment, washing procedure, live feed storage). The content and composition of bacteria and the loading of organic carbon depend on the rotifer feed used, feeding routines, rotifer densities, rinsing procedures and other optional post-harvest treatments employed.

2.3 *Artemia* nauplii

Artemia nauplii have been used for all cultured marine species of fish larvae, and *Artemia franciscana* is the most commonly used *Artemia* species. The cyst quality is variable, but because of cyst formation, the biochemical composition of *Artemia* is much more stable than that of rotifers. The basic nutritional-composition, *e.g.*, protein and lipid contents, is relatively constant from the hatching of the cysts to larval feeding. However, n-3 HUFA is virtually absent in the *Artemia* cysts, and this critical component must be carefully bioencapsulated by established enrichment techniques before the *Artemia* become adequate as live food for marine fish larvae. A specific problem in this regard is the rapid DHA catabolism in *Artemia*, because DHA is considered to be the most important fatty acid for marine fish larvae.

2.4 Juvenile *Artemia*

Larger on-grown stages of *Artemia* (2–7 d old) have been tested as live food for the relatively large larvae of grouper, and they may also be advantageous for other species. Juvenile *A. franciscana* have a higher, and therefore a more attractive, ratio of protein to lipids than *Artemia* nauplii, but they still suffer from a high DHA degradation rate. A larger prey will theoretically be more optimal for the larger larval stages, but nauplii, which are easier and less expensive to cultivate, may still be adequate. The production technology for juvenile *Artemia* that are suitable for marine fish larvae has been established and described, although there is still room for improvements. The need to consider the composition of the bacterial flora is the same as that with *Artemia* nauplii.

3 Production of rotifers

3.1 The process of rotifer production

The rotifer strain has both an asexual (amictic or parthenogenetic) and a sexual (mictic) life cycle, but males and females carrying resting eggs have only been observed on a very few occasions.

The normal production temperature and salinity is 20–22 °C and 20, respectively. Under these conditions, the average amictic females produce their first eggs 1.4 d after hatching. Thereafter, they produce 21 eggs during the following 6.7 d. The developmental time of the eggs is 0.41 d. The post-reproductive period of the rotifer is 2.4 d, giving a total average life span of 10.5 d (Korstad et al., 1989a).

3.1.1 Internal control method

A. Make decision on culture growth rate (for example 0.3 d^{-1}, the recommended range is 0.2–0.4 d^{-1}).

B. Estimate the corresponding specific food ration (SFR) using Equation 1 (m_{net} = 0.3 d^{-1} requires 1.3 mg yeast per individual per day, SFR range for the recommended mnet-range (0.2–0.4 d^{-1}) is 0.94–1.8 mg yeast per individual per day).

where m_{net} is the net specific growth rate of the rotifer cultures (d^{-1}), and SFR is the specific food ration (weight of yeast (mg) per individual per day).

$$SFR = 0.489 \cdot (26.6)^{m_{net}} \quad \text{(batch culture)} \quad \text{(Equation 1)}$$

C. Count rotifer density (RD) per unit volume.

D. Estimate the culture food ration (FR, food per unit volume and day) (FR = SFR · RD).

E. Feed the estimated ration to the culture (in one, two or several portions).

F. Repeat steps C–E with constant SFR for the next few days until the maximum sustainable feed ration (FR_{max}), which has already been determined, is reached. Feed the cultures their maximum sustainable feed ration from that day.

3.1.2 External control method

A. Make decision on culture growth rate (for example 0.3 d^{-1}, the recommended range here is 0.2–0.4 d^{-1})

B. Estimate the corresponding SFR using Equation 1.

C. Count initial rotifer density per unit volume (RD_0).

D. Estimate the initial culture food ration (FR_0, food per unit volume and per day on day 0) ($FR0 = SFR · RD_0$).

E. Feed the estimated day 0 culture ration (FR_0) to the culture (in one, two or several portions).

F. For the next day (day 1) and later (days 2, 3, ..., n), estimate the food ration for day n according to Equation 3. The daily percentage increase in FR depends on the predetermined growth rate. A growth rate of, for example, 0.3 d^{-1} requires a daily increase in the food ration (IR) of 35% per day (Equation 2).

The cultures do not need to be counted daily, but the quality of the cultures should be evaluated visually before the food is added. The exponential rate of increase in food rations (IR, (%) d^{-1}) must be kept with in the growth capabilities of the rotifers. A net growth rate range of 0.20–0.40 d^{-1} corresponds to an IR range of 22%–49% d^{-1}, as estimated by Equation 2.

$$IR = (e^{m_{net}} - 1) \cdot 100\% \qquad \text{(Equation 2)}$$

The food ration for day n (FR_n) can be calculated as.

$$FR_n = FR_0 (IR/100 + 1)^n \qquad \text{(Equation 3)}$$

where FR_0 is the food ration at the initial day ($n = 0$) and IR is the daily increase rate of the food ration.

G. If the quality of the cultures is acceptable (see below), feed the day n dose.

H. Repeat steps F–G until the maximum sustainable feed ration (FR_{max}), which has already been determined, is reached. Feed the cultures their maximum sustainable feed ration from that day.

3.2 Extremely high density rotifer mass culture

Fish larvae prefer smaller rotifers immediately after hatching and larger ones as they grow. Providing them with this size variation increases their survival (Lubzens et al., 1989). Japanese scientists have been able to produce and maintain *B. rotundiformis* in densities higher than 100,000 individuals per milliliter in 1,000 L batch cultures. The relevant breeding methods are as follows.

A. The high-intensity culture systems consist of 1 m^3 tank units in which rotifers (*B. rotundiformis*) are batch-cultured at 2–3 d intervals. Cultures are initiated at a density of 10,000 rotifers/mL, and after 2–3 d, with concentrated algae the culture density reaches 20,000–30,000 rotifers/mL. The rotifers were fed refrigerated and condensed freshwater *Chlorella*, enriched with vitamin B_{12} (Yoshimura et al., 1996, 1997; Balompapueng et al., 1997a).

B. Oxygen has to be supplied to these cultures to overcome the shortage of dissolved oxygen that results from the high amounts of food and the subsequent increase in the rotifer population. In addition, at these high densities, ammonia excreted by the rotifers becomes a significant problem.

C. The pH of the culture also increases. This, in turn, results in a higher proportion of toxic unionised ammonia of the total ammonia, which increases with increased pH, salinity and the culture temperature. Regulation of the pH at 7.0 by hydrochloric acid minimises these effects.

D. Especial nylon filtration mat is used for the removal of large amounts of suspended organic material that may also include protozoans, fungi and bacterial flocculation, which may be harmful to the fish larvae.

4 Short-term enrichment techniques to improve nutritional value

4.1 Short-term enrichment techniques

The current method involves exposure to *n*-3 HUFA-rich feed given in high concentrations for short time, normally less than 24 h. Normally, this treatment will not result in a significant growth response, but only a major change in biochemical composition. This technique was later applied for proteins as well as for vitamins.

4.2 Stability of nutritional value

A. Post-enrichment, the biochemical composition of the rotifers is a dynamic variable which specifically reflects temperature and food availability. The final process of rotifer production may involve the rinsing and storage of cultures for some time before their ultimate use as live food, and the treatment may involve starvation. The rotifers may also survive in larval tanks for some time before being eaten. This will happen because many fish larvae are not very efficient at capturing live prey when they start to search for food.

B. A practical measure against long residence times in the larval tanks is a high exchange rate of water, but this is impossible for the very early and fragile stages of many marine fish larvae. The delay between feeding and consumption will then involve starvation unless microalgae are added along with the rotifers in the larval tanks. Quantitative knowledge of the catabolic rates of essential compounds in the rotifer during starvation is useful to secure adequate post-harvest treatment and storage conditions for rotifer cultures. (Fig. 5-3-3)

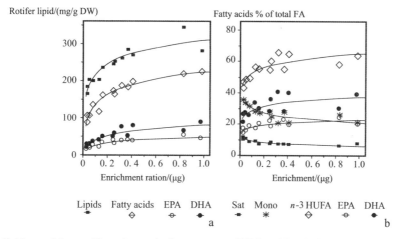

a. Total lipids, total fatty acids and quantitative contents of EPA and DHA; b. Percentage contents of important species and groups of fatty acids (right hand panel shows percentage values for feed).
Fig. 5-3-3 Lipid, fatty acid and n-3 HUFA contents and percentage fatty acid distribution as a function of the enrichment of the food ration (pooled data from two trials)

C. Both lipid per rotifer and protein per rotifer decrease rapidly with time of starvation (Fig. 5-3-4a). These are major nutrient components, and are decisive for the energy content of the rotifer and its general food value.

D. The time-courses of EPA and DHA (Fig. 5-3-4b) show an even faster reduction of these essential fatty acids. A considerable amount is lost during the first 24 h, which is well within the normal residence time of rotifers in larval tanks in the very early phase (Reitan et al., 1993). All metabolic processes will be affected by temperature, including the rate of loss of nutrients during starvation.

E. The loss rates (L) of different components are plotted as a function of the temperature during starvation in Fig. 5-3-4c and d. Its relationship with temperature is almost linear for carbon and protein per rotifer, and closer to exponential for total lipids and the other lipid components (Fig. 5-3-4d).

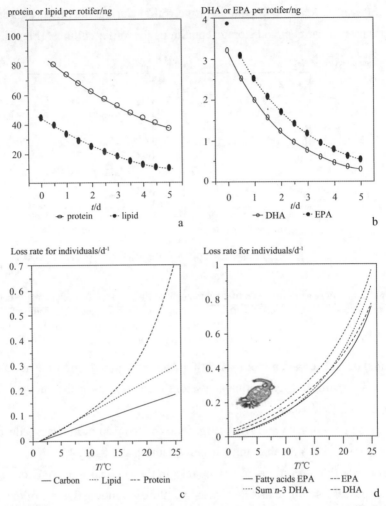

a. Losses of protein and lipid per individual as a function of starvation time at 18 ℃; b. Losses of DHA and EPA per individual as a function of starvation time at 18 ℃; c. Specific loss rates of body carbon, lipid and protein per individual as a function of starvation temperature; d. Specific loss rates of total fatty acids, sum of n-3 fatty acids, DHA and EPA per individual as a function of starvation temperature. The model curves are based on relations published by Olsen et al. (1993), and Makridis and Olsen (1999).

Fig. 5-3-4 Quantitative losses of nutritional value in *B. plicatilis* during starvation

5 Production of *Artemia*

Artemia production is different from that of rotifers because *Artemia* are hatched from resting cysts that are commercially available, and the biochemical composition and nutritional value are far more stable and reproducible in hatched nauplii than in rotifers. The challenges in making *Artemia* nutritionally adequate for marine fish larvae are primarily the manipulation n-3 HUFA content and the establishment of very strict routines for their production and use.

We now give a brief overview of some selected topics of *Artemia* technology. These involve the main challenge of adapting established international n-3 HUFA enrichment technology for

Artemia to marine fish species.

5.1 Hatching of *Artemia* cysts

Hydration of *Artemia* cysts in fresh water for 1–2 h at 25 ℃ causes the slightly shriveled chorion to regain a spherical shape, which facilitates hatching. Usually, hydrated cysts are incubated in salt water in some sort of cone-shaped tank with moderately high light and turbulence until the nauplii hatch. Then the hatching tank becomes a separatory funnel. Aeration is turned off, heavy debris sinks and is drained off to be discarded, next nauplii are drained off up to the surface layer of floating debris, which is discarded. Nauplii usually will sink quickly to the bottom of the tank, but a strong light can be used to attract them. Optimum conditions for best hatching of *Artemia* is described below: 25–30 ℃, salinity 20–25, DO > 2 mg/L, and cyst density < 5 g/L.

5.2 Feeding and growth

The Artemiidae are continuous, non-selective, obligate phagotrophic filter-feeders which start to ingest food at instar II stage using their larval antennae. The youngest individuals will then need about 20 s before the food appears in the gut, whereas larger juveniles may need as much as 3.5 min (Evjemo, 1999). The gut passage time is 24–29 min for well-fed animals.

The food ingestion rate of *A. franciscana* increases with increasing food concentration until to a maximum level, where it remains constant and independent of food concentration (Fig. 5-3-5a). The incipient limiting food concentration where the animals reach their maximum ingestion rate (I_{max}) is of the order of 5–7 mg/L. The clearance rate of the animals (mL/h per individual) is also dependent on food concentration (Fig. 5-3-5b). Unlike rotifers, the clearance rates of *Artemia* showed maximum values in the intermediate range 0.5–5 mg/L, and lower values both below and above this range.

The growth rate of *A. franciscana* during its life cycle depends on the food concentration (Fig. 5-3-5c). Well-fed animals do not show net increase in body weight earlier than 2 d after hatching, and their specific growth rate thereafter tends to increase gradually up to days 4–5. They then grow at an apparent maximum rate (m_{max} = 1.05 d^{-1}, days 4–7) up to a biomass of >150 mg dry weight per individual. Fig. 5-3-5c shows that a food concentration of 10 mg/L will support the maximum growth rate, whereas 7 mg/L is sub-optimal.

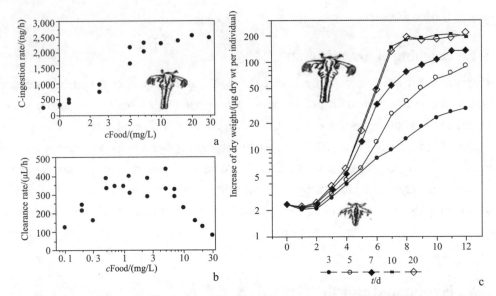

a, b. Feeding rate (a) and food clearance rate (b) of post metanauplius II and III (7 d old, 2.740 mm ± 0.11 mm length) as a function of food concentration; c. Increase of individual body dry matter with time for groups fed various concentrations of food algae. The legends express the food concentration in terms of weight of C per liter. Data are from Evjemo and Olsen (1999) and Evjemo et al. (2000)

Fig. 5-3-5   Feeding and growth characteristics of *Artemia franciscana* fed various concentrations of *Isochrysis galbana* (26–28 ℃)

5.3   Decapsulation procedure for Artemia cysts

5.3.1   Disinfection of cysts

The principal methods recommended by Van Stappen et al. (1996) for disinfection of *Artemia* cysts are:

A. Preparation of 200×10^{-6}. hypochlorite solution.

B. Soak cysts for 30 min, 50 g cysts per liter hypochlorite solution.

C. Wash cysts thoroughly in fresh water on a 125 mm screen.

D. The cysts are then ready for further hatching.

E. Decapsulation of cysts.

5.3.2   Hydration

Incubate 100 g cysts per liter in aerated water for 1 h at 25 ℃.

Collect the hydrated, rinsed cysts on a 125 μm screen.

5.3.3   Removal of capsules

A. Prepare a hypochlorite/alkaline solution (0.5 g active hypochlorite for 1 g cysts, see manual cited above for details).

B. Cool the solution to 15–20 ℃ using iced water.

C. Incubate the hydrated cysts for 5–15 min with aeration. Check the process of decapsulation using a microscope.

D. Remove the cysts from the incubation solution when they turn grey/orange (depending on the alkaline product used) or when examination under a microscope shows almost complete dissolution of the cyst shells.

5.3.4 Washing and deactivation

A. Rinse with water until no chlorine smell is detected.

B. Deactivate any traces of hypochlorite by dipping for 1 min, or in 0.1% $Na_2S_2O_3$ solution, then rinse again with water.

C. Repeat this procedure until all traces of hypochlorite have been removed.

D. The cysts may be hatched immediately or may be stored for a few days in a refrigerator (0–4 ℃). Decapsulated cysts can be used directly for marine larvae.

5.4 *n*-3 HUFA enrichment

A typical procedure for *n*-3 HUFA short-term enrichment of *Artemia* nauplii is described below.

Suitable cylindro-conical, strongly aerated tanks are filled with filtered seawater and heated to 28 ℃. The installation of a water exchange system is beneficial.

Hatched and carefully rinsed nauplii are transferred to the tanks in typical densities of 150–300 individuals per milliliter within 4–6 h after hatching.

The emulsion is dispersed and treated as recommended by the manufacturer, and is added in rations of 200 g/m^3 culture 4–6 h after the harvest of hatched nauplii. Oxygen is monitored (>2–3 mg/L).

Another similar food ration is supplied after 6–12 h, or earlier if the first ration has been consumed by the *Artemia*. Oxygen is monitored (>2–3 mg/L).

The *Artemia* nauplii are harvested using suitable harvesting gear before the enrichment diet has been completely removed (12–24 h, longer enrichment periods require further additions of food). Animals are carefully rinsed, and used as live food immediately.

Prolonged post-harvest storage will require cooling facilities.

5.5 Stability of *n*-3 fatty acids post-enrichment

Fig. 5-3-6 compares the phases of DHA enrichment (0–24 h) and DHA stability postenrichment (24–96 h) for two *Artemia* species, one of which is a Chinese strain which exhibits high DHA retention (Evjemo et al., 1997). The DHA stability of marine copepods during starvation is illustrated as a reference for comparison (Fig. 5-3-6c).

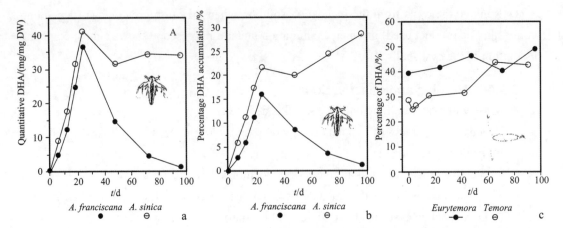

a. Time-course of quantitative DHA accumulation during 24 h enrichment (28 ℃) and later starvation for 3 d (12 ℃) in *Artemia franciscana* and *Artemia sinica*; b. Percentage DHA accumulation during 24 h enrichment (28 ℃) and later starvation for 3 d (12 ℃) in *A. franciscana* and *A. sinica*; c. Percentage DHA of total FA in two species of marine copepods maintained in starvation conditions.

Fig. 5-3-6 DHA enrichment and stability in *Artemia* strains and in marine copepods

When the enrichment diet is removed, DHA decreases exponentially in *A. franciscana*, whereas the present isolate of *A. sinica* retains DHA very efficiently. The specific loss-rate of DHA from *A. franciscana* at 12 ℃ is 1.2 d^{-1}, representing a half-time of disintegration of 14–15 h. The fact that percentage DHA decreases (Fig. 5-3-6b) implies that DHA is lost faster than the average fatty acid, and in fact faster than any other fatty acid (Evjemo et al., 2000). In *A. sinica*, however, there is an increase in percentage DHA during starvation, as in marine copepods (Fig. 5-3-6c). This means that DHA is being lost at a lower rate than the average fatty acid in this strain.

5.6 *n*-3 HUFA of *Artemia* juveniles

The technology to produce *Artemia* juveniles and adults at a very high density using cheap feed and automated feeding systems was first introduced many years ago (Sorgeloos et al., 1986). The technology has recently been adapted and fully demonstrated for grouper.

The production method described below for juveniles 3 d or older is based on a mixture of fishmeal and marine oil (modified from Olsen, 1999b). For an initial stocking density of 20 mL^{-1}:

A. Use 90 g/m^3 fish meal and 54 g/m^3 fish oil high in DHA (*e.g.*, Pronova TG 1040, DHA Selco), giving 47% lipid in the diet.

B. Feed three times a day (twice on day 0), maintain a high food level, feed more frequently for higher *Artemia* densities.

C. Use a regular water-exchange regime.

D. Use a temperature of 26–28 ℃.

E. Ensure that the oxygen concentration is >2.5 mg/L.

F. Baume degrees is above 10.

Three-day-old *A. franciscana* juveniles show a satisfactory lipid and fatty acid content (19% lipid about 19 mg/g (DW, DHA) and a protein to lipid ratio of 3.4. The survival is obviously lower and more variable than for short-termenrichment, but is still acceptable under skilled management. The biomass per prey is more than twice as high as that of short-term enriched nauplii.

High DHA catabolism is also a characteristic feature for juvenile stages of *A. franciscana*, and the nutritional problems encountered using juvenile *Artemia* are therefore more or less identical to those described for nauplii. It remains to be seen if juvenile *Artemia* will become more frequently used for feeding marine juvenile fish larvae such as tiger grouper.

6 Mass culture techniques in outdoor ponds

6.1 Copepod mass culture techniques in outdoor ponds

The live-food production units in outdoor ponds are very large. In one case, an annual production of 3 million grouper fingerings (2–3 cm) was achieved from three to four crops in Hainan Province between April and October using three 0.3 hm^2 larval rearing ponds, nine rotifer culture ponds (450 m^2) and nine copepod culture ponds (7×1 hm^2 plus 2×4 hm^2).

In the indoor system, copepods are fed to groupers from 7 mm to 2–3 cm as an *Artemia* replacement, not for first feeding. The quantity needed for the fish can be checked at night, using nets to collect the copepods in the larval rearing ponds. For 10,000 grouper fry at 1–2 cm, about 1–1.5 kg wet weight of copepods is needed and fed to the fry two to four times daily. In one case, in which 50,000 fry at 2.5 cm were copepod nauplii (about 110 μm) naturally occur in the outdoor larval-rearing ponds and are smaller than the gape of fish larvae, like giant grouper. Their presence in these ponds has enabled the successful culture of a number of grouper species.

To control competing species and those that prey on copepods, tea seed cake with 10% saponin and chemicals (bleaching powder) are used to clean pond water before starting to culture the copepods. When copepods are to be the main product, 1,000 fish are usually reared per 0.6 hm^2. Common species are red drum (*Sciaenops ocellatus*), mangrove red snapper (*Lutjanus argentimaculatus*), and tiger shrimp (*Penaeus monodon*). To provide enough food for the copepods an algal bloom must be maintained.

They are unlined, with a bottom of clay or fine sand or a mixture. Wheel paddles are used to aerate the water and to drive the water current for collection of copepods in plankton nets. Salinity of the pond water varies from 10 to 30 and is optimal at 15–20. Early-stage *Epinephelus coioides* larvae prefer to ingest copepod nauplii over rotifers.

The copepod culture ponds are dried in the sun for 1 to 2 weeks. Sea water is filtered in 100 mesh (300 μm) nets.

Chemicals and tea seed cake are then applied to clean the water.

In general, it takes about 15 d and not more than 30 d for the nauplii to appear. In most rapid cases, nauplii occurred in 3 d and copepodids in 15 d.

Usually, copepods are harvested at 3–4 h using a 100 mesh (300 μm) plankton net fixed to the wheel paddle for 20–30 min.

When the amount of harvest decreases, the harvest shifts to another pond. Generally, daily harvest continues for 7–15 d. Copepods are transported to farmers in aerating tanks by car in the morning.

After waiting for another 7–15 d, harvesting can resume.

The yield is higher from April to November in Hannan Province, when solar radiation is high. Daily yield of copepods per hectare ranges from 20 kg to 200 kg, with an average of 40–80 kg.

6.2 Rotifer mass culture techniques in outdoor ponds

6.2.1 Preliminary preparation

A. Pond condition cultivation. The breeding pond should be built close to the nursery with convenient transportation and accessible water resources, away from the wind to the sun place, in a for construction and drainage. The final area is 2,000–3,000 m^2, 80–120 cm deep pool in a rectangle form. The soil should be of argillaceous sediment quality, or the bottom and wall leakage. The number of soil pools is generally 3–5 or more, which is convenient to divide the soil pools.

B. Clear the pond with new water filtered. The excavation pool can be directly filtered into the water, and the remaining water should be applied with drugs, killing fish and crustaceans. Common cleansing drugs include the following kinds. Bleach: detected before using chlorine content, chlorine content 30% bleach, dosage of 25–30 g/m^3, general efficacy of 3–5 d disappearance. Fish rattan essence: 2 g/m^3 cubed.

The water can be filtered after cleaning the pond. The filter net is made of 160–200 mesh silk screen to prevent the mixing of harmful organisms. In the initial stage, the water was fed for 40–60 cm, and the salinity was adjusted to 10–20. Then fertilize the water to achieve eutrophication.

C. Fertilizer and water. Fertilizer used is divided into organic fertilizer and inorganic fertilizer. The apply amount depends on the circumstance of pool water. Fertilizer dosage is generally 3 g/m^3, urea 15–20 g/m^3. Chicken manure, due to its high fertilizer efficiency and strong persistence, with a dosage of 200 g/m^3, should be fermented before use. After fertilization, in general light temperature conditions, it usually takes 3 to 5 d after the color from clear to pale green or light brown. Water can be added as the color deepens.

6.2.2 Inoculate rotifers

When the natural water temperature reaches 15 ℃, pool water is more fat and can be timely inoculated with rotifers. If rotifers have been cultured in the soil pond for a year, Bamboo rake can be used to pull the bottom mud, the dormant egg stir up. the rotifers to be inoculated can also be live rotifers bought from the greenhouse nearby, with high egg rate. The inoculation was carried out when the water temperature is at its peak in sunny days. Inoculation density is generally 10 rotifers/L, or high as the case may be.

6.2.3 Training management

A. Fertilization. In the early stage of cultivation, due to large one-off fertilizer application and in the later stage, with the increase of water temperature, algae reproduction is accelerated, and nutrient salt consumption in water body needs to be replenished in time. Making up fertilizer is before noon. The dosage depends on the water quality and generally a small amount of frequent application of fertilizer.

B. Add water. In the early stage, water is mainly added, and with the propagation of algae, water is added at right. In the later stage, the material circulation in the pool with higher temperature is accelerated. As rotifers are constantly produced, some trace elements in the water are out of balance, so fresh seawater needs to be replenished. The added seawater should be filtered and preferably treated with disinfection. Water exchange is carried out by changing generally 1/5-1/3 of total water every 10 d. Water exchange can be carried out simultaneously with harvest.

6.2.4 Training management

A. Feeding. When the density of rotifers is too high to prevent the pond water from being filtered out within a short time, rotifers need to be harvested in a timely manner. In the meantime, to make the population grow continuously, rotifers can also be fed with mainly soybean milk, yeast, wheat and so on. The density of rotifers is controlled by rotifer production and the feed supplemented, which are single-celled algae in appropriate density.

B. Daily test. Check the tank wall for leakage. Check whether there is any change in the color of water. If the color of water becomes light, the density of algae species may decrease and need to be fertilized in time. If the color of water changes, it may be because of the change of dominant algae species, with little effect on rotifers. Check the density and ovulation rate of rotifers (total number of eggs in the female body). When the ovulation rate drops below 13%, rotifer population will decline, and the pond can be emptied and re-inoculated with water.

6.2.5 Methods of harvest

Make a long cylinder with 240 mesh silk screen, 6-10 m long and 40 cm in diameter. Fix the electric floating pump on the boat in the middle of the pool. Cover one end of the silk screen at the pump mouth and let it fully extend, and fix the other end at the edge of the pool and use live ligation. Check the mesh bag before use.

Generally, harvest is better in the morning. Because in the afternoon the water temperature is higher, the amount of viscous substances in the water body increases, and the dissolved oxygen in the water body is often oversaturated, thus forming a lot of foam in the bag, affecting the efficiency and product quality.

6.2.6 Discussion

Sometimes, the density and ovulation rate of rotifers decline rapidly, possibly due to high NH_3-N in the water, and water exchange measures are usually adopted. In high water temperature, breeding can be harvested quickly. Generally, the daily harvest is 1/6-1/3 of the pool amount. In addition, a large amount of rotifers should be harvested in time and fertilizer and water should be added to maintain the continuous growth of the population.

7 Microalgae

7.1 Chlorella (*Chlorella* spp.)

The classification status: Chlorophyta, Chlorophyceae, Chlorococcales, *Chlorella*.

Morphological characteristics: single, small cells, 2-12 μm. The mode of reproduction: when cells divide, protoplast is divided into 2, 4, 8 pro spores, wait until mother cell wall broken, which seems like kiss spores are released.

Ecological conditions: temperature: 10-36 ℃ (25-30 ℃); salinity: 0-45; light intensity: the optimal range is 3,000 – 10,000 lx; pH: 5.5-8.0.

Chlorella (Fig. 5-3-7) is rich in protein content, so it is healthy food for animals. In aquaculture, *Chlorella* is often used to cultivate rotifers.

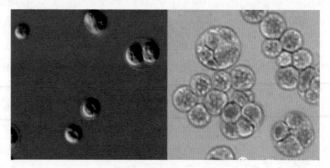

Fig. 5-3-7   *Chlorella* spp

7.2 *Nannochloris oculata*

The majority of marine chloropsis is *Nannochloropsis oculata* (Fig. 5-3-8), which is characterized by nearly spherical cells, with a diameter of 2-4 μm, and is very similar to *Chlorella*.

However, this alga has no cell wall, and when reproducing, it divides into two by longitudinal division.

Ecological condition: the suitable temperature for growth is 10–36 ℃; the optimum temperature is 25–30 ℃; suitable salinity range is 4–36;

optimal pH is 7.5–8.5; optimal light intensity is 10,000 lx.

Nannochloris oculata is mostly used in the culture of rotifers.

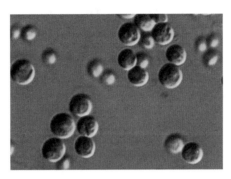

Fig. 5-3-8   *Nannochloris oculate*

7.3   Liquid formula (for green algae culture)

The formula for green algae culture is as follows:

$NaNO_3$	60 g
Urea	18 g
NaH_2PO_4	4–8 g
$FeC_6H_5O_7$	0.5–1 g
Disinfect seawater	1 m³

It is suitable for the production of *Platymonas subcordiformis*, *Pyramidomonas* sp. and so on.

7.4   Liquid formula for *Isochrysis* culture

The formula for *Isochrysis* culture is as follows:

$NaNO_3$	50–80 g
NaH_2PO_4	4–8 g
$FeC_6H_5O_7$	0.5–1 g
Vitamin B_1	100 mg
Vitamin B_{12}	0.5 mg
Disinfect seawater	1 m³

It is suitable for producing and cultivating *Isochrysis galbana* and *I. zhanjiangensis*.

7.5   Liquid formula for *Nannochloris oculata* culture liquid

The formula for *Nannochloris oculata* culture is as follows:

Ammonium sulfate	$(NH_4)_2SO_4$	100 g/m³
Calcium superphosphate	$Ca(H_2PO_4)_2 \cdot H_2O$	15–30 g/m³

Urea　　　　　　　　　　$CO(NH_2)_2$　　　　　　　　10-30 g/m^3

General methods of algae culture can also be used.

7.6 Method of disinfection and inoculation of seawater

Step1: Disinfection. Add 125 mL NaClO (valid chlorine 8%-11%) per cubic filter seawater.

Step2: Disinfection aeration. Stir and park for 6 h, and add 17 g $Na_2S_2O_3$ for neutralization.

Step3: If the residual chlorine test paper does not show blue reaction, fertilizer can be added.

Step4: Inoculation of microalgae. After stirring evenly, algae seeds can be inoculated.

In view of the importance of PUFA in live food technology of marine fish larvae breeding, PUFA, ARA, EPA and DHA contained in microalgae species are listed in Table 5-3-1, to make better use of the culture and use of these microalgae.

Table 5-3-1　The common microalgae species containing PUFA in research and production

Types of microalgae	Types of PUFA		
	ARA	EPA	DHA
Bacillariophyta			
Phaeodactylum tricornutum		√	
Nitzschia		√	
Cyclotella		√	
Odontella		√	
Chlorophyta			
Ostreococcus tauri		√	√
Micromonas pusilla		√	√

Microalgae are cultivated on a farm.

Preservation culture of *Chlorella* in seawater: Heat the filtered seawater to 70-80 ℃ for cooling reserve, or use the bleach solution (sodium hypochlorite solution) to sterilize the seawater, with sodium hyposulfite neutralizing it for the thermostatic culture box; use 2 liter flask, keep water temperature 20 ℃, light 2,500 lx, and inflate culture algae seeds. When the concentration of chlorella reaches 5×10^7 mL^{-1}, it is then transferred to other bottles and diluted to 5×10^6 mL^{-1} for seed preservation or expansion (Figs. 5-3-9-10).

Fig. 5-3-9　Microalgae cultured in fiberglass tubes

Fig. 5-3-10　Microalgae cultured in 3,000 mL flasks and 500 L sink

8　Recommendations

A. Use rotifers that are well enriched with n-3 HUFA during cultivation, and avoid using rotifers harvested from cultures with growth rates lower than half of their maximum growth rate (*e.g.*, >0.2 d^{-1}, 20 ℃).

B. The n-3 HUFA or protein content of rotifers may be additionally enhanced through short-term n-3 HUFA or protein enrichment post-harvest.

C. If rotifers are harvested while growing at a rate close to their maximum rate (>0.4 d^{-1}, 20 ℃, salinity 20), the protein content will be optimal.

D. Short-term enrichment with emulsified oil reduces the mobility and viability of rotifers and *Artemia* in sea water.

E. Use DHA-rich algae (*e.g.*, *Isochrysis galbana*) in larval tanks along with rotifers.

F. Use emulsified oil diets rich in DHA with a low ratio of EPA to DHA for short term enrichment of *n*-3 HUFA in *Artemia*, and reduce the starvation phase of post-enrichment to a minimum for the *Artemia*.

G. The lipid content of the short-term-enriched *Artemia* nauplii may be higher than the optimal. This can be counteracted by using a leaner *Artemia* grown on for 2–4 d.

References

Evjemo J O, Olsen Y. Lipid and fatty acid content in cultivated live feed organisms compared to marine copepods [J]. Hydrobiologia, 1997, 358(1–3): 159–162.

Evjemo J O, Olsen Y. Effect of food concentration on the growth and production rate of *Artemia franciscana* feeding on algae (T. iso) [J]. Journal of Experimental Marine Biology & Ecology, 1999, 242(2): 273–296.

Evjemo J O, Danielsen T L, Olsen Y. Losses of lipid, protein and *n*-3 fatty acids in enriched *Artemia franciscana* starved at different temperatures [J]. Aquaculture, 2001, 193(1): 65–80.

Evjemo J O, Reitan K I, Olsen Y. Copepods as live food organisms in the larval rearing of halibut larvae (*Hippoglossus hippoglossus* L.) with special emphasis on the nutritional value [J]. Aquaculture, 2003, 227(1): 191–210.

Evjemo J O, Vadstein O, Olsen Y. Feeding and assimilation kinetics of *Artemia franciscana* fed *Isochrysis galbana* (clone T. iso) [J]. Marine Biology (Berlin), 2000, 136(6): 1099–1109.

Hamre K, Yufera M, Ronnestad I, et al. Fishlarval nutrition and feed formulation: knowledge gaps and bottlenecks for advancesin larval rearing [J]. Rev. Aquac. 2013, 5: S26–S58.

Helland S, Oehme M, Ibieta P, et al. Effects of enriching rotifers *Brachionus* (Cayman) with protein, taurine, arginine, or phospholipids—startfeed for cod larvae *Gadus morhua* L. [C] Aquaculture Europe 2010, October 6–10. European Aquaculture Society, Porto.

Howell B R. Experiments on the rearing of larval turbot, *Scophthalmus maximus* L. [J]. Aquaculture, 1979, 18(3): 215–225.

Karlsen Ø, van der Meeren, T, Rønnestad I, et al. Copepods enhance nutritional status, growth and development in Atlantic cod (*Gadus morhua* L.) larvae—can we identify the underlying factors [J]. Peer J. 2015, 3: e902.

Kobayashi T, Nagase T, Hino A, et al. Effect of combination feeding of *Nannochloropsis* and freshwater *Chlorella* on the fatty acid composition of rotifer *Brachionus plicatilis* in a continuous culture [J]. Fish Sci., 2008, 74: 649–656.

Kobayashi T, Nagase T, Kurano N, et al. Fatty acid composition of the L-type rotifer *Brachionus plicatilis* produced by a continuous culture system under the provision of high density *Nannochloropsis* [J]. Nippon Suisan Gakkkaishi, 2005, 71: 328–334.

Korstad, J E, Olsen, Y, Vadstein, O. Life history of *Brachionus plicatilis* fed different algae [J]. Hydrobiologia, 1989a, 186/187: 43–50.

Lubzens E. Raising rotifers for use in aquaculture [J]. Hydrobiologia, 1987, 147(1): 245-255.

Lubzens E, Tandler A, Minkoff G. Rotifers as food in aquaculture [J]. Hydrobiologia, 1989, 186-187(1): 387-400.

Olsen A I. Development of production technology of juvenile *Artemia* optimal for feeding and production of Atlantic halibut fry [J]. Doktor Ingernioeravhandling, 1999.

Olsen A I, Maeland A, Waagbø R, et al. Effect of algal addition on stability of fatty acids and some water-soluble vitamins in juvenile *Artemia franciscana* [J]. Aquaculture Nutrition, 2000, 6(4): 263-273.

Olsen Y. Lipids and essential fatty acids in aquatic food webs: what can freshwater ecologists learn from mariculture? [M]//Lipids in freshwater ecosystems. New York, NY: Springer, 1999: 161-202.

Reitan K I, Rainuzzo J R, Øie G, et al. Nutritional effects of algal addition in first-feeding of turbot (*Scophthalmus maximus* L.) larvae [J]. Aquaculture, 1993, 118(3-4): 257-275.

Rothhaupt K O. Differences in particle size-dependent feeding efficiencies of closely related rotifer species [J]. Limnol. Oceanogr., 1990a, 35(1): 16-23.

Rothhaupt K O. Changes of the functional responses of the rotifers *Brachionus rubens* and *Brachionus calyciflorus* with particle sizes [J]. Limnol. Oceanogr., 1990b, 35(1): 24-32.

Shields R J, Bell J G, Luizi F S, et al. Natural copepods are superior to enriched *Artemia* nauplii as feed for larvae (*Hippoglossus hippoglossus*) in terms of survival, pigmentation and retinal morphology: relation to dietary essential fatty acids [J]. The Journal of Nutrition 1999a,129(6): 1186-1194.

Van Stappen G, Merchie G. Dhont J, et al. Artemia [M]// Manual on the Production and Use of Live Food for Aquaculture. Lavens P, Sorgeloos P. FAO Fisheries Technical Paper No. 361. 1996: 79-136.

Watanabe T, Kitajima C, Arakawa T, et al. Nutritional quality of rotifer *Brachionus plicatilis* as a living feed from the viewpoint of essential fatty acids for fish [J]. Bull. Jpn. Soc. Sci. Fish., 1978, 44: 1109-1114.

Watanabe T, Kitajima C, Fujita S. Nutritional values of live organisms used in Japan for mass propagation of fish: a review [J]. Aquaculture, 1983, 34(1-2): 115-143.

Yoshimura K, Usuki K, Yoshimatsu T, et al. Recent development of a high density mass culture system for the rotifer *Brachionus rotundiformis* Tschugunoff [J]. Hydrobiologia, 1997, 358(1-3): 139-144.

Protein nutrition for aquacultural fish

Author: Wei Yuliang
E-mail:weiyl@ysfri.ac.cn

1 Definition of protein

Protein is the major organic material in fish tissues. It constitutes 65%–75% of the total body weight on dry matter basis.

Fish nutrition research shows that the protein requirement for fish is significantly higher than that of human and the terrestrial animals. It is shown that fish need two to four folds in the diet as compared to terrestrial animals.

2 Protein requirements

Protein requirements have been obtained mainly from dose-response curves in which graded amounts of high-quality protein were fed in partially various semipurified or purified diets.

2.1 The evaluation index was mainly growth performance

Since the growth potential of aquatic animals is affected by nutrient intake, growth is often used as an important indicator to evaluate the requirement of a nutrient. The broken-line model has been the most widely used method of evaluating dose-response data in nutrient requirement studies with aquatic species (Fig. 5-4-1) (Luo et al., 2004).

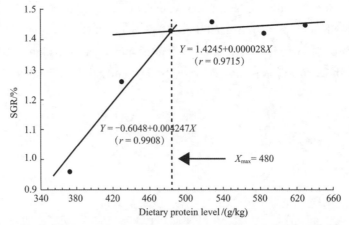

Fig. 5-4-1　Optimal dietary protein requirement of grouper *Epinephelus coioides* juveniles, based on the broken-line regression model of specific growth rate (SGR) versus dietary protein level

2.2 For most fish, protein requirements rang from 30% to 55%

The estimated protein requirements of several species of juvenile fish are summarized in Table 5-4-1 (Wilson, 2003). Most of these values have been estimated from dose-response curves, yielding the minimum amount of dietary protein which results in maximum growth.

Table 5-4-1 Estimated dietary protein requirement for maximal growth of some species of juvenile fish (as fed basis)

Species	Protein source	Estimated protein requirement/%
Atlantic salmon	Casein and gelatin	45
Channel catfish	Whole egg protein	32–36
Chinook salmon	Casein, gelatin and amino acids	40
Coho salmon	Casein	40
Common carp	Casein	31–38
Estuary grouper	Tuna muscle meal	40–50
Gilthead sea bream	Casein, fish protein concentrate and amino acids	40
Grass carp	Casein	41–43
Japanese eel	Casein and amino acids	44.5
Largemouth bass	Casein and fish protein concentrate	40
Milkfish	Casein	40
Plaice	Cod muscle	50
Puffer fish	Casein	50
Rainbow trout	Fishmeal, casein, gelatin and amino acids	40
Red sea bream	Casein	55
Smallmouth bass	Casein and fish protein concentrate	45
Snakehead	Fishmeal	52
Sockeye salmon	Casein, gelatin and amino acids	45
Striped bass	Fishmeal and soy proteinate	47
Blue tilapia	Casein and egg albumin	34
Mossambique tilapia	White fishmeal	40
Nile tilapia	Casein	30
Zillii's tilapia	Casein	35
Yellowtail	Sand eel and fishmeal	55

2.3 Factors affecting protein requirements

2.3.1 Fish species
Carnivorous is greater than omnivorous, followed by herbivorous.

2.3.2 Size and age
Generally, the protein requirements of fish decrease with increasing size and age.

2.3.3 Environmental conditions, such as water temperature
In general, the growth rate and feed intake increase as the water temperature increases. For example, chinook salmon were found to require 40% protein at 8 ℃ and 55% protein at 15 ℃. Similarly, striped bass were found to require 47% protein at 20 ℃ and about 55% protein at 24 ℃.

3 Protein sources

Protein sources are the major concern during the formulation of fish feed. They are the most expensive feed for fish and the most important factors contributing to the growth performance of cultured species.

3.1 Dietary protein serves three purposes in the nutrition of fish

A. Supply amino acids.

B. Meet requirements for functional proteins, such as enzymes, hormones and structural proteins.

C. Provide energy.

3.2 Fish meal

Until recently, fish feed was regarded as an important protein ingredient. Generally, fish meal requirement for omnivores is about 30% to 40% and for carnivores it is more than 40%.

Fish meal (Fig. 5-4-2) has been the protein source of choice in aquafeeds for many reasons.

A. High protein content.

B. Excellent amino acid profile.

C. High nutrient digestibility.

D. General lack of antinutritional factors.

Fig. 5-4-2   Fish meal

3.2.1 How can aquaculture be sustainable

Sustainable aquaculture needs to reduce the dependence of fish meal in feed industry.

If aquaculture consumes wild fish in the form of fishmeal at higher amounts than what is produced, then aquaculture is a net consumer of fish, not a net producer (Figs. 5-4-3 to 5-4-4).

Fig. 5-4-3 Input: wild fish   Fig. 5-4-4 Output: aquaculture

If the reverse is true, aquaculture is a net producer of fish. No sustainability!

3.2.2 Solving this problem

Fish meal production is finite

If fish meal is used at current rates in aquafeeds and aquaculture production also grows at current rates, aquaculture's demand for fishmeal will exceed fish meal production sooner or later.

Two possible strategies to address the problem

Firstly, the reduction of the protein in the feeds based on protein sparing effects of carbohydrates and lipids. Secondly, the use of some raw materials is to replace fish meal.

A. Optimum protein: energy ratio (P/E).

A decrease in P/E has indeed proven to be extremely efficient in improving protein utilization and decreasing nitrogenous loses in most fish (Table 5-4-2) (Wilson, 2003).

Table 5-4-2 Optimum P/E for different fish

Fish species	Dietary protein/%	Energy content/(kcal/g)	P/E/(mg/kcal)	Fish body weight/g	Evaluation index
Channel catfish	22.2	2.33	95	526	Weight gain
	28.8	3.07	94	34	Weight gain
	27	2.78	97	10	Weight gain
	27	3.14	86	266	Weight gain
	24.4	3.05	81	600	Weight gain

(*to be continued*)

Fish species	Dietary protein/%	Energy content/(kcal/g)	P/E/(mg/kcal)	Fish body weight/g	Evaluation index
Red drum	31.5	3.2	98	43	Weight gain
Tilapia nilotica	30	2.9	108	20	Weight gain
Common carp	31.5	2.9	103	50	Weight gain
Rainbow trout	33	3.6	92	90	Weight gain
	42	4.1	105	94	Weight gain

B. Use of fish meal replaces.

a. Plant protein sources—The main alternative protein sources (Fig. 5-4-5).

Cereals by-product meals includes milled/processed cereals (maize/corn, wheat, rice, barley, sorghum, oats, rye, millet, triticale, etc.).

Oilseed meals includes full-fat (soybean) and solvent extracted oilseed meals (soybean, rapeseed, cotton, groundnut/peanut, sunflower, palm kernel, copra).

Protein concentrate meals includes pea protein concentrate, lupin protein concentrate, soybean protein concentrates, rapeseed/canola protein concentrate.

Fig. 5-4-5 Plant protein sources

Apart from the partial replacement of fish meal using the plant proteins (Fig. 5-4-5), complete fish meal replaced by plant protein in fish diets had been achieved in some species, such as rainbow trout, Nile tilapia (Table 5-4-3) (Daniel, 2018), when some essential amino acids (methionine and lysine) had been supplemented to non-fish meal diet. For example, a previous study found that Nile tilapia received soybean meal and extruded full-fat soybean was able to completely replace dietary fishmeal when supplemented with DL-methionine and L-lysine.

Table 5-4-3 Complete fish meal replaced by plant protein in fish diets did not affect the animal's performances

Species studied	Plant ingredients used	Inclusion level supported	Remarks
Gilthead sea bream	Mixture of corn gluten meal, wheat gluten, extruded peas, rapeseed meal balanced with EAAs	100%	Improved the protein deposition than those of fish meal based diet

(to be continued)

Species studied	Plant ingredients used	Inclusion level supported	Remarks
Nile tilapia	Mixture of plant protein sources	100%	No adverse effect on growth performances; Around 36% of the feed production cost was reduced
Abalone	Soybean combined with either corn gluten meal or silkworm pupae meal	100%	Growth performances were not interfered
Rainbow trout	Mix of corn gluten, yellow soy protein concentrate and wheat gluten meal supplied with limiting EAAs and inorganic phosphate	100%	No apparent reduction in growth performance and feed utilization
Rainbow trout	Protein from plant protein concentrates with multiple EAA supplementations and using krill meal and the water soluble fraction of krill as feed attractant	100%	No adverse effects on feed intake or growth
Siberian sturgeon	Mix of soybean meal and wheat gluten meal with crystalline EAAs and mono-calcium phosphate	100%	No adverse effects on growth and protein utilization

b. Animal protein sources.

Aquatic animal protein meals: fish by-product meals, fish hydrolysates, silages and fermentation products.

Terrestrial animal protein meals: meat by-product meals, poultry by-product meals and blood by-product meals.

Microbial nutrient sources: yeast residue, fungi residue and algae residue.

4 Amino acids

A. Fish consume protein to obtain amino acids.

B. These amino acids are used by the various tissues to synthesize new protein.

C. A regular intake of protein or amino acids is required because amino acids are used continually by the fish, either to build new proteins (as during growth and reproduction) or to replace existing proteins (maintenance) (Fig. 5-4-6).

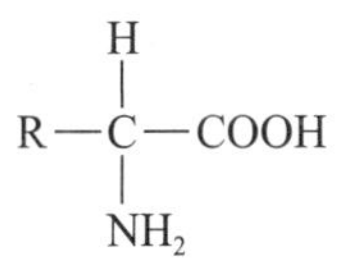

Fig. 5-4-6 The general formula of amino acid

D. Speaking scientifically, any animal as such does not need protein in the diet; they need balanced amounts of amino acids.

Essential amino acids (EAA) are defined as either those amino acids whose carbon skeletons cannot be synthesized or those that are inadequately synthesized *de novo* by the body and must be provided from the diet to meet optimal requirements.

The following 10 EAAs and 2 semi-EAAs were indispensable for fish:

A. Arginine.

B. Histidine.

C. Branched-chain amino acid: Leucine, Valine, Isoleusine.

D. Lysine.

E. Sulfur amino acid: Methionine, Cysteine (semi-EAA).

F. Threonine.

G. Aromatic amino acid: Phenylalanine, Tyrosine (semi-EAA).

H. Tryptophan.

4.1 The methodology of quantitative amino acid requirements

4.1.1 Growth studies—most amino acids

Replicate groups of fish are fed with diets containing graded levels of the test amino acid until measurable differences appear in the weight gain of the test fish. A linear increase in weight gain is normally observed with increasing amino acid intake up to a breaking point corresponding to the requirement of the specific amino acid, at which the weight gain levels off or plateaus (Fig. 5-4-7) (Mai et al., 2006).

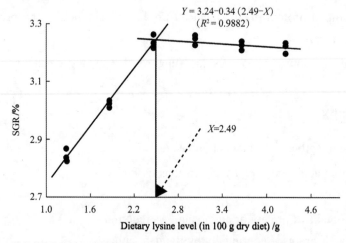

Fig. 5-4-7 Relationship between dietary lysine level and SGR of Japanese sea bass fed the six diets after 10 weeks

4.1.2 Serum or tissue amino acid studies—only a few cases

The hypothesis is that the serum or tissue content of the amino acid should remain low until the requirement for the amino acid is met and then increase to high levels when excessive amounts of the amino acid are fed (Figs. 5-4-8 to 5-4-9) (Yun et al., 2016).

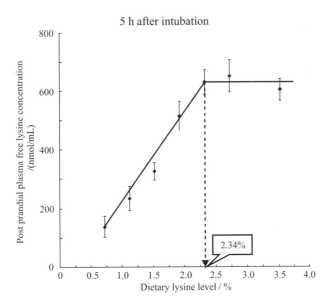

Fig. 5-4-8　Based on plasma free lysine concentrations: 2.20% and 2.34% of diet (dietary protein)

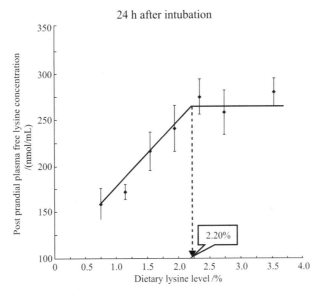

Fig. 5-4-9　Based on the growth performance: 2.23% and 2.32% of diet (dietary protein)

4.1.3 Amino acid oxidation studies—only limited success in rainbow trout

The general hypothesis is that when an amino acid is deficient in the diet, the major portion will be utilized for protein synthesis, and little will be oxidized to carbon dioxide, whereas when the quantity of an amino acid is supplied in excess, and thus is not a limiting factor for protein synthesis, most of the amino acid will be oxidized (Table 5-4-4, Fig. 5-4-10) (Walton et al., 1984).

Table 5-4-4 Oxidation of L-[14COOH]tryptophan for 20 h by rainbow trout given diets containing different levels of tryptophan

	Diet						SEM
	1	2	3	4	5	6	
Dietary tryptophan /(g/kg)	0.8	1.3	2.0	3.0	4.0	6.0	
Dose oxidized /%	3.13^a	3.31^a	4.50^a	13.40^b	22.44^c	31.98^d	1.48

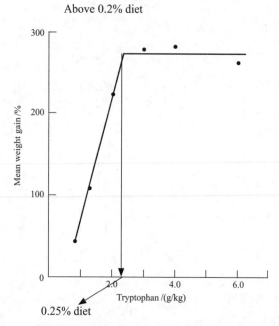

Fig. 5-4-10 Estimation of tryptophan requirement of rainbow trout

4.2 Several important amino acids in fish growth and metabolism

4.2.1 Lysine—The first limiting amino acid in ingredients (4% to 5% of dietary protein for most fish)

Dietary lysine levels critically affect fish growth performance and health. Lysine is a substrate for the synthesis of carnitine, which is required for the transport of long chain fatty acids from the cytosol into mitochondria for oxidation.

4.2.2 Methionine—A sulfur-containing amino acid (2% to 3% of dietary protein for most fish)

S-Adenosylmethionine is an important methyl group donor for creatine and spermidine synthesis. Pathways of methionine transmethylation, remethylation, transsulfuration for the synthesis of cysteine and taurine are likely present in fish.

4.2.3 Arginine—A particularly highly required amino acid for fish (3.3% to 5.9% of dietary protein for most fish)

Arginine is the precursor for the synthesis of protein, nitric oxide, urea, polyamines, proline, glutamate, creatine and agmatine, which can regulate endocrine and reproductive functions, as well as extra-endocrine signaling pathways.

4.2.4 Taurine—A sulfonic acid found in high concentrations but not incorporated into proteins (a conditionally essential amino acid for some marine fish species)

A. Osmotic regulation.

B. Antioxidation.

C. Feeding stimulation.

D. Retina development and vision.

E. Bile acid metabolism.

F. Modulation of immune response.

4.3 Amino acid interactions

4.3.1 Antagonism

A. Arginine and lysine.

B. Branched-chain amino acid: leucine, isoleucine, valine.

4.3.2 Synthesis

A. Methionine and cysteine.

Cysteine is considered dispensable because it can be synthesized by the fish from the indispensable amino acid methionine.

B. Tyrosine and phenylalanine.

Tyrosine is also considered dispensable because it can be synthesized by the fish from the indispensable amino acid phenylalanine.

5 Dietary protein digestibility, absorption and metabolism

5.1 Protein digestibility

In fish, dietary protein is digested or hydrolyzed by proteolytic enzymes and releases di/tri-peptide or free amino acids, which are absorbed from the intestinal tract and distributed by the blood to the organs and tissues (Fig. 5-4-11).

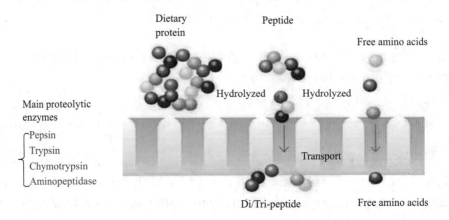

Fig. 5-4-11 Protein digestibility

5.2 Protein absorption

5.2.1 Di/Tri-peptides

Peptide transporter includes PEPT1 (SLC15A1), PEPT2 (SLC15A2), PHT1 (SLC15A4) and PHT2 (SLC15A3). The PEPT1 and PEPT2 proteins can transport all 400 different dipeptides and 8,000 different tripeptides derived from the 20 proteinogenic L-α-amino acids.

5.2.2 Free amino acids

Based on transport substrates, five transporters were proposed:

A. neutral amino acid transporters transporting all neutral amino acids, such as methionine, leucine, phenylalanine;

B. cationic amino acid transporters transporting cationic amino acids, such as lysine, histidine, arginine;

C. anionic amino acid transporters transporting glutamate and aspartate;

D. imino acid and glycine transporters transporting proline, hydroxyproline and glycine;

E. β-amino acid transporters transporting taurine, β-alanine and GABA (Fig. 5-4-12).

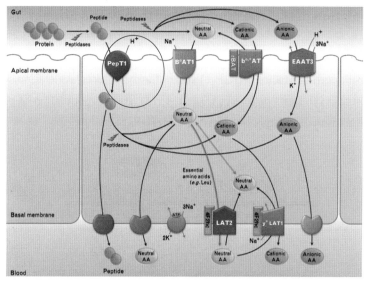

Fig. 5-4-12　The major luminal transport systems for amino acids and peptide transport systems

5.3　Protein metabolism—synthesis and degradation (catabolism)

Body protein is in a continuous process of synthesis and catabolism. This is a fundamental fact in understanding how to quantify the processes of intake, digestion, circulation, synthesis and hydrolysis, where protein/amino acids are in a state of turnover (Fig. 5-4-13).

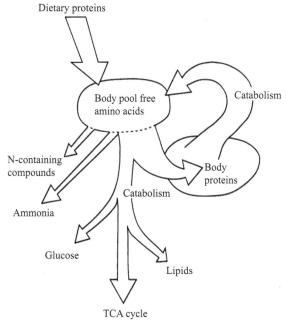

Fig. 5-4-13　Main pathways of protein and amino acid metabolism

5.4 Regulation of skeletal muscle hypertrophy and atrophy

5.4.1 Muscle hypertrophy

The insulin-like growth factor 1 (IGF-1)/Akt pathway induces skeletal muscle hypertrophy by increasing protein synthesis pathways (Fig. 5-4-14). IGF-1 modulation of protein synthesis operates via Akt/mammalian target of rapamycin (mTOR) signaling.

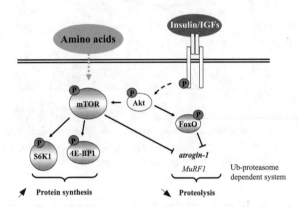

Fig. 5-4-14 mTOR signalling and control of protein synthesis and proteolysis

5.4.2 Muscle atrophy

Two markers of skeletal muscle atrophy have been characterized, namely the muscle-specific E3 ubiquitin ligases muscle RING-finger 1 (*MuRF1*) and muscle atrophy F-box (*MAFbx*; also called *atrogin-1*).

Akt phosphorylates and thus inhibits the family of FoxO transcription factors, which are required for *MuRF1* and *MAFbx* upregulation.

References

Daniel N. A review on replacing fish meal in aqua feeds using plant protein sources [J]. International Journal of Fisheries and Aquatic Studies, 2018, 6(2): 164–179.

Luo Z, Liu Y J, Mai K S, et al. Optimal dietary protein requirement of grouper *Epinephelus coioides* juveniles fed isoenergetic diets in floating net cages [J]. Aquaculture Nutrition, 2004, 10(4): 247–252.

Mai K, Zhang L, Ai Q, et al. Dietary lysine requirement of juvenile Japanese seabass, *Lateolabrax* [J] Aquaculture, 2006, 258(1–4): 535–542.

Walton M J, Coloso R M, Cowey C B, et al. The effects of dietary tryptophan levels on growth and metabolism of rainbow trout (*Salmo gairdneri*) [J]. British Journal of Nutrition, 1984,

51(2): 279–287.

Wilson R P. Amino acids and proteins [M]. Halver, J E, Hardy, R W. Fish Nutrition. 3rd ed. San Diego: Academic Press, 2003: 143–179.

Yun H, Park G, Ok I, et al. Determination of the dietary lysine requirement by measuring plasma free lysine concentrations in rainbow trout *Oncorhynchus mykiss* after dorsal aorta cannulation [J]. Fisheries and aquatic sciences, 2016, 19(1): 1–7.

Chapter 6
International science and technology cooperation in mariculture of Yellow Sea Fisheries Research Institute, Chinese Academy of Fishery Sciences

International science and technology cooperation in mariculture of Yellow Sea Fisheries Research Institute, Chinese Academy of Fishery Sciences

Author: Li Jian, Xu Jiakun, Zhao Fuwen
E-mail: xujk@ysfri.ac.cn

1 Introduction

Yellow Sea Fisheries Research Institute (YSFRI), Chinese Academy of Fishery Sciences (Fig. 6-1-1) is a comprehensive marine fishery research institute affiliated to the Ministry of Agriculture and Rural Affairs of the People's Republic of China (MARA). It was established in 1947 and is one of the oldest marine fisheries research institutions in China. YSFRI, focusing on the key tasks of sustainable utilization of marine living resources, has developed a large number of basic and forward-looking applied technologies in the fields of marine fishery resources and environment, germplasm resources and healthy aquaculture, aquatic products processing and quality control. YSFRI has made significant contributions to the development of China's marine fishery technologies and industrial economy.

Fig. 6-1-1   Landscape of YSFRI

YSFRI has carried out a wide range of international cooperation and established smooth, diversified international cooperation channels with universities, research institutions, enterprises of more than 40 countries and regions, as well as global and regional governmental and non-governmental organizations such as Food and Agriculture Organization of the United Nations (FAO), United Nations Industrial Development Organization (UNIDO), United Nations Development Programme (UNDP), World Organisation for Animal Health (OIE), North Pacific Marine Science Organization (PICES), Commission for the Conservation of Antarctic Marine Living Resources (CCAMLR), Network of Aquaculture Centres in Asia-Pacific Centres (NACA), etc.

During the past 10 years, YSFRI has undertaken more than 90 international cooperation projects at various levels, including the international cooperation projects from the Ministry of Science and Technology, the MARA, UNDP, the Seventh Framework Programme of the European Union, and the Horizon 2020. Among them, the implementation of Sino-Norwegian "Beidou" project (Fig. 6-1-2) from 1984 to 2004 laid the cornerstone for the acoustic assessment and the investigation of marine fishery resources in China; the successful introduction and breeding of turbot in 1992 promoted the development of marine fish farming industry in China; the implementation of the series of international China-Canada and China-EU cooperation projects since the 1990s has promoted the development of marine aquaculture ecological disciplines, forming a world-leading level of Integrated Multi-Trophic Aquaculture (IMTA) modes. The implementation of the "Science, Technology and Innovation Cooperation in Aquaculture with Tropical Countries along the 'Belt and Road' Project" granted by MARA will effectively promote the international fisheries technologies and industry cooperation, which will also support regional economic development.

Fig. 6-1-2 "Beidou" fishery research vessel of YSFRI

At a new historical starting point, YSFRI will adhere to the spirit of "peace and cooperation, openness and inclusiveness, mutual learning and mutual benefit". YSFRI will carry out extensive international fishery science and technology cooperation in the fields of protection and utilization of fishery resources, aquatic genetic breeding, healthy aquaculture technologies, quality and safety of aquatic products, aquatic products processing and utilization of product resources, which will contribute to the development of world fishery science and technology.

2　Cooperation platforms

2.1　"Belt and Road" Training Base for Mariculture Technologies, Ministry of Agriculture and Rural Affairs of the People's Republic of China

The "Belt and Road" Training Base for Mariculture Technologies was approved by the Ministry of Agriculture and Rural Affairs of the People's Republic of China in 2018 (Fig. 6-1-3). The base mainly provides theoretical and technical training in mariculture for fishery technical and managerial personnel from home and abroad. It also vigorously promotes and supports international exchanges and cooperation with the "Belt and Road" countries, technologically advanced countries in fishery and main international fishery organizations. Meanwhile, consultation services are also provided for fishery enterprises who are seeking for international cooperation.

Fig. 6-1-3　Unveiling Ceremony of "Belt and Road" Training Base for Mariculture Technologies of the Ministry of Agriculture and Rural Affairs of the People's Republic of China

2.2　International Fisheries Cooperation Base for Marine Fisheries, Ministry of Science and Technology of the People's Republic of China

In 2010, YSFRI was designated as the International Fisheries Cooperation Base for Marine Fisheries by the Ministry of Science and Technology of the People's Republic of China (Fig. 6-1-4). The base carries out extensive fishery science and technology cooperation in the fields of protection and utilization, etc., of marine fishery resources, aquatic genetic breeding, healthy aquaculture, aquatic product quality and safety, aquatic product processing and product resource utilization, etc., and provides effective support for the development of world fishery science and technology.

Fig. 6-1-4　YSFRI was designated as the International Fisheries Cooperation Base for Marine Fisheries by the Ministry of Science and Technology of the People's Republic of China in 2011

2.3　Demonstration and Promotion Base for the Introduction of Foreign Intelligence of "Popularization of Shrimp Breeding and Propagation Techniques", State Administration of Foreign Experts Affairs of the People's Republic of China

In October 2010, YSFRI was designated the Demonstration and Promotion Base for the Introduction of Foreign Intelligence of "Popularization of shrimp breeding and propagation techniques", State Administration of Foreign Experts Affairs of the People's Republic of China. (Fig. 6-1-5) Advanced shrimp breeding and propagation system has been established through international cooperation. YSFRI has carried out demonstration and promotion of fine shrimp varieties in coastal areas of China, which has promoted the development of shrimp farming industry in China.

Fig. 6-1-5　Dr. Shaun Moss, the Executive Director of the Oceanic Institute at Hawaii Pacific University, visited YSFRI

2.4 World Organisation for Animal Health (OIE) Reference Laboratory for White Spot Syndrome and OIE Reference Laboratory for Infectious Hypodermal and Hematopoietic Necrosis

In 2011, YSFRI was approved as the OIE reference laboratory for white spot syndrome and OIE Reference Laboratory for infectious hypodermal and hematopoietic necrosis by OIE (Fig. 6-1-6). The OIE Reference Laboratory is the technical support organization for the international animal health, veterinary public health and standards of animal product trade. It represents the highest authority and plays the leading role in the development of standards and rules for international animal disease prevention and control and animal product safety.

Fig. 6-1-6　OIE Reference Laboratory Unveiling Ceremony in 2011

2.5　China-Norway Mariculture Technologies Cooperation Research Center

In 2009, YSFRI was approved as China-Norway Mariculture Technologies Cooperation Research Center of Shandong Province by Department of Science & Technology of Shandong Province(Fig. 6-1-7). The center has close cooperation with Institute of Marine Research and other Norwegian fishery organizations in the fields of fishery resources, aquaculture and biotechnology, which has effectively promoted the development of fishery science and technology between China and Norway.

Fig. 6-1-7　Unveiling Ceremony of China-Norway Mariculture Technologies Cooperation Research Center

2.6 China-Korea Joint Research Center for Fisheries Science

China-Korea Joint Research Center for Fisheries Science was established in 2013, which is affiliated to the Chinese Academy of Fishery Sciences and National Institute of Fisheries Science of Korea. It relies on YSFRI and West Sea Fisheries Research Institute of National Institute of Fisheries Science of Korea (Figs. 6-1-8 to 6-1-9). The center carries out the cooperative research on the conservation and sustainable utilization of biological resources of Yellow Sea, carries out exchanges, mutual visits and trainings of scientific and technological personnel, and promotes the development of fisheries science and technology in the two countries.

Fig. 6-1-8 The Opening Ceremony of China-Korea Joint Research Center for Fisheries Science in 2013

Fig. 6-1-9 2018 Annual Meeting of China-Korea Joint Research Center for Fisheries Science held in South Korea

2.7 Sino-American Open Laboratory for Fish Functional Genomics

YSFRI and Auburn University reached a cooperative agreement in 2004 to establish the Sino-American Open Laboratory for Fish Functional Genomics (Fig. 6-1-10). The laboratory focuses on marine fish functional genomics, cell engineering breeding, molecular breeding, as well as marine microorganism metagenomics, which promote the functional genomic research of marine organisms into the world leading level.

Fig. 6-1-10 Sino-American Open Laboratory for Fish Functional Genomics established in 2014

2.8 China-Mexico Joint Research Center in Mariculture

YSFRI and National Institute of Fisheries and Aquaculture (INAPESCA) of the United Mexican States reached a cooperative agreement in 2019 to establish the China-Mexico Joint Research Center in Mariculture (Fig. 6-1-11). The center carries out mariculture researches between the two countries, and encourages scientists exchange, mutual visits, and training scientific and technological personnel to promote the development of fisheries science and technology.

Fig. 6-1-11   Establishment of China-Mexico Joint Research Center in Mariculture in 2019

2.9   Sino-Scottish Aquatic Invertebrate Laboratory

Sino-Scottish Aquatic Invertebrate Laboratory was jointly founded by YSFRI and the Scottish Association for Marine Science in July 2007 (Fig. 6-1-12). It focuses on science and technology development for sustainable marine aquaculture and Integrated Multi-Trophic Aquaculture (IMTA), invertebrate nutrition, shellfish biotoxins, marine biofuel extraction and ecosystem-based marine aquaculture spatial planning and management.

Fig. 6-1-12   Unveiling Ceremony of Sino-Scottish Aquatic Invertebrate Laboratory in 2007

3   Breeding techniques and modes

3.1   Large-scale breeding of mariculture seedlings

The research on the breeding of mariculture seedlings began in the early 1950s. After more than half a century of continuous exploration, artificial propagation techniques for more than

30 kinds of economic species, including fish, shrimp, crab, shellfish, algae and so on, have been established(Fig. 6-1-13), which has made important contributions to promote the development of China's mariculture industry.

Fig. 6-1-13　Industrialized artificial seedling technique

3.1.1　Chinese shrimp

Zhao Fazhen, the CAE academician, led his team to establish the industrialized artificial seedling technique according to the biological and ecological characteristics of shrimp, mainly including important water and material control such as temperature, water quality and feed in the development of sexual glands, spawning hatching and seedling breeding. The establishment of the industrialized artificial seedling technique has realized the efficient, stable and large-scale production of shrimp seedling and laid a solid foundation for China to become the world's largest shrimp farming country. In 1985, the research achievement was awarded the first prize of National Science and Technology Progress Award.

3.1.2　Turbot

Turbot was introduced from the United Kingdom by YSFRI in 1992. After seven years of scientific and technological research, the technical difficulties in domestication broodstock, collecting eggs, controlling high albino rate of seedlings and low survival rate were solved (Fig. 6-1-14 to 6-1-15). A perfect seed breeding technology system was established, which has led to the rapid development of artificial breeding technology for marine fish in China. China has also become the world's largest flounder producing country with an annual output of more than 120 thousand tons of commercial fish and an annual output value of over 10 billion *yuan*. "Research on introduction and seed production technology of turbot" won the second prize of National Science and Technology Progress Award in 2001.

Fig. 6-1-14 Collecting eggs of turbot

Fig. 6-1-15 Broodstock of turbot

3.1.3 Chinese tongue sole

YSFRI has studied and established a series of techniques on Chinese tongue sole (Fig. 6-1-16) including the genetic sexual determination, artificial gynogenesis, molecular marker assisted sexual control, pseudo males induction, and high ratio of female fish breeding. The crucial techniques of its parental fish domestication, natural spawning and reproductive regulation have been break through. The large-scale larval rearing and healthy breeding system of Chinese tongue sole have also been developed. Thus, the stable development of sole branch fish has been achieved. A 1.5 billion-a-year fishing industry has been developed and the relating outcomes have won the second prize of National Scientific and Technological Progress Award in 2010.

Fig. 6-1-16 Female (top) and male (bottom) of Chinese tongue sole

3.1.4 Yellowtail kingfish, *Seriola lalandi*

YSFRI has made great breakthrough in artificial breeding of yellowtail kingfish, several key technology problems were conquered including germplasm evaluation, wild broodstock domestication, land-sea relay culture technology, broodstock maturation and natural spawning, etc (Figs. 6-1-17 to 6-1-18). Batches of fertilized eggs have been obtained, and the development and growth pattern in early life stages, feed series and feeding strategy, seedlings cultivation key technology of yellowtail kingfish have been established. Thousands of juveniles with an average total length of 13.6 cm and average weight of 28.4 g have been produced. This achievement will provide promising species for open ocean aquaculture in China.

Fig. 6-1-17 Broodstock of yellowtail kingfish   Fig. 6-1-18   Seedlings cultivation of yellowtail kingfish

3.1.5 Grouper

By conducting the research on gender-based conversion technology for young grouper parents, the key techniques of parental reproductive regulation and nutrition enhancement and the key technology of large-scale seedling breeding, the rate we converted the female to male were above 80% in *Epinephelus septemfasciatus*, *E.moara* and *Centropristis striata* (Fig. 6-1-19), and stable functional males were obtained. The mature rate of parental fish reached above 80%, and the water temperature of parental fish first decreased and then rose, to improve the quality of eggs and sperm, the rate of fertilization, and obtain high-quality fertilized eggs, the parental fish experienced a low temperature period in winter. The survival rates of *E. akaara*, *E. moara* and *E. septemfasciatus* seedlings reached 23.6%–26.7%, 18.3%–22.5% and 14.2%–18.6%, respectively, which were 10% higher than the traditional seedling technology.

  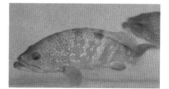

Parent fish of *Epinephelus akaara*   Male fish of *Centropristis striata*   Broodstock of *Epinephelus radiatus*

Fig. 6-1-19   Pictures of three grouper species

3.1.6 Swimming crab

The new variety swimming crab "Huangxuan No. 1" (Fig. 6-1-20) was selected successfully in 2010. Key techniques of large scale artificial breeding of swimming crab were established using farming parent crab. The mode of efficient ecological pond farming was optimized for swimming crab. The output of seedling was raised 30% by ground water overwintering and efficient ecological seedling nursing technique. Meanwhile, the seed nursery of juvenile *P. trituberculatus* and integrated multi-trophic aquaculture (crab-shrimp-shellfish-fish) were invented for best economic efficiency. "New variety of swimming crab development and large-scaled farming" won the first prize of Chinese Agriculture Science & Technology Award in 2014–2015.

Fig. 6-1-20 New variety swimming crab "Huangxuan No. 1"

3.1.7 Scallop

The new strain of Japanese scallop *Patinopecten yessoensis* named as "Yu Bei", (Fig. 6-1-21) was bred, which is characterized by pure white of the double shells, high growth rate and high heat-resistance. The genetic traits of "Yu Bei" is stable, and the color and lustre of the shell is bright. The yield rate of fresh adductor muscle rate has significantly

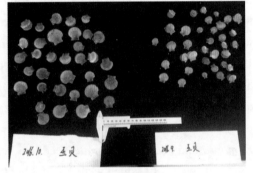

Fig. 6-1-21 New strain of Japanese scallop "Yu Bei"

increased, and the growth rate and survival rate in summer season increase by 15% and above 20%, respectively. A series of mature technical norms including seed production, intermediate culture, healthy culture, etc. of "Yu Bei", and technical means such as genetic analysis, etc., have been established. The annual production capacity of seedlings is about 200 million. "Yu Bei" has been extended and developed in large-scale culture in Changdao, Shandong Province and other places, which has promoted the industrial development of Japanese scallop. The relevant technological achievements were awarded the second prize of Science and Technology Progress in Qingdao in 2012.

3.1.8 *Sargassum thunbergii*

The technology systems for *S. thunbergii* (Fig. 6-1-22) seedling rearing with large scale under artificial conditions, and for the grow-out on the open sea have been established. Moreover, we also developed the technology for the *S. thunbergii* enhancement on the artificial reef, which has been used in the seaweed bed construction. These achievements will promote the study and application of *S. thunbergii* in aquaculture as well as other fields.

Fig. 6-1-22 *Sargassum thunbergii*

3.2 Cultivation of new varieties of mariculture

YSFRI regards the cultivation of new marine variety as a key research direction. The breeding theory and technologies have been rapidly developed, and the germplasm innovation ability has been continuously improved. With the support of the projects such as national 863 plan, national science and technology support plan, 14 new varieties have been cultivated and approved, such as Chinese shrimp "Huanghai No. 1". The cultivation and promotion of these new varieties have increased the aquaculture production and breeding survival rate, and the industrial development has been brought to a new level.

3.2.1 Chinese shrimp "Huanghai No. 1"

The Chinese shrimp *Fenneropenaeus chinensis* "Huanghai No.1" (Fig. 6-1-23) is the first artificial variety of marine aquaculture in China, which has the characteristics of fast growth, strong resistance and stable genetic structure. The average body length and body weight of "Huanghai No. 1" are 8.40% and 26.8% greater than those of those of the unselected population, respectively. "The Chinese shrimp 'Huanghai No. 1' and its healthy culture technology system" won the second prize of State Technology Invention Award.

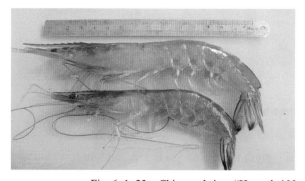

Fig. 6-1-23 Chinese shrimp "Huanghai No. 1" and the new variety certificate

3.2.2 Chinese shrimp "Huanghai No. 2"

The Chinese shrimp *F. chinensis* "Huanghai No. 2" (Fig. 6-1-24) is a new variety with many excellent characters. The growth rate of "Huanghai No. 2" is high and the harvested body weight is 30% more than that of unselected population. *F. chinensis* "Huanghai No. 2" has obvious disease resistance with the characteristics of uninfected, slow death after infection, and the death time after infection is prolonged by more than 10%.

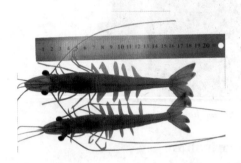 

Fig. 6-1-24 Chinese shrimp "Huanghai No. 2" and the new variety certificate

3.2.3 Chinese shrimp "Huanghai No. 3"

The Chinese shrimp *F. chinensis* "Huanghai No. 3" (Fig. 6-1-25) is the first resistant variety of marine aquaculture in China. Under the same culture conditions, the ability of "Huanghai No. 3" seeds to resist ammonia nitrogen is 21.2% higher than that of commercial seeds. The survival rate is raised by 15.2% and the harvested body weight is raised by 11.8%. The success rate of large-scale pond farming is 90%, and the yield is over 20% higher than that of commercial seeds.

Fig. 6-1-25 Chinese shrimp "Huanghai No. 3" and the new variety certificate

3.2.4 Pacific white shrimp *Litopenaeus vannamei* "Ren Hai No. 1"

The Pacific white shrimp *Litopenaeus vannamei* "Ren Hai No. 1" (Fig. 6-1-26) has the characteristics of faster growth, shorter breed period, more uniform in size, and more stable and higher survival rate. In 2012 and 2013, its productive pilot scale tests were performed in Hebei

Province, Tianjin city, and Guangdong Province, etc. Compared with the general commercial breeds, the production of the "Ren Hai No.1" increased by 20%-28%, and its survival rate improved by 10%-18%.

Fig. 6-1-26 Pacific white shrimp *Litopenaeus vannamei* " Ren Hai No. 1" and the new variety certificate

3.2.5 Swimming crab "Huangxuan No. 1"

The new variety swimming crab "Huangxuan No. 1" (Fig. 6-1-27) was selected successfully in 2010. Key techniques of large scale artificial breeding of swimming crab were established using farming parent crab. The model of efficient ecological pond farming was optimized. The output of seedling was raised 30% by ground water overwintering and efficient ecological seedling nursing technique. Meanwhile, the seed nursery of juvenile *P. trituberculatus* and integrated multi-trophic aquaculture (crab-shrimp-shellfish-fish) were invented for best economic efficiency. "New variety of swimming crab development and large-scaled farming" won the first prize of Chinese Agriculture Science & Technology Award in 2014-2015.

Fig. 6-1-27 Swimming crab "Huangxuan No. 1" and the new variety certificate

3.2.6 Turbot "Danfa"

The body weight of selected variety of "Danfa" turbot (Fig. 6-1-28) increased by 24.44%-33.2% compared with that of commercial variety. The survival rate increased by 22.5% on average, and the feed conversion rate increased by 27.9% on average. The fry production rate was 38.23% on average, and topped out at 56.3%, higher than that of unselected common fry.

The fry albinism rate of this new variety is less than 5%.

Fig. 6-1-28   Turbot "Danfa" and the new variety certificate

3.2.7   Turbot "Duo Bao No. 1"

Turbot *Scophthalmus maximus* "Duobao No. 1" (Fig. 6-1-29) has the advantages of high genetic purity, fast growth and high survival rate. Body shape has totally identical character. Compared with common cultured turbot, average weight and survival rate of "Duo Bao No. 1" increased by 36% and 25%, respectively. The genetic stability of the main economic traits reached more than 90%.

Fig. 6-1-29   Turbot "Duobao No. 1" and the new variety certificate

3.2.8   Japanese flounder "Ping You No. 1"

Japanese flounder "Ping You No. 1" (Fig. 6-1-30) is the first artificial breeding new variety of Japanese flounder in our country. It features fast growth speed and heat-resistant ability. Japanese flounder "Ping You No. 1" is a fine variety suitable for breeding in the north and south of China. In the same breeding condition, its weight increased by 30% and survival rate improved by 20% compared with the common Japanese flounder. Its breeding survival rate through the summer reached above 90%.

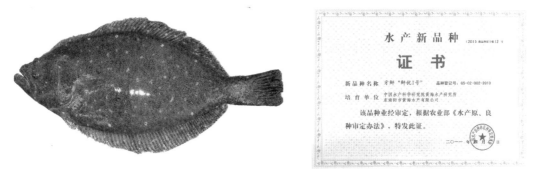

Fig. 6-1-30   Japanese flounder "PingYou No. 1" and the new variety certificate

3.2.9 Japanese flounder "Ping You No. 2"

Japanese flounder "Ping You No. 2" (Fig. 6-1-31) has the advantages of strong resistance to bacterial infection, high survival rate and high growth speed. In the same breeding condition, the growth speed of its 18-month individuals increased by 20% and survival rate improved by 20% compared with the common Japanese flounder.

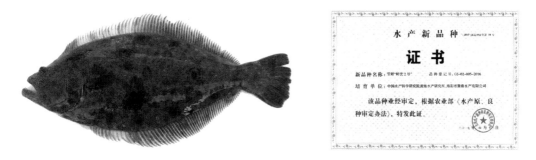

Fig. 6-1-31   Japanese flounder "Ping You No. 2" and the new variety certificate

3.2.10 "Huangguan No. 1"

The *Saccharina japonica* new cultivar "Huangguan No. 1" (Fig. 6-1-32) has a wide and flat blade with wider central band part but narrow, thicker lateral part; it is high-temperature resistant, and has a higher yield, increasing by 27% compared to the traditional cultivars. It has intermediate alginate content, balanced nutrients, and its rate of output is 81% when is shred as food, increasing by more than 20% in comparison with other cultivars.

Fig. 6-1-32 New cultivar "Huangguan No. 1"and the new variety certificate

3.2.11 Chinese shrimp "Huanghai No. 5"

The Chinese shrimp *F. chinensis* "Huanghai No. 5" (Fig. 6-1-33) was successfully bred to improve white spot syndrome virus (WSSV) resistance and growth rate after six generations of multi-trait selection. Under the same culture conditions, the WSSV resistance and growth rate of "Huanghai No. 5" improved by an average of 30.10% and 26.52%, respectively, compared with unselected Chinese shrimp larvae. "Huanghai No. 5" was ratified by the National Certification Committee for Aquatic Varieties of China in 2017.

Fig. 6-1-33 Chinese shrimp "Huanghai No. 5" and the new variety certificate

3.2.12 Ridgetail white prawn "Huangyu No. 1"

The ridgetail white prawn *Exopalaemon carinicauda* "Huangyu No. 1" (Fig. 6-1-34) is a new variety obtained from population selection of six generations based on body length and weight traits. Under the same culture conditions, the body length and weight of 3 months of "Huangyu No. 1" increased by 12.62% and 18.40% respectively, compared with the wild population. In 2017, *E. carinicauda* "Huangyu No. 1" was approved by the National Certification Committee for Aquatic Varieties of China.

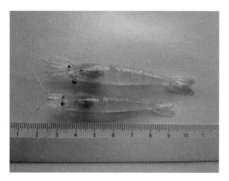 

Fig. 6-1-34　Ridgetail white prawn "Huangyu No. 1" and the new variety certificate

3.2.13　Sea cucumber "Shenyou No. 1"

"Shenyou No. 1" (Fig. 6-1-35) is a new variety of sea cucumber (*Apostichopus japonicus*) ratified by the National Certification Committee for Aquatic Varieties of China in 2017. It was selected for 4 generations by means of mass selective breeding. The survival rate at 6-month-old after Vibrio splendidus infection is 11.68% higher than that of unselected sea cucumber. Using pond culture type, the body weight and the survival rate are 24.46% and 23.52% higher than that of unselected sea cucumber.

Fig. 6-1-35　Sea cucumber "Shenyou No. 1" and the new variety certificate

3.2.14　Swimming crab "Huangxuan No. 2"

Compared with those unselective breeding individuals, the survival rate of swimming crab "Huangxuan 2" (Fig. 6-1-36) increases by 31.2% on average, and the average weight increases by 18.8%; compared with "Huangxuan No. 1", the survival rate of "Huangxuan No. 2" increases by 10.7% on average, and there is no significant difference in body weight. The resistance of "Huangxuan No. 2" to low-salinity stress is significantly improved.

Fig. 6-1-36   Swimming crab "Huangxuan No. 2" and the new variety certificate

3.3   Marine aquaculture disease prevention and control

Aiming at resolving main disease problems, the laboratory focuses on the key research directions including early warning of marine aquaculture disease, disease occurrence and epidemic mechanism, immunity and disease resistance mechanism and management of aquaculture health. Breakthroughs have been made in virology of aquaculture animal, interaction of pathogens and host cells, histopathology of aquaculture animals, molecular detection techniques for pathogens, non-specific immunity of shrimp and shellfish and their regulation techniques.

3.3.1   Highly sensitive and rapid detection technology of aquatic animal pathogens

The serial key technical problems of restricting large-scale application of rapid detection techniques for aquatic animal pathogens were solved by YSFRI researchers through development of core technologies of rapid extraction of nucleic acids, long-term-storage of reagents at room temperature and immobilization of nucleic acid dyes. The kits for more than 20 aquatic animal pathogens have been developed for on-site rapid detection (Fig. 6-1-37). The kits have been widely used in research institutes and farms in 15 provinces of China, and feedbacks of the kit application are quite positive. The detection kits are also used by customers of foreign countries such as Egypt, Indonesia and Thailand.

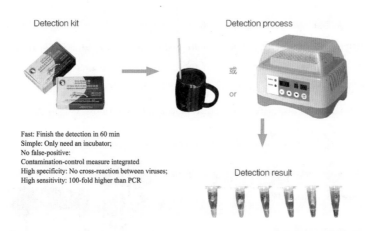

Fig. 6-1-37　Rapid detection kits

3.3.2　Chinese herbal medicine-Zhong Ren Yi Hao

In order to control the diseases such as shrimp acute hepatopancreatic necrosis disease (AHPND), white spot syndrome (WSSV) and fish *Cryptocaryon irritans* disease (CID), we developed pathogen rapid detection kit, green disinfector, antagonistic probiotic, Chinese herbal medicine (Fig. 6-1-38), efficient drug delivery technology, aquaculture water harmful microbes quantitative control technology, VR viewer for underwater environment, which have achieved integrated and standardized application. Using these technologies to prevent the outbreak of disease, clinical cure rate is more than 80% and survival rate increases greatly.

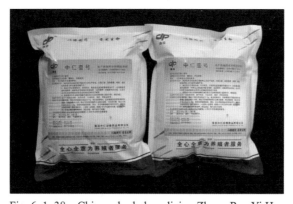

Fig. 6-1-38　Chinese herbal medicine-Zhong Ren Yi Hao

3.3.3　Remote diagnosis system for aquatic diseases

Remote diagnosis system for aquatic diseases (Fig. 6-1-39) is established by YSFRI and can greatly improve the timeliness and accuracy of aquatic diseases diagnosis. The system consists of hardware platform and software platform, which are divided into one central station and several terminal stations. The central station is the core of the whole system to ensure remote consultation normal operation and all functions realization. The terminal station is generally located in technical service centers or aquaculture enterprises in China or abroad. The remote disease diagnosis or consultation can be carried out through the Internet between the central and terminal stations. The transmission of video, audio, micrographs and other images can be used by the experts in central station to diagnose the disease. The software platform of this system includes center station website, aquatic biology database, aquatic diseases database, aquatic drugs database and aquatic knowledge library. These databases support the network and mobile phone access, search and query.

Fig. 6-1-39 Remote diagnosis system for aquatic diseases

3.4 Mariculture modes

In terms of marine aquaculture modes, YSFRI has been exploring new modes for sustainable development of marine aquaculture suitable for China's national conditions, including land-based industrialization, seawater ponds, tidal flats, shallow seas, marine pastures and deep-water cages. A new concept of ecosystem and carrying capacity-based aquaculture was proposed.

3.4.1 Inshore integrated multi-trophic aquaculture (IMTA) modes

The inshore IMTA (Fig. 6-1-40) is based on ecological principles including eco-balance and bio-symbiosis, which rationally matches groups with multiple trophic-level and different bio-functions to achieve effective recycling of nutrients in the system. The IMTA helps to improve carrying capacity and output efficiency, and to reduce the negative impact on the environment. According to the inshore ecological mariculture mode and the results of carrying capacity of the technology research group, YSFRI has constructed and implemented many IMTA modes including fish-shellfish-seaweed IMTA mode, shellfish-seaweed IMTA mode, abalone-sea cucumber-seaweed IMTA mode, etc. in the coastal waters of Shandong and Liaoning provinces. These IMTA modes have been promoted for industrialization, providing theoretical basis for establishing an adaptive management strategy based on the ecosystem level and exploring the environmentally friendly modern mariculture industry with principles of "high efficiency, high quality, ecology, health and safety". In 2016, the FAO and the NACA promoted the Sanggou Bay Integrated Culture Mode as one of the 12 successful examples of sustainable intensive aquaculture in the Asia-Pacific region.

Fig. 6-1-40 Inshore integrated multi-trophic aquaculture modes

3.4.2 Ecological seawater pond IMTA mode

Eco-friendly seawater pond IMTA (Fig. 6-1-41) has been developed according to the complementary multi-level utilization principle. Various mariculture species at different trophic levels, such as shrimp, crab, fish, clam and algae, were selected to construct a manmade ecosystem in a seawater pond, and the specific techniques including aeration by microporous system, nutrients supplement and water quality manipulation were performed during the process of culture. The total production of such aquaculture mode was significantly increased without additional land input. It has significantly improved the production efficiency and has obvious economic and social benefits. For example, in area of Rizhao of Shandong Province and Ningbo of Zhejiang Province, where the shrimp-crab-clam-fish mode was applied, the production of the Chinese shrimp were increased to about 1,500 kg/hm^2, and 900-1,200 kg/hm^2 for crab, 4,500-7,500 kg/hm^2 for clams, and 150 kg/hm^2 for fish. The annual total output value was 300,000 RMB/hm^2.

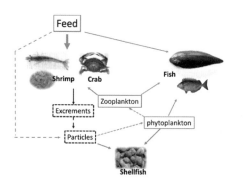 

Fig. 6-1-41 Ecological seawater pond integrated multi-trophic aquaculture mode

3.4.3 Modern marine ranching

In the early 1980s, YSFRI began to explore and practice marine ranching and accumulated plenty of materials and experiences. Until now, the techniques such as artificial reef engineering technology, proliferation and release, multi-nutrient level comprehensive three-dimensional breeding mode have been integrated (Fig. 6-1-42), and six marine ranching demonstration areas have been established in Laizhou Bay, Xuejia Island in Qingdao, Shuangdao Bay in Weihai, Beidaihe in Qinhuangdao, Zhangzidao in Dalian and other sea areas, with a total scale of more than 3300 hm^2. YSFRI has completed the planning and design of six large-scale marine pastures such as Nanri Island in Fujian, Changhai County in Dalian, Fangcheng Port in Guangxi, Long Island and Laizhou Bay in Shandong, providing strong technical support for the construction and development of China's marine ranching.

Fig. 6-1-42　Modern marine ranching

3.4.4 Offshore cage culture

According to China's sea conditions and the habits of different farmed fish, YSFRI has successively developed three kinds of net cage products, including HDPE round net cage, wind & wave resistant metal net cage (Fig. 6-1-43) and special net cage for flounder fish breeding (Fig. 6-1-44). We successfully solved the technical problem that the open sea cannot be cultivated because of the deep water, big wave and rapid flow. The overall performance of the cage developed has reached or exceeded the level of the same kind of imported products abroad and led the domestic market.

Fig. 6-1-43　Wind&wave resistant metal net cage　　Fig. 6-1-44　Special net cage for flounder fish breeding

3.4.5　Energy saving and environmentally friendly recirculating aquaculture system (RAS)

Focusing on energy saving, environmental protection and seafood safety, the key equipments and facilities have been invented and developed for aquaculture wastewater treatment, and the RAS (Figs. 6-1-45 to 6-1-46) has been constructed. The new RAS which includes filtration, biological purification, degassing, sterilization and oxygenation procedures, has the advantages of high removal efficiency, eco-friendly, water saving, energy saving and controllable, which promoted the scale and level of the industrialized circulating water aquaculture. The technology has been applied to more than 20 coastal provinces and cities, and the application area is up to 170,000 m^2.

Fig. 6-1-45　Cultivation of red fin oriental dolphin in RAS system　　Fig. 6-1-46　RAS system

3.4.6 Industrialized shrimp aquaculture

The aim of industrialized shrimp aquaculture is to effectively utilize the natural resources, such as land, water, light and heat, to reduce the adverse effects of seasonal and climate changes on shrimp production, and to improve the feed utilization rate and the aquaculture success rate (Fig. 6-1-47). In such mode, shrimp are cultured in a closed room or a greenhouse, equipped with water supply, drainage, aeration, warming, and other specific supporting systems such as bio-filters and disinfection. The culture system is relatively human-controllable and the water quality is stable. It is a kind of healthy and efficient shrimp aquaculture mode with high production. The culture can be carried out throughout the year with 3-4 crops per year and the average production is 5-8 kg/m^2/crop.

Fig. 6-1-47 Industrialized shrimp aquaculture